Cycling Pathways

Cycling Pathways

The Politics and Governance of Dutch Cycling Infrastructure, 1920-2020

Henk-Jan Dekker

Routledge
Taylor & Francis Group
LONDON AND NEW YORK

The publication of this book was made possible through the generous support of Rijkswaterstaat, Eindhoven University of Technology, School of Innovation Sciences, Stichting Unger-Van Brero Fonds, the J.E. Jurriaanse Stichting and Stichting Historie der Techniek.

First published in 2022 by Amsterdam University Press Ltd.

Published 2025 by Routledge
4 Park Square, Milton Park, Abingdon, Oxon OX14 4RN
605 Third Avenue, New York, NY 10158

Routledge is an imprint of the Taylor & Francis Group, an informa business

ISBN: 9789463728478 (hbk)
ISBN: 9781041177876 (pbk)
ISBN: 9781003693611 (ebk)
NUR 696

Cover illustration: Utrecht, Lucasbolwerk, c. 1955
Source: F.F. van der Werf, Collectie Het Utrechts Archief, no. 502431

Cover design: Coördesign, Leiden

DOI 10.5117/9789463728478

For Product Safety Concerns and Information please contact our EU representative:
GPSR@taylorandfrancis.com
Taylor & Francis Verlag GmbH, Kaufingerstraße 24, 80331 München, Germany

Acknowledgements

Researching and writing this book was not unlike one of my weekend bike rides: long, full of (mental) ups and downs, harder and harder as the end approaches, and, in the past year and a half, also pretty solitary at times. Making it to the end of such a ride often requires a bit of tailwind or shelter behind fellow cyclists. Likewise, to finish this PhD, many people have helped me along the way. Without them I could not have written this book.

First and foremost, I would like to express my deepest appreciation to my supervisors Ruth Oldenziel and Frank Veraart. In many decisive ways they have shaped this project. They started studying cycling at a time when cycling was not taken seriously as a topic of historical research, and they are still highly interested in it. I could not have wished for supervisors more genuinely invested in this project. They were always willing and enthusiastic when reading my drafts and thinking through the issues. Their many insightful suggestions improved this book immeasurably. Without their patience, unwavering support, and belief in my work throughout the process, this study would simply not have been possible.

I also gratefully acknowledge the help of people who generously took the time to read and comment on my many drafts: Peter Norton, Arwen Mohun, Frank Schipper, Bert Toussaint, Rick Lindeman, and especially Adri Albert de la Bruhèze, who all provided valuable insights from different perspectives. Special thanks go to Frank, for getting the project started and familiarizing me with the academic world and the PhD process.

I have gained a deeper understanding of the more recent history of cycling in the Netherlands thanks to interviews with former policymakers and activists who graciously contributed their time. I am especially grateful to Jan Ploeger, who generously shared his extensive knowledge of cycling policy history and library of policy documents, which have benefited this book enormously.

Many thanks also to the archivists whose assistance made this project possible. I would particularly like to thank ANWB archivist Leo van Luxemburg for being so helpful and accommodating in providing access to archives that were not open to the public, like many state archives I have visited.

Special thanks go to the funders of the TUe Bicycle Challenges project (2016-2021): Eindhoven University of Technology, the Dutch Public Works administration Rijkswaterstaat, and Pon Holding. The publication of this book was made possible through the generous financial support of Rijkswaterstaat, the Stichting Unger-Van Brero and the J.E. Jurriaanse Stichting.

I am also extremely grateful to my fellow cycling PhDs in Eindhoven. Patrick Bek, Brett Petzer, and Sun Qi, with whom I shared an office, and who formed, together with Matthew Bruno and George Liu, a great support group of cycling researchers. They read many versions of my chapters in our work-in-progress sessions, and with their various disciplinary backgrounds their work has greatly enriched my understanding of cycling, both past and present. Patrick's historical bottom-up perspective on workers' mobility complemented my policy perspective greatly and I have benefited much from our many conversations. Matthew's sharp analyses of contemporary cycling governance and innovation led to fruitful dialogues and collaborations involving my historical approach. Jan's historical work on the bicycle-train combination was invaluable for contextualizing many of the developments in my book. Brett's admirably critical work on space and bike-share was a source of inspiration. I had many fascinating office discussions with Qi about her comparative perspective on e-bikes. Finally, George's work on the built environment also has many areas of overlap with my work. Together they created an extremely stimulating environment to work in. Many of them also brought knowledge and experience of cycling in non-Dutch contexts to the table, which was eye-opening for me. I feel very lucky to have worked alongside them for the past four and a half years and cannot imagine having gone through this process without their companionship.

All my Eindhoven colleagues in the Technology, Innovation, and Society group provided a welcoming and supportive environment in which I have enjoyed working. My History Lab colleagues Harry Lintsen, Erik van der Vleuten, Ruth Oldenziel, Mila Davids, Frank Veraart, Karena Kalmbach, Jan Korsten, Eric Berkers, and Ginevra Sanvitale deserve a special mention. Thanks also to my fellow TIS PhD students with whom I shared a hallway or floor and many lunches: Aleid, Ana, Aziz, Darja, Edgar, Ginevra, Luc, Natascha, Nick, Peter, and Tanja. A special word for the various teammates in our football team, as I always enjoyed our lunch-break games so much. Thanks Leon, Daan, Emilio, Erik, Jolanda, Frank, Luc, Oktay, and Richard.

For the past year and a half, I have also worked part-time as assistant book review editor for the journal *Technology & Culture*. Many thanks to Ruth Oldenziel, Hermione Giffard, and Dick van Lente for the enjoyable teamwork and the opportunity to get to know the academic world outside my research project.

Finally, I would like to thank my family for supporting me throughout this long process and putting up with me whenever I was quietly struggling through the difficult stretches of writing a book.

Maartensdijk, July 2021
Henk-Jan Dekker

Table of Contents

Part II
Divergence: How Dutch Cycling Policy and Practice Persevered, 1950s-1970s

Part III
Dutch Model: How Urban Cycling Became a National Political Demand after 1970

List of Abbreviations

Archives

AE	Archief Eemland
BHIC	Brabants Historisch Informatie Centrum
DA	Drents Archief
GA	Gelders Archief
GAA	Gemeentearchief Amsterdam
GADH	Gemeentearchief Den Haag
GAZ	Gemeenterachief Zeist
HCO	Historisch Centrum Overijssel
HGA	Haags Gemeentearchief
HUA	Het Utrechts Archief
NA	Nationaal Archief
NHA	Noord-Hollands Archief
RHCE	Regionaal Historisch Centrum Eindhoven

Organizations

ANWB	Algemene Nederlandsche Wielrijders-Bond *Dutch Cyclists' Association* [Nowadays Royal Dutch Touring Club ANWB]
BOVAG	Bond van Automobielhandelaren en Garagehouders *Federation of Car Dealers and Garage Owners*
COW	Commissie van Overleg voor de Wegen *Consultation on Roads Committee*
CRM	Ministerie van Cultuur, Recreatie en Maatschappelijk Werk *Ministry of Culture, Recreation and Social Work*
CROW	Centrum voor Regelgeving en Onderzoek in de Grond-, Water- en Wegenbouw en de Verkeerstechniek *Centre for Regulations and Research in Civil and Traffic Engineering*
EMO	Eindhoven met Omstreken *Cycling Path Organization Eindhoven and surroundings*
ENFB	Echte Nederlandse Fietsersbond *Real Dutch Cyclists' Union*
ENWB	Enige Echte Nederlandse Wielrijdersbond *Only Real Dutch Cyclists' Union*
FB	Fietsersbond *Cyclists' Union*

FNRV	Federatie van Nederlandse Rijwielpadverenigingen *Federation of Dutch Cycling Path Organizations*
IPO	Interprovinciaal Overleg *Inter-Provincial Consultation*
KNLC	Koninklijk Nederlands Landbouwcomité *Royal Dutch Agriculture Committee*
NEVAS	Nederlandsche Vereniging voor Autosnelwegen *Dutch Highway Organization*
NIVS	Nederlands Instituut voor Volkshuisvesting en Stedebouw *Dutch Institute for Public Housing and Urban Development*
NTFU	Nederlandse ToerFiets Unie *Dutch Touring Bike Federation*
NUCA	Nederlandsche Unie van Chauffeurs en overig Automobielpersoneel *Dutch Union of Drivers and other Automotive Personnel*
PPD	Provinciale Planologische Dienst *Provincial Spatial Planning Office*
PWS	Provinciale Waterstaat *Provincial Public Works Administration*
RAI	Rijwiel- en Automobiel Industrie *Car and Bicycle Industry*
RWS	Rijkswaterstaat *National Public Works Administration*
SWOV	Stichting Wetenschappelijk Onderzoek Verkeersveiligheid *Road Safety Research Foundation*
UMO	Utrecht met Omstreken *Cycling Path Organization Utrecht and surroundings*
VNG	Vereniging van Nederlandse Gemeenten *Association of Dutch Municipalities*
VVN	Veilig Verkeer Nederland *Safe Traffic the Netherlands*
VVV	Vereniging voor Vreemdelingenverkeer *Dutch Tourist Information*

Other Abbreviations

BDU	Brede Doeluitkering *Broad Special-Purpose Grant*
GDU	Gebundelde Doeluitkering *Combined Special-Purpose Grant*
RONA	Richtlijnen Ontwerp Niet-Autosnelwegen *Design Guidelines Non-Motorways*
VERDI	Verkeer en Vervoer: Regionaal, Decentraal, Integraal *Traffic and Transport: Regional, Decentral, Integral*

List of Figures and Tables

Figures

Tables

Introduction

In 1930, B. Duyts, an accountant from the Dutch village of Loosdrecht, decided he had had enough. For years he had petitioned the municipality to improve the cycling route he used every day for getting to work, to no avail. There was an improvised path, but it was in poor condition, alongside a sand road; heavy lorries carrying tree trunks made it unsuitable for cycling. A well-paved separate cycling path, inaccessible to motorized vehicles, would make his commute safer and more pleasant. If the government would not build it, Duyts decided he could do it himself. He organized some neighbors and together they asked the town for a small subsidy to purchase paving material and wooden poles. They then improved the path by paving it and used the poles to separate it from the road, preventing cars and lorries from swerving onto the cycling path.[1] Duyts even placed signs at the beginning of the path banning motorized vehicles.

If Duyts thought this would be the end of his problems, he was wrong. Confusion broke out among local municipalities, the province of Noord-Holland, and provincial public works. Did Duyts have the authority as a private citizen to declare a part of this road an exclusive cycling path? Could this new cycling path be recognized as official? And if so, which government agency would be responsible for its maintenance? His municipality declined to take up this task. This infuriated Duyts, who argued that mobility was a right that was being denied to him and the other cyclists who used his street, Raaweg: like all citizens, they paid numerous taxes "and so like other citizens they could claim a right to a decent road rather than a quagmire."[2] The town advised Duyts to find someone willing to maintain the path. This exasperated him: had he not done more than his fair share already by improving the path? Now it was up to local authorities to take responsibility and do their part in facilitating cycling traffic. As he wrote: "After all the trouble I have gone to already, it is not clear to me what further steps I could take to set up an arrangement regarding maintenance." After all, Duyts stressed "that I, together with the Raaweg residents, have carried out repairs only as an absolute necessity" – could the authorities not show a little more initiative and willingness to help commuting cyclists?[3]

1 Archives Noord-Holland (NHA), Archive Provincial Public Works Noord-Holland (553), inv. no. 307.

2 NHA 553, inv. no. 307, letter B. Duyts to mayor and aldermen Hilversum, October 8, 1930.

3 NHA 553, inv. no. 307, letter B. Duyts, undated, around January 1931.

Citizens like Duyts played an important role in Dutch cycling governance. They were actively involved by organizing interest and action groups, lobbying the government to acknowledge their rights. While legally on the same footing, not all citizens of the state are treated the same way. Some groups have a higher status than others. Historically, policymakers and engineers in many countries accorded motorized travelers as a group higher status and privileges than cyclists. It is in this sense that the word citizen is used throughout *Cycling Pathways: The Politics and Governance of Dutch Cycling Infrastructure, 1920-2020*. In the Netherlands, as I will show, the political appreciation of cyclists as citizens with rights (to infrastructure, safety, comfort, and so on) was stronger than elsewhere. Had Duyts lived today, he could have commuted by bicycle almost anywhere along the many cycling-friendly and government-maintained roads in the Netherlands. Over nearly a hundred years, an extensive cycling network has been constructed in the Netherlands, for both recreational and utilitarian purposes. Today, constructing cycling infrastructure is considered a public good and a state task. As the 1930 episode shows, that was not always the case. What changed? How did providing for cyclists become a state task? Which actors shaped cycling policies and how did they do so? *Cycling Pathways* uncovers this governance process to shed light on the status of Dutch cyclists, both on the road and in policymaking from the 1920s to today.

Over the past few years, cycling has become considered the key to a sustainable and livable future for cities. International experts and scientists like urban planners John Pucher and Ralph Buehler praise Dutch cycling as a success story – even a cyclists' paradise – and an example of a pathway to a more sustainable transport system.[4] Dutch engineers and planners share their knowledge about planning for cyclists all over the world.[5] As they are among the world's most experienced cycling planners, this seems to make sense. On closer inspection, however, it is rather surprising. If Dutch experts are to help engineers in other countries, this presupposes that they know how to progress from these humble beginnings to a fully-fledged cycling system or network. This assumes historical knowledge about what led to this outcome. However, little has been written on the emergence of the Netherlands as a "cyclists' paradise."

How can the Dutch export their cycling knowledge to policymakers across the globe when so little is known about which government policies formed

4 John Pucher and Ralph Buehler, "Making Cycling Irresistible: Lessons from The Netherlands, Denmark and Germany," *Transport Reviews* 28, no. 4 (2008): 495-528.

5 For instance, through the Dutch Cycling Embassy: https://www.dutchcycling.nl/en/.

the Dutch cycling system? What role did the engineering community and social movements play? Should we ascribe a central role to non-governmental actors that tried to put cycling on the agenda? These questions need answers if we want to explain how the Netherlands became a nation of cyclists. And yet, historians have hardly focused on what is often considered one of the quintessential Dutch features. In fact, cycling is so mundane in the Netherlands that its history is not considered a serious issue of study. This project fills a knowledge gap important to both an international audience interested in boosting cycling levels and a Dutch audience wanting to understand how cycling became so deeply ingrained in everyday practices.

Studies of the history of urban cycling in the Netherlands are available, but none have covered the role of provincial and national governmental actors. While cycling activism has been studied, the role of citizens in social movements building expertise and political leverage requires further elaboration. *Cycling Pathways* addresses this gap by providing long overdue insights on mobility governance as a product of the interplay between engineering norms and ideology, politics, social movements, and users. This process of the mutual shaping of daily mobility in the Netherlands over the past century is central to this study. It shows the significant long-term consequences of mobility policies on aspects ranging from street design to the modal split and spatial planning. One of the insights of history – and mobility history in particular – is that decisions taken in the past often have far-reaching and unforeseen consequences. This mechanism is often referred to as *path dependence*: when decisions are built into concrete or asphalt, the results become hard to remove or modify.[6] A reflection on such policy choices made by various actors in the past, and the reasons behind them, can help us understand the present and shed some light on what (not) to do in the future.

The bigger question *Cycling Pathways* raises is: Why did the Netherlands become such a successful cycling country? Multiple factors explain this. As I explain in more detail below, Adri Albert de la Bruhèze, Frank Veraart, Martin Emanuel, and Ruth Oldenziel have proposed five key variables: spatial structure and morphology, the availability of mobility alternatives, the cultural status of cycling, government traffic policy, and the influence of social movements.[7] Given my ambition to discuss a long time

6 James Mahoney, "Path Dependence in Historical Sociology," *Theory and Society* 29, no. 4 (2000): 507-48.

7 Ruth Oldenziel, Martin Emanuel, Adri Albert de la Bruhèze, and Frank Veraart, eds., *Cycling Cities: The European Experience* (Eindhoven: Foundation for the History of Technology and

period (1920-2020), I limit myself to a focus on two interrelated factors: government traffic policy and the influence of organized citizens in social movements. Together, these shape the division of road space – in particular, cycling infrastructure. While cycling infrastructure may not be a sufficient condition for a successful cycling country, research suggests that it is a necessary condition in today's settings.[8] In high-automobility contexts, a combination of traffic-calming and traffic-separating measures are needed to make cycling broadly accessible to citizens. Historically, the distribution of road space and the provision of cycling space – sometimes seen as neutral and technical problems – are in fact political projects. The car's dominant place on our streets is the outcome of a contested and long-term process in which other modes of transport were denied access to the road. Politicians, policymakers, and engineers have governed – and still govern – this process, as have user-citizens organized in social movements and the media. Despite opposition, in many countries pro-car interests have managed to bar cyclists and pedestrians almost completely from the road; in many cases there is also little to no investment in separate infrastructure for these road users. The story of the Netherlands is different – *Cycling Pathways* addresses why.

My main research question is: to what extent has Dutch cycling governance since 1880 contributed to the success of the Netherlands as a cycling country? Two aspects of this research question, cycling and governance, need further elaboration and form persistent themes. The first is the term *cycling* itself. It suggests a uniformity that does not exist. Cyclists form a heterogenous group, consisting of (partly overlapping) different usage types. Utilitarian cycling (from commuters to school going kids) is a different activity from recreational or sports cycling; urban cycling can be distinguished from suburban or rural practices. Age, gender, and social class also form dividing lines. Historically, not all these groups have received equal attention from policymakers, engineers, and activists. By studying policies for these different groups, made at multiple government levels over more than a century, this project goes further than existing cycling and mobility history.

To operationalize my main question, I ask a number of sub-questions. First, which actors were involved in cycling governance throughout the research period? Second, how did they frame cycling as a public good?

Rachel Carson Center for Environment and Society, 2016).

8 Paul Schepers et al., "The Dutch Road to a High Level of Cycling Safety," *Safety Science* 92 (2017): 264-73; John Pucher and Ralph Buehler, "Safer Cycling Through Improved Infrastructure," *American Journal of Public Health* 106, no. 12 (2016): 2089-91.

Third, how did these actors interact to shape cycling policies? And finally, what were the main cycling policies resulting from this coalition? These questions, requiring extensive archival research, form the main empirical contribution of the project. To answer these questions, a series of roughly chronological chapters will discuss the actor coalitions around specific types of cycling, and the cycling policies and funding streams they organized.

Introducing Cycling Governance

As the opening story of the Dutch cyclist and accountant suggests, the political appreciation of cycling has changed enormously over the past century. Moreover, social advocacy and state support for cycling developed in different ways in different countries, leading to vastly different material infrastructures. Together with other developments – the rise of automobility, competition with public transit, growing commuter distances, and changing cultural appreciation for cycling – politics and governance are key factors that shape cycling. But how do these political transformations take place at the level of everyday governance and policymaking? That is the question I study here. To demand a state role in cycling governance, cycling advocates need to make the case for cycling infrastructure as a public good. After agreeing that some form of cycling infrastructure provision is a state task, there are still multiple steps between that acknowledgement and realizing infrastructure. Engineering norms and the coordination and division of responsibilities between different state bodies must be agreed. This process, from abstract political claim-making to implementing concrete policies, forms the topic of this book. The concepts of governance and public good that will help me answer these questions on a theoretical level are elaborated below.

Governance is a broad concept that has generated a vast amount of scholarship in recent decades. The issue of governance has been inspired by the declining role of nation-states and their governments since the 1980s and the growing role of supranational organizations, market parties, and civil society. Governance theories use the term *steering* to allude to a society-centered view, as opposed to a state-centered one. This is apt for the analysis of the political and administrative processes that have shaped Dutch cycling, in which a key argument is that non-state actors played an important role. According to public administration scholar Walter Kickert, at the most basic level, governance is "the mutual steering relations between

State and society."[9] A more elaborate definition, applied specifically to science and technology, comes from sociologists Susana Borrás and Jakob Edler, who consider governance "the way in which societal and state actors intentionally interact in order to transform [socio-technical and innovation] systems, by regulating issues of societal concern, defining the processes and direction of how technological artefacts and innovations are produced, and shaping how these are introduced, absorbed, diffused, and used within society and economy."[10] Given its relatively recent rise to prominence and frequent theoretical application to developments since the 1980s, why is governance a suitable concept with which to analyze a history stretching back to the early 1900s? According to Kickert, "the concept of governance may be a novel theoretical invention of modern political and administrative sciences but the empirical phenomenon existed before."[11] Similarly, political scientist Jeremy Richardson reminds us that "nonhierarchical styles of government are not at all new. Governments of all persuasions have always consulted and often bargained with a range of private actors in the formulation and implementation of public policy."[12] Two other prominent scholars of governance, Mark Bevir and Rod Rhodes, have also noted that the idea of "a monolithic state in control of itself" is a myth. It obscures "the reality of diverse state practices that escaped the control of the center because they arose from the contingent beliefs and actions of diverse actors at the boundary of state and civil society. The state is never monolithic and it always negotiates with others."[13] This is especially true of the Netherlands, as Kickert argues that "steering by a strong central State has hardly ever existed in the Netherlands. Governance has almost always been a matter of deliberation, persuasion, and compromise."[14] In retrospect, the powerful role the state and its institutions could assume in the 1950s and 1960s seems to have been the exception rather than the rule.[15]

9 Walter J.M. Kickert, *The History of Governance in the Netherlands: Continuity and Exceptions* (The Hague: Elsevier Overheid, 2004), 10.

10 Susana Borrás and Jakob Edler, "Introduction: On Governance, Systems and Change," in *The Governance of Socio-technical Systems: Explaining Change*, eds. Susana Borrás, and Jakob Edler (Cheltenham, UK: Edward Elgar Publishing, 2014), 14.

11 Kickert, *The History of Governance in the Netherlands*, 7.

12 Jeremy Richardson, "New Governance or Old Governance? A Policy Style Perspective," in *The Oxford Handbook of Governance*, ed. David Levi-Faur (Oxford: Oxford University Press, 2012), 312.

13 Mark Bevir and R. Rhodes, *The State as Cultural Practice* (Oxford: Oxford University Press, 2010), 93.

14 Kickert, *The History of Governance in the Netherlands*, 8.

15 Walter J.M. Kickert, "Beneath Consensual Corporatism: Traditions of Governance in the Netherlands," *Public Administration* 81, no. 1 (2003): 121-23.

How then did providing cycling infrastructure become a governmental task? As Part I, "Roots," shows, this was never a foregone conclusion. What the state deemed part of its tasks varied over time and depended on what was regarded as a public good – a point also observed by sociologists.[16] According to sociologist Rhys Williams: "Everyone is in favor of the public good, but just what constitutes that public good – or more accurately, whose public good is to be promoted – is a matter of political contention."[17] With regard to mobility, historians have convincingly shown how policymakers across the world saw the promotion and facilitation of automobility as a public good requiring extensive state intervention and funding. Cyclists also demanded infrastructure and provisions. The status of cycling infrastructure as a public good, however, has been far more precarious throughout the twentieth century. Cyclists and cyclists' organizations have had to fight for this status. To quote Williams again: "Movements use particular constructions of the public good to frame public politics in ways that benefit their agenda. By talking about the public good in a particular way, movements simultaneously legitimate their involvement and solutions, while casting aspersions on their opponents' positions. The discursive struggle is part of all public politics; a vision of the public good is a valuable tool in this process."[18] Indeed, countless arguments for constructing cycling infrastructure as beneficial to society as a whole and cyclists in particular came to the fore in the Netherlands. Together, they presented a case that was sufficiently convincing to justify state funding and coordination.

My approach consists of tracing two long-term processes. The first is this (discursive) struggle to frame cycling as a public good. It pertains to the first two sub-questions: who is involved in cycling policy, and how do these key stakeholders frame cycling and consequently the government's role? *Cycling Pathways* shows how state engineers, national and local politicians, and citizens (organized in social movements), throughout the history of cycling regarded particular types of cycling (recreational) as a public good, while ignoring others (utilitarian). The second process is the actual governance or steering process. This addresses the third and fourth sub-questions: what policy coalitions were established to govern cycling, and what policies did

16 Kevin Morrell, "Governance and the Public Good," *Public Administration* 87, no. 3 (2009): 538-56.

17 Rhys H. Williams, "Constructing the Public Good: Social Movements and Cultural Resources," *Social Problems* 42, no. 1 (1995): 124-44, 124.

18 Ibid., 126.

they introduce? In other words, once some state role in cycling became accepted, how was this support translated into concrete policies? According to cycling historian James Longhurst, we can also term this the distinction between politics and policies: politics are "the rhetoric and mechanisms by which groups and individuals come to power," whereas policy refers to "the deliberations, decisions, and actions of government in pursuit of a shared goal or public good."[19] To analyze the policies put in place, I take inspiration from political sciences, that is, the Policy Arrangement Approach.[20] From the perspective of this theory's proponents, the concepts in the policy arrangement approach serve as a heuristic framework that sensitizes us to certain processes and actors in governance. Accordingly, the four elements of the approach recur implicitly throughout each chapter but do not shape the actual structure of the chapters.

The proponents of the Policy Arrangement Approach define policy arrangements as "the temporary stabilization of the organization and substance of a policy domain at a specific level of policy making."[21] Specific policy arrangements can explain how long-term processes of political change led to actual policy measures: it is the link that may clarify the functioning of a given policy network through four analytical lenses. One concerns the content or substance (the discourse) within the network. The other three concern organization, specifically focusing on the actors, their resources and power, and the rules of the game.

Substance is analyzed in terms of *policy discourse*, which refers to "concepts, ideas, views, *buzzwords* and the like, which give meaning to a

19 James Longhurst, *Bike Battles: A History of Sharing the American Road* (Seattle/London: University of Washington Press, 2015), 12.

20 Scholars studying the development of Dutch nature governance pioneered the Policy Arrangement Approach to governance. There are parallels between nature and cycling governance: both are policy areas dealing with issues that are local in scale, but still subject to rules and processes formed on a provincial, national, and international scale. In addition, local attempts to place nature protection on the political agenda resemble the cycling activists' struggle in many ways: often a relatively small and powerless group was up against large vested interests (the agricultural lobby in the case of nature protection, the car lobby in the case of cycling activism). Several scholars have fruitfully applied this approach to Dutch environmental policy. See: Jan van Tatenhove, Bas Arts, and Pieter Leroy, eds., *Political Modernisation and the Environment: The Renewal of Environmental Policy Arrangements*, Environment & Policy, vol. 24 (Dordrecht: Kluwer, 2000); in particular "Policy Arrangements," 53-69; also Bas Arts and Pieter Leroy, eds., *Institutional Dynamics in Environmental Governance*, Environment & Policy, vol. 47 (Dordrecht: Springer, 2006). On applying the theory: Rikke Arnouts, "Regional Nature Governance in the Netherlands: Four Decades of Governance Modes and Shifts in the Utrechtse Heuvelrug and Midden-Brabant" (PhD diss., Wageningen University, 2010).

21 Arts, Tatenhove, and Leroy, "Policy Arrangements," 54.

policy domain."[22] Often, one discourse will dominate a policy arrangement challenged by competing discourses. These discourses can have varying levels of specificity: "Consistency, elaboration, operationalization, and coherence, however, are not essential features of discourse." On the contrary, "the vaguer a policy discourse ... the more open it is to different interpretations, the greater its mobilizing capability, and the more impressive its consensus-building ability."[23]

Scholars analyze actors through the lens of so-called *policy coalitions*. These consist "of a number of players who share interpretations of a policy discourse or resources, in the context of the rules of the game."[24] *Power and resources* refer to the actors' access to resources that can help them influence political decisions or frame public debates and agendas. Power can also be seen as a more hidden phenomenon that creates a hierarchy of dependent relations among actors. Importantly, knowledge can also serve as a source of power and legitimacy. Finally, the *rules of the game* refer to the more informal, unwritten norms and political culture governing the interaction of stakeholders. Within these policy arrangements, change in time occurs either through conscious actions by actors within the network (agency) or through the influence of external factors beyond their control (structure).

The authors working with the policy arrangements approach prefer to approach the agency-structure issue through Giddens's concept of structuration.[25] This concept tries to balance agency and structure, without giving primacy to either. *Duality of structure* is the idea that actors cannot act outside certain structures, but that they simultaneously shape and (re) produce them. As Marleen Buizer states in her book on local initiatives in nature policy: "Duality of structure means that there is a reciprocal relationship between actor and structure in which actors are neither powerless subjects of structure, nor powerful enough to change structure according to their wishes."[26] Actors within the cycling governance coalition made contingent choices, but also operated in a landscape of (path-dependent) constraints. We need to take a long-term perspective to understand this.

22 Ibid., 56. Italics in original.

23 Ibid., 64.

24 Ibid., 57.

25 Anthony Giddens, *The Constitution of Society: Outline of the Theory of Structuration* (Cambridge: Polity Press, 1984). The debate on agency and structure is fundamental in history. For a collection of essays, see Merritt Roe Smith and Leo Marx, eds., *Does Technology Drive History? The Dilemma of Technological Determinism* (Cambridge, MA/London: MIT Press, 1994).

26 Marleen Buizer, "Worlds Apart: Interactions Between Local Policy Initiatives and Established Policy" (PhD diss., Wageningen University, 2008), 19.

While it is possible to concentrate purely on the survival of urban cycling after 1970, studying the cycling policies of the 1950s and 1960s helps us to understand the structures in which activists and engineers operated in Dutch cities from the 1970s onwards. We cannot, in turn, understand the 1950s and 1960s without investigating the roots of cycling policies in the 1920s. Earlier choices, often contingent and sometimes with unintended consequences, shaped the long-term development of cycling and cycling governance. One concept often invoked to study this process of foreclosure is path dependency.

The long timescale of this project allows me to investigate the long-term effects of certain policy choices, and the possible feedback mechanisms that create path dependent processes. As a policy domain, mobility and spatial planning produce mobility patterns and physical infrastructures like cycle paths which are so deeply embedded that changing them is both difficult and expensive.[27] Initial choices, albeit contingent and small, can in the long run have significant and almost unavoidable consequences.[28] According to historian James Mahoney, path dependency is often used loosely, becoming little more than the truism that "earlier events affect later possibilities and foreclose certain options."[29] For a process to be properly called path-dependent, it has to be very hard to change at a later stage because of earlier choices. Positive feedback processes increase the cost of unmaking earlier decisions.[30]

So, the question is, can we explain the endurance of cycling with such a path-dependent process through a historical analysis? Did the large presence of cyclists in the Netherlands early on in history set in motion a positive cycle when planners provided for cyclists, leading to higher levels of cycling that in turn encouraged more pro-cycling policies?[31] The opposite may also be true. Where cycling levels are low, a vicious circle emerges: planners who believe

27 Anique Hommels, *Unbuilding Cities: Obduracy in Urban Sociotechnical Change* (Cambridge, MA: MIT Press, 2005); Anique Hommels, "Studying Obduracy in the City: Toward a Productive Fusion between Technology Studies and Urban Studies," *Science, Technology, & Human Values* 30, no. 3 (2005): 323-51.

28 The classic example is the QWERTY keyboard: once a contingent choice, changing it now is nearly impossible: Paul A. David, "Clio and the Economics of QWERTY," *American Economic Review* 75, no. 2 (1985): 332-37.

29 Mahoney, "Path Dependence in Historical Sociology," 509.

30 Paul Pierson, *Politics in Time: History, Institutions, and Social Analysis* (Princeton: Princeton University Press, 2004), 21.

31 Martin Emanuel, "Moments of Unsustainability: Planning, Path Dependence, and Cycling in Stockholm," in *Cycling and Recycling: Histories of Sustainable Practices*, eds. Ruth Oldenziel and Helmuth Trischler (New York/Oxford: Berghahn, 2016), 101-21.

cycling is on the decline do not invest in it; cycling subsequently becomes less attractive and the number of cyclists drops, which is used to justify withholding further support. Put differently, a self-reinforcing process in environments with both high and low cycling levels may become so ingrained in existing institutions that it is too costly to change. At the same time, emphasizing the constraints historical factors placed on actors may obscure that there is always room for agency as in Giddens's concept of structuration.

This historical study allows us to test ideas about path dependency. As I will show, the early stages of systematic roadbuilding in the 1920s and 1930s included a strong emphasis on parallel but separate cycling paths and fierce debates about cyclists' rightful place and the relationship between drivers and cyclists. Each group had different standpoints, and different outcomes were possible. As we will see, ultimately, traffic separation was the contingent solution. What path-dependent properties did this choice develop over the following decades? Path dependency theory suggests the need for a close look at the early, formative stage of Dutch roadbuilding policies (a "critical juncture" in theoretical terms), since so many of the later choices and developments could only take place within the limited range of possibilities left by these choices. It is also an open question whether engineers and policymakers were (or felt) compelled to continue catering to cyclists in the 1950s and 1960s, when the facilitation of automobility occupied the political center stage. Is there indeed a path dependency pattern here which made cycling relatively secure in the Netherlands, or was there a real danger of cycling disappearing almost entirely as it did in other European countries at this time?

To close this section on governance, some introductory remarks about the historical traditions and context of Dutch politics are in order. First, the Dutch political system has a decentralized tradition of local autonomy: lower provincial and municipal governments have a significant amount of independence. Twentieth-century national policymakers could not exert a great deal of influence on urban mobility planning. Taxation, which is largely centralized, is one crucial exception. The national government controls the distribution of funding as one of its most consequential means of steering policy at a local level. Financial decentralization only took place in the 1990s when the national government awarded an infrastructure budget to lower governments as a lump sum, which local communities could spend as they saw fit. Until that time, the Ministry of Public Works decided how much money each province received as a road budget. Further division meant that lower government levels could still spend this road budget on cycling facilities. These road budgets were limited and had to be supplemented with occasional subsidies. While the national government took a strong role in the highway network from the early twentieth century, its focus on cycling came and went. When the state paid attention, it generated a significant boost for local governments. As it turns out, historically the structural, long-term, and sustained attention necessary to create good cycling conditions first came from the provinces and municipalities.[32]

A second characterization political scientists and historians often apply to the Netherlands is that of a "consensus-driven, neo-corporatist system."[33] Society is seen to consist of different groups (*corpora*) whom certain people or organizations represent. Distinct from state (or fascist) corporatism, neo-corporatism is "explicitly a democratic model of bottom-up interest representation, hence the adjective 'neo'."[34] Characteristic of this model is that "the State recognizes a limited number of interest associations, involves them in the decision-making and commits them, grants them privileges and delegates the execution of certain public tasks to them."[35] Indeed, as governance scholars Rudy Andeweg and Galen Irwin note: "The existence of strong, well-organized interest groups is an important precondition for a corporatist model of policy-making to work."[36] This recognition is more or less informal, although in the Dutch context it has become common for the government to subsidize action groups – a measure of public recognition. State officials deliberate with these representatives of social groups to reach policy solutions which take into account as many interests as possible.[37]

32 Rudy B. Andeweg and Galen A. Irwin, *Governance and Politics of the Netherlands*, 2nd edition (Basingstoke: Palgrave Macmillan, 2005).

33 Caspar F. van den Berg, "Dynamics in the Dutch Policy Advisory System: Externalization, Politicization and the Legacy of Pillarization," *Policy Sciences* 50, no. 1 (2017): 63-84, here 64; Frits van der Meer, Jos Raadschelders, and Toon Kerkhoff, "Van nachtwakersstaat naar waarborgstaat: proliferatie en vervlechting van het Nederlandse openbaar bestuur in de lange twintigste eeuw (1880-2005)," in *Duizend jaar openbaar bestuur in Nederland: van patrimoniaal bestuur naar waarborgstaat*, eds. Pieter Wagenaar, Toon Kerkhoff, and Mark Rutgers (Bussum: Coutinho, 2011), 221-90.

34 Kickert, *The History of Governance in the Netherlands*, 38.

35 Ibid.

36 Andeweg and Irwin, *Governance and Politics of the Netherlands*, 150.

37 Adriejan van Veen, "From Neo-Corporatism to Regulatory Governance: Interests, Expertise, and Power in Dutch Extraparliamentary Governance, c. 1900-2018," in *Information and Power in History: Towards a Global Approach*, eds. Ida Nijenhuis Nijenhuis, Marijke van Faassen, Ronald Sluijter, Joris Gijsenbergh, and Wim de Jong (London: Routledge, 2020), 85-102; Erik Schrijvers and Stefan Couperus, "Voorbij verkiezing en parlement. Alternatieve representatie in Nederland na 1870," in *Omstreden democratie. Over de problemen van een succesverhaal*, eds. Remieg Aerts and Peter de Goede (Amsterdam: Boom, 2013), 67-88; Ido de Haan, "Verplaatste democratie? Politieke representatie in functionele organen," in *Omstreden democratie*, 89-107.

This search for consensus and compromise is often seen as a hallmark of Dutch political culture. Indeed, the political development of consensus and compromise in the cycling policy domain is analyzed in this book. At the same time, one consequence of this system where state officials and interest groups work closely together is "the fragmentation of policy-making into various 'Iron Triangles', 'policy networks' or 'policy communities'." Relatively isolated policy networks tend to focus "on a particular policy area without much attention to the full range of government activities."[38] Relatively informal arrangements, personal contacts, and unwritten rules are often quite central to these networks.[39]

Third, the consensus-driven nature of Dutch politics is often analyzed as the "polder model."[40] The term refers to the low-lying nation's historical battle against water. The outside threat required people to work together, although the term's applicability to this policy domain is contested.[41] Political science scholarship uses the concept to analyze economic policy and the negotiations between state, employers, and unions about wage and industrial policies. The question of whether the tensions between labor and capital are characterized by consensus-seeking rather than conflict is a hotly contested argument among political scientists. Nevertheless, we can say that "in a broader political science sense, neo-corporatism as State co-operation with interest groups certainly did exist in the Netherlands." This is in line with an age-old tradition of "pragmatism, tolerance and consensus."[42] Accommodating different interests by minutely negotiating policy choices and distributing costs and benefits more or less equally over society are characteristic of Dutch politics.

Regarding the workings of the national state bureaucracy, political scientists describe the Dutch governance system as heavily compartmentalized. Most ministries, in cooperation with society's interest groups, form sectoral policy domains which struggle to communicate with other domains.[43] For infrastructure policy in general, this means that communication between

38 Andeweg and Irwin, *Governance and Politics of the Netherlands*, 158.

39 Rivke Jaffe and Martijn Koster, "The Myth of Formality in the Global North: Informality-as-Innovation in Dutch Governance," *International Journal of Urban and Regional Research* 43, no. 3 (2019): 536-68.

40 A classic work is Arend Lijphart, *The Politics of Accommodation: Pluralism and Democracy in the Netherlands* (Berkeley: University of California Press, 1968).

41 Milja van Tielhof, "Op zoek naar het poldermodel in de waterstaatsgeschiedenis," *Tijdschrift voor geschiedenis* 122, no. 2 (2009): 148-61.

42 Kickert, *The History of Governance in the Netherlands*, 12.

43 Andeweg and Irwin, *Governance and Politics of the Netherlands*, 149-63.

mobility, spatial planning, and recreation domains (all connected to cycling in different ways) did not always occur. In practice, this meant that different ministries governed and funded different types of cycling infrastructure. There are downsides and upsides to this dispersed governance system. Below the national level, the province is generally regarded as far less important than either the national state or the municipality. Both in terms of budget and personnel, the provincial level is quite small compared to the national and local level.[44] That said, the province plays a key role in mobility policy: traffic formed the largest item in the provincial budget until about 1980, and has remained a significant item since. The provinces also maintain their own provincial public work departments. While for many policy fields the province might be more or less ignored, in mobility policy it has had a significant role historically. In the first half of the twentieth century, national policymakers increasingly believed the province or region was the appropriate level for cycling governance – though the national state did consider automobility its mandate. Since the 1950s, provinces indeed took up this task to varying degrees and have remained significant in this domain. Applying these political science insights, I explore the multi-level nature of cycling governance.

Taking Stock of Cycling History

How does cycling feature in historical scholarship? Historians of innovation have extensively studied the governance of (mobility) technologies like the bicycle as the friction between the experts or system-builders on the one hand and users on the other. In the seminal *Making Europe* book series, which analyzes European history of technology, this is one of the key lines of conflict.[45] Historians have discredited the popular narrative of omnipotent inventors who can shape the world exactly as they envision. The manifold

44 Theo A.J. Toonen, "Dutch Provinces and the Struggle for the Meso," in *The Rise of Meso Government in Europe*, ed. L.J. Sharpe (London: Sage, 1993), 117-53; Michiel S. de Vries, "Institutional Fleecing: The Slow Death of Dutch Provinces," *Public Organization Review* 4, no. 4 (2004): 295-315.
45 See first three volumes: Ruth Oldenziel and Mikael Hård, *Consumers, Tinkerers, Rebels: The People Who Shaped Europe*, vol. 1, Making Europe (New York/Basingstoke: Palgrave Macmillan, 2013); Martin Kohlrausch and Helmuth Trischler, *Building Europe on Expertise: Innovators, Organizers, Networkers*, Making Europe, vol. 2 (Basingstoke: Palgrave Macmillan, 2014); Per Högselius, Arne Kaijer, and Erik van der Vleuten, *Europe's Infrastructure Transition: Economy, War, Nature*, Making Europe, vol. 3. (Basingstoke: Palgrave Macmillan, 2015); the other volumes are also important. On the entire series: Eda Kranakis, "Writing Technology into History," *Technology and Culture* 62, no. 1 (2021): 212-40.

ways users shape technological objects and systems are well documented. Innovation sociologists Trevor Pinch and Nelly Oudhoorn's *How Users Matter* is one of the premier works in this tradition.[46] By tinkering with objects, by using them in different ways than intended, or even by not applying certain technologies, technology users can exert their influence. A significant amount of this literature focuses on how users shape the design and construction of particular technical objects like bicycles or cars.[47] Wiebe Bijker famously used the bicycle as an example of his SCOT approach.[48] This book does not focus on the bicycle as a technical object, but on how cycling citizens organized in social movements shaped cycling as a socio-technical system.

Users and non-state actors play a major role in shaping this socio-technical cycling system. Much institutional, legal, and financial power, however, lies with engineers and experts. Historians of technology have extensively studied technocracy – the idea that engineers and experts' supposedly neutral and value-free scientific approach can solve all kinds of (social) problems.[49] There is a consensus that engineers' approach to technology is not neutral: at no stage in the engineering process can we speak of a purely technical approach. According to Michel Callon, engineers should always also be seen as sociologists whose technological work is driven by a certain view of society.[50] Globally, automobility dominated their view of mobility for much of the twentieth century.[51] When the changing cultural climate of

46 Nelly Oudshoorn and Trevor Pinch, eds., *How Users Matter: The Co-Construction of Users and Technologies* (London: MIT Press, 2003).

47 Ronald Kline and Trevor Pinch, "Users as Agents of Technological Change: The Social Construction of the Automobile in the Rural United States," *Technology and Culture* 37, no. 4 (1996): 763-95.

48 Wiebe E. Bijker, *Of Bicycles, Bakelites, and Bulbs: Toward a Theory of Sociotechnical Change* (Cambridge, MA: MIT Press, 1995).

49 On technocracy in roadbuilding, see Johan W. Schot and Vincent Lagendijk, "Technocratic Internationalism in the Interwar Years: Building Europe on Motorways and Electricity Networks," *Journal of Modern European History* 6, no. 2 (2008): 196-217; Dick van Lente and Johan W. Schot, "Techniek als politiek: ingenieurs en de vormgeving van de Nederlandse samenleving," in *Techniek in Nederland in de twintigste eeuw – Deel VII – Techniek en modernisering, balans van de twintigste eeuw*, eds. Adri A. Albert de la Bruhèze, Harry Lintsen, Arie Rip and Johan Schot (Zutphen: Walburg Pers/Stichting Historie der Techniek, 2003). On the extensive European network of road engineers and planners, see Frank Schipper, "Driving Europe: Building Europe on Roads in the Twentieth Century" (PhD diss., TU Eindhoven, 2008).

50 Michel Callon, "Society in the Making: The Study of Technology as a Tool for Sociological Analysis," in *The Social Construction of Technological Systems: New Directions in the Sociology and History of Technology*, eds. Wiebe E. Bijker, Thomas P. Hughes, and Trevor Pinch (Cambridge, MA: MIT Press, 1987), 83-103.

51 On conflicts between users and system-builders, see Peter Norton, *Fighting Traffic: The Dawn of the Motor Age in the American City* (Cambridge, MA: MIT Press, 2008).

the late 1960s and 1970s made cyclists more daring in voicing their demands, the system-builders had to acknowledge the gap between their vision of mobility and that of a large group of users. How did Dutch cyclists manage to address this knowledge gap between engineering models and daily practice?

Cycling Pathways studies the clash between road engineers and road users, specifically cyclists. Most traffic experts simply knew very little about how cyclists behaved, why they made certain (route and mode) choices, and had no insight into how they could improve cycling conditions and safety. Historian David Arnold has called traffic a manifestation of the "everyday state."[52] Writing on British colonial policies in India, Arnold notes that "the world of urban planning was often far removed from the reality of life on the streets and the fractured modernity it represented."[53] In the Dutch context, particularly since the 1960s, when critique of technocratic governance resonated and a more egalitarian social structure emerged – known by political scientists as the era of de-pillarization – this allowed lay persons like cyclists to influence urban planners and engineers. In other national contexts where this was not possible, cyclists could still use the bicycle in subversive ways to evade government control but did not have the power to change the planning and design of streets and road space.[54]

Until a few decades ago, historians writing on transport focused on one modality in one specific country, often centered around trains, cars, and airplanes; they dealt with the inventors and manufacturers of these machines and were less concerned with how these transport technologies shaped people's lives and the political processes involved. Since then, the focus has shifted when new approaches to technology (SCOT, Actor-Network Theory, and others) made an inroad into transport history – as did the mobility studies by John Urry and Mimi Sheller.[55] This led mobility historian Gijs Mom to call for a move from this traditional transport history to a mobility history, or an "integrated transport history," which examines the interaction between mobility modes and applies new methodological approaches.[56]

52 David Arnold, "The Problem of Traffic: The Street-Life of Modernity in Late-Colonial India," *Modern Asian Studies* 46, no. 1 (2012): 120.

53 Ibid., 130.

54 David Arnold and Erich DeWald, "Cycles of Empowerment? The Bicycle and Everyday Technology in Colonial India and Vietnam," *Comparative Studies in Society and History* 53, no. 4 (2011): 971-96.

55 Mimi Sheller and John Urry, "The New Mobilities Paradigm," *Environment and Planning A: Economy and Space* 38, no. 2 (2006): 207-26; Peter Norton, "Urban Transport and Mobility," *Technology and Culture* 61, no. 4 (2020): 1197-211.

56 State of the art in transport history: Gijs Mom, "What Kind of Transport History Did We Get?: Half a Century of JTH and the Future of the Field," *The Journal of Transport History* 24, no. 2 (2003):

The approach opened up the possibility of studying non-motorized forms of transport like walking and cycling that scholars had ignored. Indeed, marginalized mobility groups like cyclists and pedestrians have received a boost in recent years, not least because of the rising interest in transport or mobility justice issues.[57]

In 2008, Gijs Mom and Ruud Filarski wrote an authoritative and thoroughly documented history of Dutch mobility. Yet the authors still focused on car use and the railroads, the development of their infrastructures, and the conflict between motorized and public transport in the Interbellum.[58] While acknowledging the strong role of "slow modes" of transport in the Netherlands, they did not analyze their emergence.[59] They characterized the emergence of the Dutch automobile system in the first half of the twentieth century as the outcome of an interplay of multiple actors with different goals and visions rather than a systematically planned effort. Mom and Filarski also argued that the car did not oust other forms of mobility in the Netherlands like in the US, where automobility took over pedestrianism, cycling, and public transit. The Dutch government did support railway

121-38; Gijs Mom, "The Crisis of Transport History: A Critique, and a Vista," *Mobility in History* 6, no. 1 (2015): 7-19. For the debate on what a new mobility history should entail: Kevin Hannam, Mimi Sheller, and John Urry, "Editorial: Mobilities, Immobilities and Moorings," *Mobilities* 1, no. 1 (2006): 1-22; Colin Michael Divall, Peter Lyth, and Gijs Mom, "Towards a Paradigm Shift? A Decade of Transport and Mobility History," in *Mobility in History: The State of the Art in the History of Transport, Traffic and Mobility*, eds. Gijs Mom, Gordon Pirie, and Laurent Tissot (Neuchatel: Alphil, 2009), 13-40; Colin Divall, "Mobilizing the History of Technology," *Technology and Culture* 51, no. 4 (2010): 938-60; Peter Merriman, "Mobilities, Crises, and Turns: Some Comments on Dissensus, Comparative Studies, and Spatial Histories," *Mobility in History* 6, no. 1 (2015): 20-34; Gijs Mom, "Mao or Merriman? On Pitjantjatjara and Other Mobilities – A Response," *Mobility in History* 6, no. 1 (2015): 35-39.

57 Karel Martens, *Transport Justice: Designing fair transportation systems* (London: Routledge, 2016); Aaron Golub, Melody L. Hoffmann, Adonia E. Lugo, and Gerardo F. Sandoval, *Bicycle Justice and Urban Transformation: Biking for all?* (London/New York: Routledge, 2017); Mimi Sheller, *Mobility Justice: The Politics of Movement in an Age of Extremes* (London: Verso, 2018); Mimi Sheller, "Theorising Mobility Justice," *Tempo Social* 30, no. 2 (2018): 17-34.

58 Gijs Mom and Ruud Filarski, *Van transport naar mobiliteit: de mobiliteitsexplosie [1895-2005]* (Zutphen: Walburg Pers /Stichting Historie der Techniek, 2008); Ruud Filarski and Gijs Mom, *Van transport naar mobiliteit: de transportrevolutie [1800-1900]* (Zutphen: Walburg Pers /Stichting Historie der Techniek, 2008); Ruud Filarski, "The Emergence of the Bus Industry: Dutch Transport Policy During the Interwar Years," *Transfers* 1, no. 2 (2011): 61-82. As follow-up, a large international research network studied car mobility and truck transport vs public transportation networks and the coordination problems for national governments in the Interbellum. The bicycle did not have a prominent place, despite its importance: Bert Toussaint, "Using the Usable Past: Reflections and Practices in the Netherlands," in *Transport Policy: Learning Lessons from History*, eds. Colin Divall, Julian Hine, and Colin Pooley (Farnham: Ashgate, 2016), 21-22.

59 Mom and Filarski, *De mobiliteitsexplosie*, 397.

transport and inland navigation, which otherwise might not have survived.[60] Does this argument also apply to cycling?

A key point of contention concerned the issue of governance: who was responsible or had authority over parts of the system? This question also recurs in scholarship on the history of spatial and urban planning in the Netherlands. In the 1920s, planners claimed the region was the appropriate scale for planning. Municipalities, however, jealously guarded their autonomy and did not work with their neighbors. The national government was reluctant to legally enforce this regional cooperation.[61] According to planning historians Andreas Faludi and Arnold van der Valk, a measure of uniformity existed in Dutch planning, despite the absence of top-down hierarchical governance: "rule and order is not imposed from above, it pervades the Dutch way of doing things."[62] They argue that ideas and informal documents are as important as the legal status of official plans.[63] Thus, it is crucial to heed the planning discourse. Extending this analysis to cycling means asking how engineers discussed cycling and its place in governance. In other words: who was regarded as responsible for governing cycling and who was regulating or facilitating cycling? Was this lack of a centralized, top-down approach also typical of cycling? And if so, can this uniformity be created through shared policy beliefs?

Cycling Pathways is not the first work to discuss cycling history. Although a relatively new field, growing numbers of studies are emerging in the past decade. The existing cycling history gravitates towards cultural approaches (cycling as a form of leisure and sociability) and focuses on the late nineteenth and early twentieth century.[64] This might have to do with an "elitist bias" in the sources, as when cycling becomes a mundane, everyday activity, it is harder to find traces in archives and published sources.[65]

60 Ibid., 390-93.

61 Hans van der Cammen and Len de Klerk, *Ruimtelijke ordening: van grachtengordel tot Vinex-wijk* (Utrecht: Spectrum, 2003), 118-19.

62 Andreas Faludi and Arnold van der Valk, *Rule and Order: Dutch Planning Doctrine in the Twentieth Century* (Dordrecht: Springer, 1994), 8.

63 Ibid., 61.

64 For overviews, Harry Oosterhuis and Manuel Stoffers, "'Ons populairste vervoersmiddel.' De Nederlandse fietshistoriografie in internationaal perspectief," *BMGN – Low Countries Historical Review* 124, no. 3 (2009): 390-418; Harry Oosterhuis, "Bicycle Research between Bicycle Policies and Bicycle Culture," *Mobility in History* 5 (2014): 20-36; Harry Oosterhuis, "Ingebakken gewoonte of buitenissige liefhebberij?" *Sociologie* 11, no. 1 (2015): 3-30; Harry Oosterhuis, "Cycling, Modernity and National Culture," *Social History* 41, no. 3 (2016): 233-48.

65 Oosterhuis, "Cycling, Modernity and National Culture," 235. For an interesting case on cycle paths for workers in north Sweden's forests, see: Anna-Maria Rautio and Lars Östlund,

For certain countries, notably Italy, France, Spain, and Belgium, cycling historiography extends beyond this early focus but then concentrates on vibrant professional cycling cultures (related to national identity) without discussing the long history of everyday utility cycling.[66] As sociologists Colin Pooley, Jean Turnbull, and Mags Adams argue in their book on everyday twentieth-century mobility in the UK, this is unjustified. Everyday mobility might be a mundane activity, it is also a highly important one given the amount of time we devote to it, its frequency, and what it costs the environment (pollution), the economy (congestion), the government (costs of policies), society and culture (stress, alienation, lack of leisure), and individuals (time, health).[67] Cycling has been a key utilitarian mode of transport since at least the 1920s and requires the long view which I provide here. The rise, fall, and resurgence of cycling is a century-long story. Unlike most studies of cycling history, this book takes a perspective that stretches from the late nineteenth century to the present.

Scholars discussing utilitarian cycling in the twentieth century emphasize its marginalization by policymakers favoring the car. They demonstrate the

"'Starvation Strings' and the Public Good: Development of a Swedish Bike Trail Network in the Early Twentieth Century," *Journal of Transport History* 33, no. 1 (2012): 42-63.

66 For France: Hugh Dauncey, *French Cycling: A Social and Cultural History*, (Liverpool: Liverpool University Press, 2012); Hugh Dauncey and Geoff Hare, eds., *The Tour de France 1903-2003: A Century of Sporting Structures, Meanings and Values* (London: Routledge/Cass, 2003); Hugh Dauncey and Geoff Hare, "The Tour de France: a Pre-Modern Contest in a Post-Modern Context," *The International Journal of the History of Sport* 20, no. 2 (2003): 1-29, and Philippe Gaboriau, "The Tour de France and cycling's Belle Epoque," 57-78. Another publication by Gaboriau pays utilitarian cycling some attention: Philippe Gaboriau, "Les trois âges du vélo en France," *Vingtième Siècle. Revue d'histoire* 29, no. 1 (1991): 17-34. For Belgium: Stijn Knuts and Pascal Delheye, "Identiteiten in koers: Roeselaarse wielrenners als kopmannen van lokale, regionale en (sub) nationale identiteiten, 1900-1960," *Belgisch Tijdschrift voor Nieuwste Geschiedenis/ Revue Belge de Histoire Contemporaine* 41, no. 1-2 (2011): 167-214; Stijn Knuts and Pascal Delheye, "Cycling in the City? Belgian Cyclists Conquering Urban Spaces, 1860-1900," *International Journal of the History of Sport* 29, no. 14 (2012): 1942-62; Stijn Knuts and Pascal Delheye, "Connecting City and Countryside? Faces of Cycling Mobility in Belgium, 1890-1914," *Dutch Crossing* 37, no. 3 (2013): 240-59. For Italy: Stefano Pivato, "The Bicycle as a Political Symbol, Italy, 1885-1955," *International Journal of the History of Sport* 7, no. 2 (1990): 173-87; Stefano Pivato, "Italian Cycling and the Creation of a Catholic Hero. The Bartali Myth," *International Journal of the History of Sport* 13, no. 1 (1996): 128-38. For Spain: Bernat López, "Sport, Media, Politics and Nationalism on the Eve of the Spanish Civil War: The First Vuelta Ciclista a España (1935)," *The International Journal of the History of Sport* 27, no. 4 (2010): 635-57; Bernat López, "The Failed Vuelta Ciclista a España of 1913 and the Launching of the Volta a Catalunya (1911-1913): Centre Versus Periphery in the Struggle for the Governance of Cycling in Early Twentieth-Century Spain," *Sport in History* 30, no. 4 (2010): 547-69.

67 Colin Pooley, Jean Turnbull, and Mags Adams, *A Mobile Century? Changes in Everyday Mobility in Britain in the Twentieth Century* (Aldershot: Ashgate, 2005), 9-10.

various national contests with regard to political traditions and engineering policies. Studies on cycling in the US show this marginalization in convincing ways.[68] Historians like Peter Norton in *Fighting Traffic*, James Longhurst in *Bike Battles*, or Zack Furness in *One Less Car* show how American traffic policy forcefully pushed cycling aside – a clear contrast with the more accommodating and consensus-seeking approach to road space distribution in the Netherlands. Germany chose the middle path, establishing multiple initiatives around cycling path construction in the first half of the twentieth century, but not encouraging it to coalesce into a national culture as it did in the Netherlands.[69] Denmark's political traditions and cycling culture seem closest to those of the Netherlands, but a national account beyond Copenhagen is still lacking.[70] For other countries, publications on cycling history are few and

68 On Canada and The United States: Glen Norcliffe, *The Ride to Modernity: The Bicycle in Canada, 1869-1900* (Toronto: University of Toronto Press, 2001); Norton, *Fighting Traffic*; Zack Furness, *One Less Car: Bicycling and the Politics of Automobility* (Philadelphia: Temple University Press, 2010); Christopher W. Wells, *Car Country: An Environmental History* (Seattle: University of Washington Press, 2012); Lorenz Finison, *Boston's Cycling Craze, 1880-1900: A Story of Race, Sport, and Society* (Amherst: University of Massachusetts Press, 2014); Longhurst, *Bike Battles*; Evan Friss, *The Cycling City: Bicycles and Urban America in the 1890s* (Chicago: University of Chicago Press, 2015); Evan Friss, *On Bicycles: A 200-year History of Cycling in New York City* (New York: Columbia University Press, 2019). Longhurst covers American cycling, which he argues was resilient enough to survive the strongly car-centred policies of the 1900s. See also Bruce Epperson, "The Great Schism: Federal Bicycle Safety Regulation and the Unraveling of American Bicycle Planning," *Transportation Law Journal* 37, no. 2 (2010): 73-118; Bruce Epperson, *Bicycles in American Highway Planning: The Critical Years of Policy-Making 1969-1991* (Jefferson, NC: McFarland & Company, 2014).

69 Verein Deutscher Fahrrad-Industrieller, *Festschrift zum Vierzigjährigen Bestehen des Vereins Deutscher Fahrrad-Industrieller e.V. 1888-1928* (Berlin: Verein Deutscher Fahrrad-Industrieller, 1928), 88-98; Burkhard Horn, "Vom Niedergang eines Massenverkehrsmittels – Zur Geschichte der Städtischen Radverkehrsplanung" (Master Thesis, Gesamthochschule Kassel, 1990); Volker Briese, *Besondere Wege für Radfahrer. Zur Geschichte des Radwegebaus in Deutschland von den Anfangen bis 1940* (Unpublished manuscript, Paderborn, 1993); Thomas Fläschner, "Stahlroß auf dem Aussterbe-Etat: zur Geschichte des Fahrrades und seiner Verdrängung in den 50er-Jahren," *Eckstein – Journal für Geschichte* 9 (2000): 4-22; Anne-Katrin Ebert, *Radelnde Nationen: Die Geschichte des Fahrrads in Deutschland und den Niederlanden bis 1940* (Frankfurt: Campus Verlag, 2010); Thomas Fläschner, "Zum Gebrauch des Rades gezwungen: Fahrradkultur in Saarbrücken zwischen 1885 und 1939," in *Saarbrücken in Fahrt: 125 Jahre Automobil an der Saar*, eds. Hans-Christian Herrmann and Ruth Bauer, Veröffentlichungen des Stadtarchivs Saarbrücken, vol. 1 (Marpingen-Alsweiler: Edition Schaumberg, 2011), 184-221. I am very grateful to Thomas Fläschner and Burkhard Horn for sharing their work.

70 Trine Agervig Carstensen et al., "The Spatio-Temporal Development of Copenhagen's Bicycle Infrastructure, 1912-2013," *Geografisk Tidsskrift-Danish Journal of Geography* 115, no. 2 (2015): 142-56; Martin Emanuel, "Making a Bicycle City: Infrastructure and Cycling in Copenhagen since 1880," *Urban History* 46, no. 3 (2019): 493-517; Walther Knudsen and Thomas Krag, *På cykel i 100 år – Dansk Cyklist Forbund 1905-2005* (Copenhagen: Dansk Cyklist Forbund, 2005); Marie

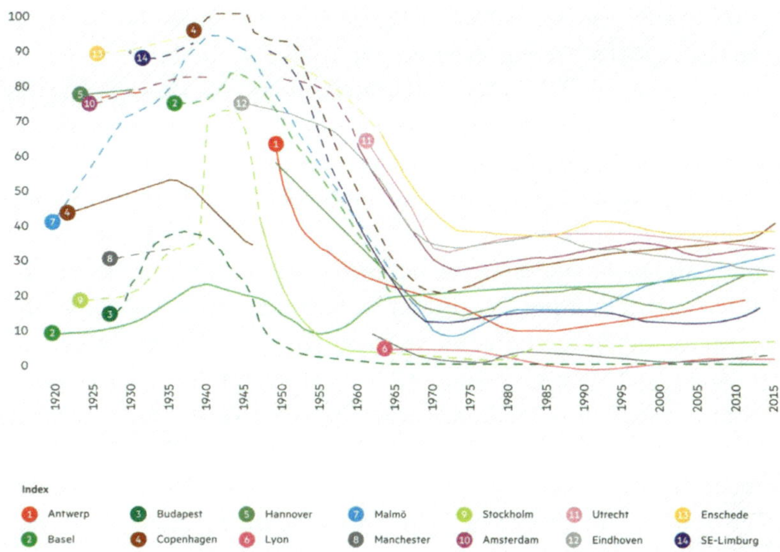

Figure 1 Cycling trends as part of the modal split in fourteen European cities/areas from 1920 to 2015, in percentages. By Frank Veraart in Cycling Cities: The European Experience, *13.*

far between.[71] Most scholarship explains why once thriving cycling cultures disappeared in the mid-1900s. The belief in the modernity of driving and the outdatedness of cycling typically explain why policymakers neglected cyclists. Yet the larger question is why cycling levels, while following international trends, remained so much higher in the Netherlands, even when automobility levels eventually became as high there as elsewhere (see figure 1).

Explaining the historical development of cycling is a complex affair involving multiple causal factors. Rather than providing simplistic mono-causal explanations, as the American bicycle historian Bruce Epperson has emphasized, "clear cause-and-effect-explanations have proven elusive" and there is more to explaining cycling levels historically than one might think.[72] Simple explanations pointing to the (flat) morphology of the Dutch

Kåstrup, "Hverdagens beskedne demokrati – Analyser af cyklen som symbol på danskhed" (Master thesis, University of Copenhagen, 2007).

71 On Ireland: Erika Hanna, "Seeing Like a Cyclist: Visibility and Mobility in Modern Dublin, c. 1930-1980," *Urban History* 42, no. 2 (2015): 273-89. On Sweden: Rautio and Östlund, "'Starvation Strings' and the Public Good." For Finland: Tiina Männistö-Funk, "The Crossroads of Technology and Tradition: Vernacular Bicycles in Rural Finland, 1880-1910," *Technology and Culture* 52, no. 4 (2011): 733-56.

72 Epperson, *Bicycles in American Highway Planning*, 59. Also arguing that geographical conditions only partly explain cycling levels, Trine Agervig Carstensen and Anne-Katrin Ebert,

landscape and the shorter distances traveled are not sufficient either: even between Dutch cities, cycling levels vary considerably. In their review of the historiography of Dutch cycling, historians Harry Oosterhuis and Manuel Stoffers note the cycling scholarship's suggestions that policy choices made decades earlier often explain differences in present cycling levels. Additionally, the interaction between policy and cultural perceptions of cycling are key explanatory factors.[73]

The first and most rigorous explanation of why cycling levels diverged so much internationally after very similar rates in the early 1900s, comes from the single most important source on which my project builds: Adri Albert de la Bruhèze and Frank Veraart's pioneering work on cycling history, *Fietsverkeer in praktijk en beleid in de twintigste eeuw* ("Cycling Traffic in Practice and Policy in the Twentieth Century," published in 1999). They study the historical evolution of cycling levels and practices in nine European cities over a century, to explain the high Dutch level of cycling: Amsterdam, Eindhoven, Enschede, and the South-East-Limburg agglomeration within the Netherlands, as well as Antwerp, Manchester, Copenhagen, Hannover, and Basel. Through the painstaking collection of statistical data and other qualitative methods, the authors reveal very high cycling levels until roughly the 1950s, when they dropped sharply. In some countries cycling almost disappeared, while in others such as the Netherlands, cycling levels were still respectable even at their lowest point, around 1970: some 20-25 percent in most Dutch cities. Since the 1970s, cycling levels have stabilized or even risen again.

According to Albert de la Bruhèze and Veraart, there are four major causal factors: (1) spatial structure: suburbanization and increased commuting shape people's mobility options. Increasing distances to work limit the potential for cycling if public transit is lacking and cars are the only alternative. In countries with good public transit, these long distances can still be negotiated using bike-train combinations, though this requires investment in bicycle parking facilities at stations; (2) availability of mobility alternatives such as public transit and the car: wage growth influences car ownership and use, as does car tax and the status of car ownership; (3) cycling's place in traffic policy: to what extent authorities are interested in cycling at all, and if so, whether and how they promote it; and (4) cultural perceptions of the bicycle: important in that cycling and driving have different connotations: if cycling is perceived as low-status or old-fashioned, this will limit the

"Cycling Cultures in Northern Europe: From 'Golden Age' to 'Renaissance'," in *Cycling and Sustainability*, ed. John Parkin (Bingley: Emerald, 2012), 23-58.

73 Oosterhuis and Stoffers, "'Ons populairste vervoersmiddel'," 400.

amount of people willing to cycle, as well as authorities' support. This is what happened historically in many Western countries and is now happening across the globe in growing economies where the car is a sign of status and progress.[74] A recently added fifth actor is the role of social movements in demanding more space for cyclists and their (Dutch) success in working with the government to achieve this. As widespread as social unrest over car-centric cities was in the 1970s, relevant research is still in its infancy. Harald Engler notes in a recent article on German protest movements "resistance towards the car-friendly city ... is still under-researched."[75]

The addition of social movements as a factor came from a major new impetus in cycling history research when a wider group of researchers led by Ruth Oldenziel took up Albert de la Bruhèze and Veraart's work and worked with them to create a new, updated, and expanded publication in English: *Cycling Cities: The European Experience* (2016). Along with the extended explanatory framework, new cities were added: Utrecht, Enschede, Malmö, Stockholm, Budapest, and Lyon.[76] Since then, more cities have shown an interest in their own cycling history and commissioned further books. To date these include Rotterdam, The Hague, Arnhem/Nijmegen, and Maastricht, as well as a growing number beyond the Netherlands, such as Munich and Johannesburg.[77] Together, these researchers have done invaluable work exploring neglected forms of everyday mobility through

74 Ruth Oldenziel, "Mode of the Past or Promise for the Future? Cycling in China and the Sustainability Challenge, 1955-Present," *Zeithistorische Forschungen/Studies in Contemporary History* 14, no. 3 (2017): 465-86. In a colonial context, cycling had a subversive potential as low-technology transport mode that escaped regulations. See Arnold and DeWald, "Cycles of Empowerment?"

75 Harald Engler, "Social Movement and the Failure of Car-friendly City Projects: East and West Berlin (1970s and 1980s)," *Journal of Transport History* 41, no. 3 (2020): 353-380, here 354. For German resistance to car-centric cities: Astrid Mignon Kirchhof, "For a Decent Quality of Life": Environmental Groups in East and West Berlin," *Journal of Urban History* 41, no. 4 (2015): 625-46; Annika Levels, "(Re-)claiming Urban Streets: The Conflicting (Auto)mobilities of Cycling and Driving in Berlin and New York," *The Journal of Transport History* 41, no. 3 (2020): 381-401.

76 Oldenziel et al., *Cycling Cities: The European Experience.*

77 Eric Berkers and Ruth Oldenziel, *Cycling Cities: The Arnhem and Nijmegen Experience* (Eindhoven: Foundation for the History of Technology, 2017); Eric Berkers, Frans Botma, and Ruth Oldenziel, *Cycling Cities: The Hague Experience* (Eindhoven: Foundation for the History of Technology, 2018); Adri A. Albert de la Bruhèze and Ruth Oldenziel, *Cycling Cities: The Munich Experience* (Eindhoven: Foundation for the History of Technology, 2018); Erik Berkers, Frank Schipper, Patrick Bek and Ruth Oldenziel, *Cycling Cities: The Rotterdam Experience* (Eindhoven: Foundation for the History of Technology, 2019); Njogu Morgan, *Cycling Cities: The Johannesburg Experience* (Eindhoven: Foundation for the History of Technology, 2019); Eric Berkers, "Fietsgebruik en -beleid in Maastricht en Parkstad in historisch perspectief," (Eindhoven: TU Eindhoven/Stichting Historie der Techniek, 2017).

comparative analysis across cities worldwide. The focus is on cities – and as this is a relatively new field of research, many questions remain.

Taking a national approach, my research asks, to what extent has Dutch cycling governance since 1920 contributed to the success of the Netherlands as a cycling country? Building on the governance and politics of cycling scholarship, which takes the city as its point of departure, I also investigate how policymakers at provincial and national levels have influenced cycling policies in the Netherlands. Historian Anne-Katrin Ebert also identified this gap in her review of *Cycling Cities,* stating that "the historical intertwining of urban and national policies is not investigated in any systematic way in this volume."[78] I respond to Ebert's call by investigating the intertwining of this multi-level dynamic of governance. I also take into account the important role of non-state actors in shaping policy. Like *Cycling Cities*, but unlike many other studies of cycling history, I study a timescale from the early 1900s to the present. My goal is to elucidate the dynamics between different governance levels and thus enrich the picture these authors sketched of cycling governance at the city level. My work on the intricacies of Dutch cycling governance examines two of the *Cycling Cities* project's explanatory factors: cycling's place in traffic policy and the role of social movements. The concept cycling governance captures both. Given the important historical role of social movements in Dutch cycling history, I consider these factors together rather than in isolation. When relevant for my analysis, I discuss other explanatory factors in the *Cycling Cities* comparative model. Engineers, policymakers, and politicians' cultural perception of cycling partly explains their political choices.[79] Traffic policies also include other transport modes linked to cycling policies. Finally, spatial planning is closely related to traffic planning and will feature in certain case studies.[80] There is still much work to do, and this book can only cover part of that ground.[81]

78 Anne-Katrin Ebert, "Review of: Ruth Oldenziel, Martin Emanuel, Adri Albert de la Bruhèze, Frank Veraart eds., *Cycling Cities: The European Experience*," *The Journal of Transport History* 39, no. 1 (2018): 128-30.

79 A good example is Martin Emanuel's "Constructing the Cyclist: Ideology and Representations in Urban Traffic Planning in Stockholm, 1930-70," *The Journal of Transport History* 33, no. 1 (2012): 67-91.

80 On the integration of spatial and transport planning: Gijs Mom, "The Emancipation of the Urban View: Dutch Spatial Planning in an International Context (1920-1950)," in *Builders and Planners: A History of Land-Use and Infrastructure Planning in the Netherlands*, eds. Jos Arts, et al. (Delft: Uitgeverij Eburon, 2016), 73-137; Gijs Mom, "The West and the Rest: The Green Heart and the Breakthrough of Spatial Planning (1950-1982)," in *Builders and Planners*, 143-99.

81 See forthcoming books by Patrick Bek and Jan Ploeger, who both provide crucial new perspectives on commuter patterns and the spatial aspects of mobility.

Other Dutch cycling works deserve a mention in this introduction. Particularly important are Albert de la Bruhèze and Oldenziel's articles on Dutch bicycle tax and cycling path construction.[82] They apply the contextualized approach also used here, by looking at cycling policy in the context of mobility policy and the governance of other modes of transport, through comparisons with car policies. Crucial in the bicycle tax debate was the alleged status of the bicycle as a luxury good, or the very definition of what constituted a public good, as in the American debate on bicycle taxes around 1900 that James Longhurst has rescued from oblivion.[83] For Longhurst, these debates raging continuously, from the introduction of the bicycle in the late 1860s up to the present, revolved around seeing the street and road space as an "exhaustible resource" and should be considered "attempts to support or delegitimize competing interest-group claims to an exhaustible resource."[84] In the Dutch case, Albert de la Bruhèze and Oldenziel emphasize the class dimensions of this struggle. Upper-class drivers literally marginalized the working-class cyclist.[85] Cyclists – of all classes and types – were forced to pay a bicycle tax that was used to build roads (for a small elite) on which these cyclists were no longer welcome.

This analysis is closely linked to the Foucauldian analysis of cycling by sociologist Jennifer Bonham and Peter Cox, who consider cycling paths a way to marginalize cyclists. These paths force cyclists to squeeze into a mobility system that is planned from the driver's perspective and only grudgingly gives in to other road users' demands.[86] Albert de la Bruhèze and Oldenziel emphasize how traffic separation favored car users in the interwar period when cyclists were dominant and cars in the minority, but nevertheless, "policymakers, traffic engineers, and urban planners were all

82 Adri A. Albert de la Bruhèze and Ruth Oldenziel, "Who Pays, Who Benefits? Bicycle Taxes as Policy Tool, 1890-2012," in *Cycling and Recycling: Histories of Sustainable Practices* eds. Ruth Oldenziel and Helmuth Trischler (New York/London: Berghahn, 2015), 73-100. Ruth Oldenziel and Adri A. Albert de la Bruhèze, "Contested Spaces: Bicycle Lanes in Urban Europe, 1900-1995," *Transfers* 1, no. 2 (2011): 29-49. On the Dutch bicycle tax between 1926 and 1941, see F.H.M. Grapperhaus, *Over de loden last van het koperen fietsplaatje. De Nederlandse rijwielbelasting 1924-1941* (Franeker: Uitgeverij Van Wijnen, 2006); Anja Lelieveld, "Het plaatje met of zonder gaatje," *De oude fiets* 1992, no. 2 (1992): 1-8; Otto Beaujon, "De Franse fietsbelasting," *De oude fiets* 1997, no. 3 (1997): 4-6.

83 James Longhurst, "The Sidepath Not Taken: Bicycles, Taxes, and the Rhetoric of the Public Good in the 1890s," *Journal of Policy History* 25, no. 4 (2013): 557-86; Longhurst, *Bike Battles*.

84 Longhurst, *Bike Battles*, 20.

85 Albert de la Bruhèze and Oldenziel, "Who Pays, Who Benefits?," 74-75.

86 Jennifer Bonham and Peter Cox, "The Disruptive Traveller? A Foucauldian Analysis of Cycleways," *Road and Transport Research* 19, no. 2 (2010): 42-54.

convinced, even though bicycle use was booming in most cities, that cars would inevitably be the dominant mode of transport in the future."[87] The analysis of marginalized non-motorized road users is also studied historically from the pedestrian perspective.[88]

To explain cycling's unique position in the Netherlands, historian Ebert, who compared Dutch and German cycling history in her important study *Radelnde Nationen*, points to the role of the tourist organization ANWB (*Algemene Nederlandsche Wielrijders-Bond* – "Dutch Cyclists' Association"), an association that started as a cyclists' club in 1883, before becoming an interest group for both car drivers and cyclists around 1920, then evolving into a defender of drivers in the present day. She argues that because the ANWB was the only cycling organization in the Netherlands, it could convincingly claim to represent all cyclists. This gave it the political clout that the divided German organizations never achieved. During the formative period of Dutch roadbuilding, in the 1920s and 1930s, the organization strongly advocated traffic separation as being in the interest of both drivers and cyclists. According to Ebert, the existence of the bicycle tax was a key bargaining chip in these debates, and the powerful ANWB used it to claim investments in cycling paths.[89] This important argument will be discussed in Chapter 2.

Cycling's cultural status is also important in *Cycling Cities*, yet strikingly, while cycling is seen as typically Dutch, cultural-historical studies of this phenomenon are still lacking. The only major contribution is an article by sociologist Giselinde Kuipers, who coined the phrase "inconspicuous

87 Oldenziel and Albert de la Bruhèze, "Contested Spaces," 37.

88 For an introduction to the topic, Colin Pooley et al., "Introduction: historical perspectives on pedestrians and the city," *Urban History*, 1-7. For case studies: Martin Emanuel, "Controlling Walking in Stockholm during the Inter-War Period," *Urban History* 48, no. 2 (2021): 248-65; Peter Norton, "Persistent Pedestrianism: Urban Walking in Motor Age America, 1920s-1960s," *Urban History* 48, no. 2 (2021): 266-89; Peter Norton, "Street Rivals: Jaywalking and the Invention of the Motor Age Street," *Technology and Culture* 48, no. 2 (2007): 331-59; Joe Moran, "Crossing the Road in Britain, 1931-1976," *The Historical Journal* 49, no. 2 (2006): 477-96; Muhammad M. Ishaque and Robert B. Noland, "Making Roads Safe for Pedestrians or Keeping Them Out of the Way?," *The Journal of Transport History* 27, no. 1 (2006): 115-37. See also Martin Emanuel, Frank Schipper, and Ruth Oldenziel, eds., *A U-Turn to the Future: Sustainable Urban Mobility since 1850* (New York: Berghahn Books, 2020).

89 Anne-Katrin Ebert, "When Cycling Gets Political: Building Cycling Paths in Germany and the Netherlands, 1910-40," *Journal of Transport History* 33, no. 1 (2012): 123-26; Frank Veraart, "Fietspaden. Van gerieflijke paden tot fietsbeleid," *Verkeerskunde* 60, no. 5 (2009): 23-24. On the ANWB's involvement in cycling path construction, see Hans Buiter, "Als een spin in het web: de ANWB en de Nederlandse rijwielpadenverenigingen," *Tijdschrift voor historische geografie* 1, no. 2 (2016): 76-86.

consumption" to explain why the Dutch did not immediately adopt higher-status cars when these became affordable.[90] By continuing to cycle, Dutch citizens, so goes the argument, displayed qualities like being down to earth and modest, which are valued more than conspicuous displays of wealth. This is also why the Dutch royal family and leading politicians like to have their photo taken on bicycles. I will not go into detail here, but it will become clear how deeply ingrained cycling was and is in Dutch mobility practices and cultural representations. Cycling's relationship with national identity surely deserves further research. Similarly, whether people have mobility alternatives affects cycling levels – like the late diffusion of car ownership in the Netherlands, coupled with limited public transit. These factors will only feature tangentially in my research.

My research contributes to cycling historiography by furthering our understanding of the long-term development of cycling governance, that is, the way state actors and non-state actors like user advocacy groups and social movements shaped cycling infrastructure and traffic policies in the Netherlands. Uniquely, I adopt a multi-level rather than an urban perspective. Albert de la Bruhèze and Veraart have already acknowledged the role of traffic policy. Their project was – by design – limited to urban governance. This implies that urban cycling, a successful local form of transport, is also governed locally. Particularly in a small, densely networked country like the Netherlands, the question is whether regional or even national governance has also shaped cycling. Investigating these national and provincial policymakers and engineers presents a new multi-level perspective on Dutch cycling governance. National government's control of transport budgets is only one example of why it matters to study higher government levels. In addition, *Cycling Cities*' 2016 edition added the major claim that social movements were crucial in putting urban cycling on the agenda. There is hardly any scholarship on the origins and methods of these movements, nor their relations with the Dutch government. Scholars focus on other forms of Dutch activism but mobility- and cycling-related activism have not been given their due.

Unlike most studies (except *Cycling Cities*), the present work broadens the chronological scope to long-term developments. While many studies only focus on the late nineteenth century or cycling's revival since the 1970s,

90 Giselinde Kuipers, "De fiets van Hare Majesteit: Over nationale habitus en sociologische vergelijking," *Sociologie* 6, no. 3 (2010): 3-26; Giselinde Kuipers, "The Rise and Decline of National Habitus: Dutch Cycling Culture and the Shaping of National Similarity," *European Journal of Social Theory* 16, no. 1 (2013): 17-35.

this study looks at the crucial decades in between to explain the persistence and resilience of a large and vibrant cycling culture – the breeding ground for the cycling activism that in turn set off cycling's urban revival in the Netherlands.

My argument is first that cycling in the Netherlands has received a larger space in traffic policy than in other countries, partly because interest groups and social movements played a greater governance role than elsewhere. This might be attributed to the Dutch polder model, a compromise and consensus-seeking political tradition. Second, I argue that national and provincial politicians and engineers have also been crucial at various times. The new archival material I explore corroborates and refines claims about the influence of social movements. Analyzing the political dynamics between governmental agencies and non-governmental organizations, as well as the role of social movements of cycling, is crucial to understanding the historical success of Dutch cycling as a daily practice. Where Albert de la Bruhèze and Veraart primarily studied how urban utilitarian cycling was governed, this study substantially broadens the perspective to show how different types of cycling (urban, suburban, interurban, and rural, as well as utilitarian and recreational) were shaped at different levels at different times. This also allows us to see how much of the cycling governance up to the 1970s ignored urban cycling. To do all this, we need to scour the archives for new sources.

Locating Cycling Governance: Sources and Methods

Reconstructing the dual process of cycling advocacy and governance requires archival and published sources. For cycling advocacy, the main arguments are typically expressed in journals, newspapers, and if successful, end up in Parliament and other representative bodies. These published sources, as well as the records of Parliament form one important source. Second, the ensuing governance process, the implementation of cycling policies, involves negotiation between different state actors as well as between state and non-state actors. Correspondence between these groups as well as records of meetings allow us to observe this process up close. *Cycling Pathways* relies on extensive archival research, often in archives not earlier consulted, or not specifically focused on cycling policies.[91]

91 Citations from Dutch archival and policy documents have been translated into English by the author.

To trace the sources methodologically, the first step was to identify potentially relevant archives, given the scant previous research on provincial and national cycling politics. Using archive inventories, I located many documents in multiple Dutch archives, as shown in the full list in the bibliography. Given the focus on a variety of actors and their interaction, the main sources are institutional archives of national, provincial, and local public works departments, as well as those of non-governmental organizations like the ANWB and Cyclists' Union. In line with Dutch political traditions, these NGOs closely cooperated with governmental actors in state committees and through correspondence. The records of these meetings and letters between key actors provide valuable insight into their policy beliefs about cycling and cycling governance.

A main source of information was the Dutch National Archives (NA). This is where the records of Dutch ministries including Public Works are kept. Particularly useful were records of the Committee of Consultation on Roads. Given the consultation culture in the Netherlands, these committees with many different stakeholders, provide great insight into the considerations behind mobility and cycling policy. Provincial and municipal archives are also of the utmost importance. Given the tendency to assign cycling governance to lower levels of government, these archives contain a lot of material on the actual construction of cycling infrastructure. Provincial archives are a particularly rich source of information as provinces developed plans and often served as intermediaries between municipal and national government, trying to smoothen procedures and reduce friction. Municipal archives are the best place to find evidence of the street-by-street transformation of cities over the twentieth century – first into roads for cars, then after the 1970s into spaces where cyclists, pedestrians, and other non-motorized user groups regained some space. The archives of activist groups and social movements are in different locations. Those of the Cyclists' Union are at the International Institute of Social History (IISH), although regional archives have the information on local branches and Dooievaar (see Chapter 6). Other action groups or NGOs have their records at the IISH or the National Archives. The ANWB has its own extensive in-house archive, which I have also consulted.

Other sources are parliamentary documents and records.[92] Cycling featured regularly in Dutch parliament debates at various times. One such

92 Parliamentary documents are indicated with their official dossier numbers, parliamentary records as *Handelingen* [I or II for upper house/lower house, followed by parliamentary year, meeting number, date, page]. These records are digitized and accessible at https://zoek.officielebekendmakingen.nl/uitgebreidzoeken/historisch.

occasion was in the 1920s for creating road policies linked to the bicycle tax. The rising popularity of mopeds in the 1950s was another example. Since the 1970s, many politicians – and the social groups pushing them – have put pressure on the state to invest more in local cycling policies. Besides plenary meetings, MPs meet in smaller committees dedicated to specific topics with the minister and high officials. The Permanent Committee for Public Works (*Vaste Commissie voor Verkeer en Waterstaat*) is a great source for more in-depth discussions of political trade-offs in mobility policy than the plenary meetings and I use them extensively, particularly in Chapter 7.

Besides unpublished archive material, journals and newspapers are a key source of information. The ANWB's trade journals, leading in the field, contain many articles by top Dutch engineers. I have studied all the volumes of the ANWB's roadbuilding journal *Wegen* ("Roads," 1925-present) and its traffic engineering journal *Verkeerskunde* ("Traffic Engineering," published under different names since the early 1950s). While car infrastructure clearly dominates these journals, many pages are dedicated to cycling. In addition, I consulted the numerous Cyclists' Union journals, published nationally and by local branches. I have found many details otherwise lost in the archives through articles digitized in the unsurpassed national newspaper database Delpher. It contains some 120 million pages, including 15 million in newspapers from 1618 to 2005 (representing 15 percent of published newspapers), as well as hundreds of different magazines.

Semi-structured background interviews also played a significant role.[93] These were conducted between 2018 and 2020 with (former) activists, consultants, and engineers to gain new insights and triangulate archival findings.

I used the source material methodologically in a process akin to coding. While studying archival documents, I made extensive notes of documents and passages relevant to my research purposes, especially governance processes. My qualitative analysis did not involve software, but the material was marked according to categories, such as material pertaining to national, provincial, or local policymakers, non-governmental actors, or design norms, finances, and so on. The outcome of this analysis is a narrative that attempts to reconstruct the governance process behind Dutch cycling infrastructure. The thematic development "follows the actors," as in Actor-Network Theory. This means critically and reflexively constructing, but not replicating

93 Organized with Matthew Bruno and Letícia Lindenberg Lemos. See forthcoming article: "Mobility Protests in the Netherlands of the 1970s: Activism, Innovation, and Transitions," *Environmental Innovation and Societal Transitions.*

their interests and biases. The focus is therefore mainly on national and provincial politics. This approach is at odds with the dominant development in historical research, which increasingly values transnational narratives over national histories. In mobility history, the transnational circulation of roadbuilding expertise and lobbying are for instance well documented and crucial to understanding national road networks. The focus on cycling in international networks such as the road engineering association PIARC is much more limited. This research explores and tries to understand the unique Dutch cycling trajectory. While some transnational dimensions exist, comparative approaches, still adopting the nation as point of departure, can further clarify the Dutch case. However, as cycling history at the urban level is now reasonably well understood, the next step is to ask to what extent regional and national governance also impacted this local mode of transport, before extending even further to transnational developments. As this step has proven to be more than enough for one project, it is therefore practical to keep to the national level for the time being.

The book discusses the period between 1880 and 2020 in three sections. Part I covers 1880 to 1950, when the bicycle transformed from an elite recreational to a cheap mass utilitarian vehicle. Part II covers the crucial decades between 1950 and 1970, when cycling all but disappeared in many countries but remained quite firmly entrenched in the Netherlands. Part III covers the revival of urban cycling since the 1970s. Throughout, whenever possible, I make comparisons with cycling policies outside the Netherlands. The conclusions summarize the empirical findings in terms of the governance theories introduced here.

Chapter 1, "Citizen Power: From Bourgeois Clubs to Governance Groundbreakers" – like Chapter 6 – puts social actors at the forefront. The crucial role of non-governmental tourist organization ANWB in framing cycling as an all-Dutch activity and its subsequent lobbying for cycling infrastructure heralded the arrival of Dutch cycling culture. I use new material from the ANWB archives to corroborate – and nuance – the scholarship's main conclusions far more extensively than elsewhere. As a system-builder, the ANWB established close ties to government agencies, coordinated private attempts at recreational cycling path construction, and lobbied to use the proceeds of the bicycle tax for constructing non-urban cycling infrastructure. This last aspect provided the reason for the national government to become more involved in cycling governance, as discussed in Chapter 2, "A Contested Compromise: National Government Supports Commuter Cycling." The central principle of Dutch road engineering was traffic separation (between cars and bicycles) on major routes outside cities. This conviction existed to an extent already around 1920, reinforced by the bicycle tax. Though the effect was unintentional, this tax boosted cycling infrastructure. Urban cycling was in mixed spaces, largely beyond the control of non-local policymakers. Two particular aspects of cycling governance here are new. First, using parliamentary records, I examined whether politicians considered cycling as contributing to ordinary Dutch citizens' mobility and quality of life. Second, using engineering committee records, I was able to determine how the political debate about cycling's status and the bicycle tax was translated into engineering norms and practices for cycling infrastructure.

After World War II, as Part II argues, two crucial decades followed in which the Dutch trajectory diverged from international trends. While the 1950s and 1960s saw the demise of cycling in other countries due to uncompromising pro-car policies, the Netherlands kept some space for bicycles. The chapters question how governance practices in the Netherlands enabled cycling infrastructure and cycling (engineering) knowledge to endure alongside growing car use. Chapters 4 and 5 primarily deal with utilitarian cycling outside the city, while Chapter 3, "A Right to Recreation: Provincial Policymakers Design Cycling Networks," focuses on recreational cycling infrastructure. Connected to the story about providing non-governmental recreational cycling infrastructure in the first chapter ("Citizen Power"), it discusses the growing provincial involvement in recreational cycling, another episode in Dutch cycling governance which has received little attention so far. At a time when policy attention centered on cars, it was an important area where cycling governance persisted. Chapter 4, "Popular or Outdated? National Policymakers' Ambivalence about Bicycles," then asks how the cycling policies initiated in the 1920s and implemented throughout the 1930s lived on in the 1950s and 1960s, seen in the scholarship as the period of car-centered planning. Path dependency is a key topic in this chapter. In Chapter 5, "An Accident of History: How Mopeds Boosted Dutch Cycling Infrastructure," a new vehicle enters the fray, the moped – fundamentally challenging the notion of traffic separation through separate infrastructures as well as the distinction between recreational and utilitarian cycling paths. Popular across Europe, its controversial classification as a type of bicycle meant it had to use existing Dutch cycling infrastructure. What were the consequences of this decision for Dutch cycling (infrastructure)?

Part III shifts to urban cycling governance. Since national and provincial policymakers and non-governmental actors rarely interfered in urban cycling before the 1970s, the earlier chapters in this book have less to say about urban cycling. With increasing pressure on this type of cycling due to rising automobility threatening historic city centers, and the growth of new (urban) social movements, a new coalition emerged whose goal was to promote urban cycling again. The multi-level governance of this process, which activists played a significant role in shaping, is central to part III.

Chapter 6, "Citizen Expertise: Urban Activism Shapes Local Cycling Policy in the 1970s," discusses the origins of these social movements and their action strategies. How did these groups function in dialogue with a relatively accessible civil service at city level? And how successful were user-developed and lay expertise methods in transforming city streets that had become exclusively car spaces? While Chapter 6 shows how citizens forced these developments on the government, Chapter 7, "Catching Up: The State Acknowledges Urban Cycling as Public Good, 1975-1990," investigates in more detail how cycling policies responded and came about within a complex governmental bureaucracy. Consisting of multiple ministries as well as powerful local actors, the decentralized Dutch system created both opportunities and friction for politicians and engineers. Finally, Chapter 8, "Self-Evident: Mainstreaming Cycling Policy and Practice since 1990," brings this long-term story to the present by summarizing the past thirty years of cycling governance. Providing a bird's eye perspective on recent developments, placed in the context of a long time span of cycling governance, it completes our journey through more than a century of Dutch cycling policies and practices and connects the past to the present.

Part I

Roots: How Commuter and Recreational Cycling Became a Dutch Public Good, 1880s-1940s

Cycling was a highly popular activity throughout the industrializing world in the first half of the twentieth century, until its international trajectories diverged. Many cities, as historians Adri Albert de la Bruhèze and Frank Veraart show, set their sights on automobility and shunned cycling, thereby marginalizing urban cyclists in traffic policy.[1] These cyclists often formed a majority, yet had to give up their space to a minority of car drivers. This also occurred in the Netherlands, but a political compromise gave Dutch cyclists back some space in the form of separated cycling paths. Primarily serving suburban commuter cycling, cycling paths were strongly and enduringly embedded in policymaking. Although studies on road policies and the tourist organization ANWB illustrate Dutch engineers' and politicians' particular attention to non-urban cycling, both commuter and recreational, which resulted in a resilient cycling culture that survived the rise of the car after the 1950s, this commitment to providing infrastructure for some groups of cyclists is not yet fully understood; nor do we understand why it was stronger in the Netherlands than in other countries. Can alternative policies explain Dutch cycling's resilience and prominent role in mobility throughout the twentieth century?

Between the two world wars, a policy coalition of lobby groups and national engineers embedded cycling in Dutch politics and road standardization more forcefully than elsewhere. Non-governmental organizations played a significant role: for decades, citizen initiative was the sole driving force behind cycling governance. These citizen initiatives became intertwined with national and local government and remained relevant when state involvement in road policies grew after 1920. The first chapter shows how Dutch cycling clubs – like those elsewhere in the industrial world – were the first to initiate cycling policies. Supported by additional archival research, it brings together the existing scholarship on mobility and political science from a cycling advocacy perspective. An account of Dutch cycling policies would be incomplete and misleading without discussing their non-governmental origins. Uniquely, Dutch clubs retained a strong position, allowing them to find the middle ground between automobility and cycling practices in government policy.

1 Adri A. Albert de la Bruhèze and Frank Veraart, *Fietsverkeer in praktijk en beleid in de twintigste eeuw: overeenkomsten en verschillen in fietsgebruik in Amsterdam, Eindhoven, Enschede, Zuidoost-Limburg, Antwerpen, Manchester, Kopenhagen, Hannover en Basel* (Den Haag: Ministerie van Verkeer en Waterstaat, Directoraat-Generaal Rijkswaterstaat, Hoofdkantoor van de Waterstaat, Directie Kennis / Stichting Historie der Techniek, 1999).

To explain how Public Works engineers and lobbyists reached this compromise, typical of the Dutch political "polder model" culture, Chapter 2 discusses the Dutch national government's involvement in cycling policy in the 1920s and 1930s. International research shows how many thousands of (working-class) cyclists in many countries battled with a small but exceedingly powerful group of upper-class car drivers with commercial interests, who were lobbying for car-centric traffic policies. The class-driven car coalition often marginalized cycling and made life harder for (working-class) cyclists. This political struggle had a different outcome in the Netherlands, not least because the Dutch (upper) middle class also kept on cycling. National politicians, engineers, and lobby groups reached a political compromise over bicycle taxation, initially to offset the national deficit, then diverted to build highways and cycling paths alongside national and provincial roads. Most road funding came from cyclists' taxes. The cycling paths they got in return were a peace offering, a contingent development with long-lasting effects. The system of traffic separation still shapes Dutch infrastructure today. For much of the twentieth century, however, this cycling infrastructure typically ended at the city limits as provincial and national engineers had no authority within the urban built-up area. This traffic separation model created path dependencies shaping cycling infrastructure outside cities, and eventually, after the 1970s, within Dutch cities as well. From an international perspective, Dutch cycling politics found a compromise in the fight for road space and funding. Lobby groups played an important role in Dutch consensus-driven political culture and influenced this outcome.[2] The proportion of Dutch cyclists remained high because cars had their own road space, not obstructing or endangering cyclists.[3] This continuous commitment to an engineering model of traffic separation – and with it the building of separate cycling lanes – is one reason why cycling survived alongside car growth. By investigating the long-term consequences of this model and distinguishing the urban from the rural and suburban, I demonstrate the specific implementation of this traffic model in time and place by different stakeholders.

The political contest over cycling infrastructure took place in a context of dramatic changes in mobility. Internationally, first bicycles in the 1890s,

2 For intermediary groups' role in diffusion of technologies: Adri A. Albert de la Bruhèze and Ruth Oldenziel, "Theorizing the Mediation Junction for Technology and Consumption," in *Manufacturing Technology, Manufacturing Consumers: The Making of Dutch Consumer Society*, eds. Adri A. Albert de la Bruhèze and Ruth Oldenziel (Amsterdam: Aksant, 2009), 9-39.
3 Ibid., 17.

then cars in the 1950s, became increasingly affordable. And we should not overlook the role of public transit to explain why citizens did or did not opt for individual mobility modes like cycling. In Belgium, a well-developed network of local railways formed a powerful competitor for the bicycle, much more so than for their Dutch neighbors. Belgian workers had a mobility alternative to commute.[4] By contrast, in the Netherlands, the underdeveloped and expensive public transit along with the slow adoption of the car, the ANWB's cultural promotion of the bicycle, and the political status of (non-urban) cycling, explain why for many people the premier mobility option was the bicycle.

Cycling advocates may have touted the wide diffusion and use of bicycles already around 1900, but bicycles only gradually became a truly mass phenomenon. In 1899, only 2 percent of the Dutch population owned a bicycle,[5] before cycling spread socially in the next decade.[6] The ANWB, originally a cycling advocacy group, calculated that there was 1 bicycle for every 53 Dutch citizens in 1899, jumping to 13 in 1908, and 2 by 1940. The cheapest bicycles still cost a monthly median income in 1910, but cheap imports, competition, and increasing homegrown production made bicycles ever more affordable.[7]

Dutch bicycle ownership followed international trends at first. By contrast, the nation's car ownership – even when engineers and politicians claimed a spectacular future increase in car use and lobbied for a quick and comprehensive redesign of the road network – lagged behind other European countries. In 1930, only 9 in 1,000 citizens in the Netherlands had a car, fewer than in Switzerland and Belgium. In the 1930s, when German car ownership rose quickly from 8 to 30 cars per 1,000 residents, the increase in Dutch car ownership was modest, from 9 to 11 cars – even when compared to the rise

4 Greet de Block, David de Kool, and Bruno de Meulder, "Paradise Regained? Crossing Borders Between Planning Concepts in the Netherlands and Belgium (1830-2012)," in *Builders and Planners: A History of Land-Use and Infrastructure Planning in the Netherlands*, eds. Jos Arts et al. (Delft: Uitgeverij Eburon, 2016), 17-67.

5 Sue-Yen Tjong Tjin Tai, Frank Veraart, and Mila Davids, "How the Netherlands Became a Bicycle Nation: Users, Firms and Intermediaries, 1860-1940," *Business History* 57, no. 2 (2015): 265.

6 ANWB archive, inv. no. 1660. Dutch bicycle tax statistics (1899-1908) show 2-, 1-, and half-guilder tariffs, exemptions based on house rental values, and a greater number of lower categories, demonstrating that cycling was spreading to different classes. The lowest tax category increased by 775 percent, versus 387 percent for the middle and 165 percent for the full rate.

7 Tjong Tjin Tai, Veraart, and Davids, "How the Netherlands Became a Bicycle Nation," 268. In 1935, bicycles cost on average 30 guilders, half the price they were in 1922: J.B.D. Derksen and A. Rombouts, "The Demand for Bicycles in the Netherlands," *Econometrica* 5, no. 3 (1937): 299.

in another small country like Belgium (from 12 to 19).[8] Driving was obviously still a highly exclusive elite prerogative – despite the disproportional attention devoted to cars by engineers and the media. Cyclists were less fortunate. Still, in the first half of the twentieth century, Dutch national and provincial policymakers came to recognize cyclists' rights – largely for opportunistic reasons. The provision of cycling infrastructure, advocated for by the ANWB and justified by the bicycle tax, became solidly embedded in Dutch engineering norms, laying the groundwork for a path dependent process. In this first part, I set out to show how cycling put down roots between 1880 and 1950.

8 Ruud Filarski in cooperation with Gijs Mom, *Shaping Transport Policy: Two Centuries of Struggle between the Public and Private Sector: A Comparative Perspective* (The Hague: SDU Uitgevers, 2011), 91. They suggest one reason was the relatively high car tax.

1 Citizen Power: from Bourgeois Clubs to Governance Groundbreakers

"Cycle tracks will abound in Utopia, sometimes following beside the great high roads, but oftener taking their own more agreeable line amidst woods and crops and pastures." With this quote from H.G. Wells' 1905 novel *A Modern Utopia*, Dutch civil servant Jan Dirk Christiaan van Dokkum characterized a dream that he thought had become reality in the Dutch Gooi region around 1920.[1] Located southeast of Amsterdam and north of Utrecht, it was a wealthy region with an attractive landscape of waterways, sand dunes, heath, and woodlands. Many shared his sentiment: when Dutch historian Pieter Geyl lived in London in the 1920s, he had only taken to walking; back in the Netherlands, he talked about his love for cycling around Utrecht, where he became a professor in the 1930s. Initially, the historian had not planned to buy a bicycle, however, his colleague Willem Kernkamp said: "Not buying a bicycle? That is one of the greatest pleasures one can have here!" After he and his wife duly bought bicycles which they "used almost only for rides outside the city," Geyl had to admit "it is almost impossible to say how invigorating [*verkwikkend*] this has been for me."[2] He enjoyed touring alongside the canals and lakes.

Enjoying the Dutch landscape by bicycle was possible thanks to the many cycling paths – free of cars, pedestrians, horses, and other obstacles – allowing cyclists to appreciate formerly inaccessible nature. Such infrastructure was not the work of the government, however. In 1920, the state did not yet consider constructing cycling infrastructure of any type a public task. Bourgeois citizen organizations were the ones constructing recreational cycling infrastructure. Well-connected politically, yet with only minimal help from local authorities, these clubs built extensive networks of cycling paths for their leisure. Intended for bourgeois recreation, however, the paths unintentionally also provided shorter connections for commuter cyclists – a fact the leisure clubs came to eagerly exploit in the following decades when making their case for public funding.

1 J.D.C. van Dokkum and D.J. van der Ven, *Ons mooie Nederland: Het Gooi* (Amsterdam: J.M. Meulenhoff, 1921), 84-97.

2 Pieter Geyl, *Ik die zo weinig in mijn verleden leef: Autobiografie 1887-1940*, edited and annotated by Wim Berkelaar, Leen Dorsman, and Pieter van Hees (Amsterdam: Wereldbibliotheek, 2009), 291.

Historians Frank Veraart and Anne-Katrin Ebert have asked how the ANWB managed to put cycling on the political agenda. Based on the organization's records and archival material from the government's perspective, I expand on the ANWB's well-researched role building and supporting the automobile system for its cycling advocacy. My archival exploration confirms the ANWB's political role on governmental committees, its gatekeeping role in producing knowledge about cycling governance. Untapped archival material sheds light on the network-building role of ANWB-affiliated cycling path organizations. These constituted an important and successful citizen governance model, which (provincial) government agencies only incorporated in the mid-twentieth century. Since the 1880s, citizen organizations had clearly been key actors, shaping Dutch cycling policy for some forty years before the government formulated its policy in the 1920s. Thereafter, in line with the Dutch polder model tradition, the organizations retained a major role in cycling governance by working closely with government engineers.

1.1 Dutch Cycling Clubs' Unique Position in an International Context

When young bourgeois men all over the industrializing world popularized cycling as a recreational activity in the 1870s and 1880s, they organized (urban) cycling clubs. As the scholarship shows, apart from providing sociability, the clubs soon transformed into lobby organizations, addressing poor or muddy road surfaces, and created cycling paths, maps, routes, and road signs.[3] After a few decades, however, new car organizations sidelined cycling clubs, culturally and politically: the cycling club founders switched to the new big thing, the car, and left cycling behind as outdated. The car replaced the bicycle as the symbol of modernity. Countries whose cycling clubs were split along class lines had a weaker power base to compete with the new and powerful car lobby.[4] Scholars have shown this trend for German, as well as American, British, Italian, and South African clubs.[5] The

3 Frank Veraart, "Geschiedenis van de fiets in Nederland 1870-1940: Van sportmiddel naar massavervoermiddel" (Master Thesis, TU Eindhoven, 1995); M.F.A. Linders-Rooijendijk, "Gebaande wegen voor mobiliteit en vrijetijdsbesteding: I. De ANWB als vrijwillige associatie 1883-1937" (PhD diss., Katholieke Universiteit Brabant, 1989).

4 For a synthetic analysis of the international scholarship: Oldenziel and Hård, *Consumers, Tinkerers, Rebels: The People Who Shaped Europe*, 1.

5 Carlos Héctor Caracciolo, "Bicicleta, circulación vial y espacio público en la Italia Fascista," *Historia Critica* 39 (2009): 20-42; Peter Cox, "'A Denial of our Boasted Civilization': Cyclists' View

Netherlands took a different path. True, its cycling policies started with a cycling club (the ANWB), shaping cycling infrastructure for forty years. Like elsewhere, the Dutch government's Public Works (*Rijkswaterstaat*, or RWS) only took a limited interest in roadbuilding before the 1920s.[6] Unlike other countries, however, the Dutch organization remained a powerful advocate for cycling throughout the first half of the twentieth century, as we will see.

Following the early international trend, at first the many local Dutch cycling clubs merged into a national tourist organization ANWB in 1883. The name stands for General Dutch Cyclists' Union (*Algemene Nederlandsche Wielrijders Bond*). The organization's rapidly broadening scope soon made this name obsolete, and it was changed to *Koninklijke Nederlandse Toeristenbond* ANWB ("Royal Dutch Tourist Union ANWB").[7] Like many similar clubs, its members were young bourgeois and urban cyclists keen to promote recreational cycling in the countryside. Initially, their interests also included cycle races. In 1898, the Dutch club decided to scrap racing and focus on all forms of tourism, by bicycle, car, boat, or on foot. In their recent history of the organization, historians Hans Buiter and Peter Staal characterize it in the years 1883 to 1900 as a "boys' club"; 1900 to 1919 was when the club broadened into a tourist organization, and 1919 to 1949 was the period when it lobbied to shape new infrastructure (for bicycles and cars).[8] Over these different periods the ANWB developed from a single-issue bicycle club before 1900 to an organization advocating multiple forms of tourism (walking, cycling, driving) in the 1910s and 1920s. Meanwhile, the ANWB built up engineering expertise and increasingly became interested in a (utilitarian) roads network as well. After 1920, the organization therefore was an all-purpose lobbying organization which tried to balance recreational and utilitarian mobility needs.

on Conflicts over Road Use in Britain, 1926-1935," *Transfers* 2, no. 3 (2012): 4-30; Anne-Katrin Ebert, *Radelnde Nationen: Die Geschichte des Fahrrads in Deutschland und den Niederlanden bis 1940* (Frankfurt: Campus Verlag, 2010); Lorenz Finison, *Boston's Cycling Craze, 1880-1900: A Story of Race, Sport, and Society* (Amherst: University of Massachusetts Press, 2014); Evan Friss, *The Cycling City: Bicycles and Urban America in the 1890s* (Chicago: University of Chicago Press, 2015); Longhurst, "The Sidepath Not Taken: Bicycles, Taxes, and the Rhetoric of the Public Good in the 1890s," 557-86; Longhurst, *Bike Battles*; Njogu Morgan, *Cycling Cities: The Johannesburg Experience* (Eindhoven: Foundation for the History of Technology, 2019).

6 Mom and Filarski, *De mobiliteitsexplosie*, 91.

7 They kept the name ANWB because of brand recognition. The original but short-lived name *Het Nederlandsche Vélocipèdisten-Bond* changed in 1885 to the more Dutch-sounding *Algemene Nederlandsche Wielrijders-Bond*. Linders-Rooijendijk, "Gebaande wegen I" 31, 110.

8 Hans Buiter and Peter Staal, *Het avontuur van de ANWB: 135 jaar onderweg* (Bussum: Uitgeverij THOTH, 2018), 5.

A key question is how the ANWB conducted its political lobbying, since it played such a crucial governance role through its cycling advocacy. Like many European tourist and cycling clubs, the Dutch organization took up the improvement of road quality and traffic legislation earlier than state engineers.[9] The expertise gained in the process gave the organization the power to function as governance partner. The ANWB eventually became so important that people sometimes considered it a fifteenth ministry (alongside the official fourteen Dutch ministries).[10] Multiple commemorative volumes reflect the organization's sense of self-importance and it has generated more studies than most other aspects of Dutch cycling history.[11]

Historians approach the ANWB from three perspectives. M.F.A. Linders-Rooijendijk has written the organizational history of the ANWB.[12] She traces it evolving from being dedicated exclusively to cycling in the nineteenth century, to a highly professionalized organization claiming to protect the interests of all forms of tourism.[13] Ebert's cultural-historical perspective highlights the values the ANWB aimed to promote through mobility and tourism, studying its (discursive) attempts at nation-building.[14] The ANWB hoped to integrate the Dutch nation culturally and physically into one liberal-bourgeois network. In a socially divided society (the "pillars" of Protestants, Catholics, and socialists), the liberal ANWB emphatically avoided an explicit ideological profile. In line with Ebert, innovation historians

9 Emanuel, "Making a Bicycle City," 499; Catherine Bertho-Lavenir, "How the Motor Car Conquered the Road," in *Cultures of Control*, ed. Miriam R. Levin (Amsterdam: Harwood Academic Publishers, 2000), 117, 130-32. On French associations similar to the ANWB, Sébastien Gardon, "Gouverner la circulation urbaine: Des villes françaises face à l'automobile (années 1910 – années 1960)" (PhD diss., Université Lumière – Lyon II, 2009). https://tel.archives-ouvertes.fr/tel-01540257.

10 Mom and Filarski, *De mobiliteitsexplosie*, 83, 394. Before 1940, there were less ministries, but when this nickname was introduced it had grown to 14.

11 *Veertig jaar. A.N.W.B., Toeristenbond voor Nederland, 1883-1923*, (S.l.: S.n., 1923); *Een halve eeuw A.N.W.B. 1883-1933. Gedenkboek, uitgegeven ter gelegenheid van het 50-jarig jubileum van den A.N.W.B. – Toeristenbond voor Nederland, en aan dezen Bond en zijn bestuurders aangeboden door J. Jonker Roelants, Mr. H.A.M. Roelants en H.L. Jonker Roelants* (Schiedam: Roelants, 1934); D. Schaap, *Een eeuw wijzer 1883-1983. 100 jaar Koninklijke Nederlandse Toeristenbond ANWB* (Den Haag: ANWB, 1983); Buiter and Staal, *Het avontuur van de ANWB*.

12 M.F.A. Linders-Rooijendijk, *Gebaande wegen: I, and II. De ANWB van vereniging naar instituut, 1937-1983* (Den Haag: ANWB, 1992).

13 Linders-Rooijendijk, "Gebaande wegen I," 9.

14 Anne-Katrin Ebert, "Cycling Towards the Nation: The Use of the Bicycle in Germany and the Netherlands, 1880-1940," *European Review of History: Revue europeenne d'histoire* 11, no. 3 (2004): 347-64.

Gijs Mom, Johan Schot, and Pieter-Eloy Staal, consider its mission as a (paternalistic) "civilizing role": the ANWB promoted an etiquette for new cyclists and drivers, to educate them as responsible road users.[15] In the tradition of innovation studies, this third perspective considers the ANWB's system-building role for the Netherlands' (utilitarian) roads and cycling networks. Expanding on Linders-Rooijendijk and Ebert's focus on tourism and recreational cycling activities, these historians characterize the ANWB as "the main system builder ... of the Dutch automobile system" and highlight its mediating role in creating the demand for cars by promoting their usefulness as transport vehicles.[16] Mom et al. describe the ANWB leaders as intermediary actors, also for cycling, "who not only position themselves as representatives of what they see as 'the user,'... but who negotiate the alignment between the products and their use."[17] Adri Albert de la Bruhèze and Ruth Oldenziel theorize this crucial position as the "mediation junction" in shaping innovations.[18] By organizing conferences, publishing trade journals, and joining governmental committees, the tourist and lobby organization ANWB was at the heart of Dutch mobility governance. It had a strong hand in shaping not just policy but also educating (local) traffic experts. The scholarship explains this success in terms of culture and politics.

Culturally, many cycling clubs in other countries framed cycling more exclusively than the Dutch ANWB. In other countries, their single-issue advocacy made it harder to claim state support for cycling as a public good. Cycling clubs in the United States and Canada, for example, had a limited upper-class membership. According to geographer Glen Norcliffe, the Montreal Bicycle Club (founded in 1878) created an image that always implied an inside and outside group.[19] Similarly, Charles Pratt, patent lawyer for major bicycle producer Pope, who founded the Boston Cycling Club (1878) and the League of American Wheelmen (LAW, 1880), reserved membership for

15 Ebert, *Radelnde Nationen*; Gijs Mom, Johan Schot, and Peter-Eloy Staal, "Civilizing Motorized Adventure: Automotive Technology, User Culture, and the Dutch Touring Club as Mediator," in *Manufacturing Technology, Manufacturing Consumers: The Making of Dutch Consumer Society*, eds. Adri A. Albert de la Bruhèze and Ruth Oldenziel (Amsterdam: Aksant, 2009), 139-58.

16 Mom, Schot, and Staal, "Civilizing Motorized Adventure," 150-51.

17 Ibid., 139.

18 Ruth Oldenziel, Adri A. Albert de la Bruhèze, and Onno de Wit, "Europe's Mediation Junction: Technology and Consumer Society in the 20th Century," *History and Technology* 21, no. 1 (2005): 107-39; Albert de la Bruhèze and Oldenziel, "Theorizing the Mediation Junction for Technology and Consumption," 9-39.

19 Glen Norcliffe, "Associations, Modernity and the Insider-Citizens of a Victorian Highwheel Bicycle Club," *Journal of Historical Sociology* 19, no. 2 (2006): 121-50.

white, upper-class men.[20] Historian Lorenz Finison details how clubs like the LAW excluded people of color.[21] This is not to say that excluded groups did not ride bicycles. On the contrary. Non-affiliated cyclists, however, were not represented politically, rendering them invisible to policymakers – as the archives prove. In Germany, like in the Netherlands, the cycling club founders were British, as part of the transnational network of early European cycling culture.[22] Ebert describes the variety of cycling clubs in Germany, with membership differentiated by region and class. Once working-class citizens could afford bicycles, they established their own club, Solidarität, when other social classes considered cycling vulgar.[23] Oldenziel and Mikael Hård show that the rapid decline of cycling's status as modern and desirable happened across the industrializing world.[24] This had consequences for funding and state support.

Ebert's comparison with Germany highlights how the ANWB developed a more inclusive and successful framing of (recreational) cycling as nationally valuable.[25] Earlier, Veraart had pointed out that, partly to pre-empt rival organizations, the ANWB board's strategy was to claim it represented all cyclists.[26] In pillarized Dutch society, with Protestant, Catholic, and social democrat clubs, the ANWB knowingly presented itself as a broad and ideologically neutral organization. The reality was different: cycling became more affordable only gradually throughout the first half of the twentieth century – and not all cyclists were ANWB members. The political success of this claim, however, meant that in the Netherlands, only one cycling organization had a monopoly on representing cyclists until the 1970s. According to Ebert, the ANWB board members were typically liberal bourgeois men, who framed cycling as an activity for the entire country – regardless of class or gender. Cycling would unite the (middle-class) nation and create patriotic, disciplined, and healthy citizens.[27] She argues that this cultural framing of cycling in the Netherlands was part nation-building and positively

20 Ibid., 131-32. The many cycling clubs often catered to select groups. Friss, *On Bicycles*, 43.

21 Finison, *Boston's Cycling Craze*.

22 Ebert, *Radelnde Nationen*, 150.

23 Anne-Katrin Ebert, "Cycling Towards the Nation: The Use of the Bicycle in Germany and the Netherlands, 1880-1940," *European Review of History: Revue europeenne d'histoire* 11, no. 3 (2004): 357-62.

24 Oldenziel and Hård, *Consumers, Tinkerers, Rebels*, 1, 124-59.

25 Ebert, *Radelnde Nationen*. For English summaries, Ebert, "Cycling towards the nation," 347-64, and "When Cycling Gets Political," 115-37.

26 Plenty of local clubs emerged in cycling's early days but none could rival the ANWB: Veraart, "Geschiedenis van de fiets in Nederland," 30, 137-38.

27 Ebert, *Radelnde Nationen*, 363.

affected policymaking; it created a willingness among experts to facilitate cycling. In elaborating Ebert's point, sociologist Giselinde Kuipers argues the Netherlands had the right conditions for successful cultural embedding: a small, homogeneous, and highly integrated country, predominantly upper middle class yet with minor class differences. A national dislike for ostentation, moreover, had created a culture of "conspicuous non-consumption."[28] The ANWB contributed to the bicycle's representation as typically Dutch through its journals and other media expressions. The broad cultural framing of Dutch cycling mattered politically as a background condition for putting cycling on the political agenda as a public good. Class and regional tensions hindered the establishment of this status in other countries.

Politically, many international cycling clubs failed to gain long-lasting power and influence in decision making. When elites shifted their allegiance from the bicycle to the car, cycling clubs could not compete with motorist clubs. In the United States, the nineteenth-century bicycle craze disappeared rather suddenly. As soon as the bicycle spread beyond the exclusive circle, it no longer functioned as a status symbol and modern lifestyle. The car soon engaged everyone's interest. American cycling had few powerful advocates and dwindled in the early 1900s.[29] Similarly, British clubs' strong class dimensions created power conflicts between elite drivers and middle- or working-class cyclists. As cycling sociologist Peter Cox shows, the British Cyclists' Touring Club (CTC) was exclusively for cyclists, who had to fight car drivers for road space. Having to contend with engineers' single-minded privileging of cars, the CTC "was transforming from a polite, middle-class club to a campaigning organization."[30] Significantly, the British CTC strongly opposed separate cycling lanes, considering them an attack on the cyclists' right to the road.[31] In Italy, policymakers rarely embraced cycling infrastructure provision as a state task, despite lobbying by cycling clubs.[32] This brought them into direct conflict with drivers, in contrast to the Dutch case, where interest groups and the government worked out a compromise through the traffic separation model. In the short run, it was controversial. In the long run, it was effective.

In line with Buiter and Staal's periodization, in the period after 1920, the tourist and lobby organization ANWB succeeded as mobility interest group

28 Kuipers, "The Rise and Decline of National Habitus," 24.

29 Friss, *The Cycling City*, 3-4, 186ff; Longhurst, *Bike Battles*, 83.

30 Cox, "A Denial of our Boasted Civilization," 11.

31 Albert de la Bruhèze and Veraart, *Fietsverkeer in praktijk en beleid*, 133. Cox, "'A Denial of our Boasted Civilization'," 16-17.

32 Caracciolo, "Bicicleta, circulación vial y espacio público en la Italia Fascista," 29-30.

and government intermediary for multiple reasons. The organization built up an extensive body of knowledge on mobility and infrastructure issues in the Netherlands between 1880 and 1920. In return for this specialized expertise, the Dutch government granted the organization a major role in shaping mobility policy. The government's inclusion of the ANWB was not exceptional for Dutch politics: its culture allowed interest groups, representing certain parts of society, a significant role in governance. Consensus-seeking, the so-called polder model, as political scientists have pointed out, was the nation's key political value.

In marked contrast to other nations, moreover, the ANWB also became the main automobile club in the Netherlands. The US car lobby shaped the American highway code, while the cyclists' League of American Wheelmen had no influence.[33] Over time, the car boosters within the ANWB became more dominant, yet cycling advocacy – particularly for leisure purposes – remained a large and visible part of the organization. By claiming to represent both cyclists and drivers, the conflict between these groups did not play out in open power politics but in a more diffuse process of compromises, as I detail in Chapter 2.

Third, the young men who founded the ANWB in the 1880s became mature citizens, holding prominent positions in administration, law, and the economy.[34] Like their Danish counterparts, the ANWB officials' elite status ensured a close relationship with state officials.[35] Edo Bergsma, in particular, ANWB chairman for half a century between 1884 and 1937, formed a link between civil society and the government: active in many organizations and long-time mayor of the city Enschede (1896-1932), he strengthened the ANWB's position in the policy network through his personal contacts and influence. Equally his successor as chairman, Henri Bloemers, held many public functions including mayor. He was also president of the Dutch Institute for Public Housing and Urban Development (NIVS) and nature conservation groups.[36] Although the leading ANWB figures were not traffic experts and had to establish their credentials in this field, they were well connected within the Dutch state at multiple levels. This interweaving of state and civil society was a key characteristic of Dutch cycling governance and facilitated access to policymaking circles. Consequently, the Dutch

33 Longhurst, *Bike Battles*, 237.

34 Buiter and Staal, *Het avontuur van de ANWB*, 39.

35 Emanuel, "Making a Bicycle City," 500-501.

36 P.R.A. van Iddekinge, "Bloemers, Henri Peter Johan (1880-1947)," in *Biografisch Woordenboek van Nederland*. http://resources.huygens.knaw.nl/bwn1880-2000/lemmata/bwn3/bloemers [accessed August 18, 2017].

cycling policy coalition's non-governmental stakeholders stood close to government policymakers and engineers.

Through new research in state archives, I highlight two of the ANWB's cycling governance roles: governing cycling with a coalition of non-governmental actors and, in parallel, putting cycling on the political agenda, particularly after 1920. Section 1.2 shows how this coalition and affiliated cycling path organizations built recreational cycling infrastructure in rural areas with little help from the government. As state engineers showed more interest in cars and roads in the 1920s, the ANWB also tried to carve out a role in this new governmental sphere. It became a key stakeholder by working closely with the government to shape cycling governance and defend cycling's interests. The organization framed cycling broadly as a recreational but also very much utilitarian activity for everyone – a recurring strategy and distinction throughout the twentieth century. Although several historians have studied many elements of these roles, this chapter, "Citizen Power," brings the activities into an integrated analytical framework, filling in the blanks and drawing a more comprehensive picture of independent initiatives' major role in Dutch cycling governance between 1880 and 1950.

1.2 Citizens Building Recreational Cycling Paths

Until the 1950s, non-governmental actors took responsibility for building a tourist cycling infrastructure in the Netherlands – the paths that led civil servant Jan van Dokkum to wax so lyrical about cycling infrastructure in 1921. Governmental involvement was limited to occasional subsidies. The ANWB promoted domestic tourism by advocating a cycling network. As Ebert argues: "Bicycle path construction was neither an environmental nor a social project, but was brought about by the efforts of a social elite that used bicycle tourism to strengthen national identity and increase social restraint."[37] Before 1920, the Dutch state made no concerted effort to build cycling lanes. The only paths it built were in response to citizen petitioners – usually recreational cyclists demanding comfortable paths, or access where sand roads or the lack of roads prevented them from enjoying nature. Recreational cyclists also needed road signs, maps, guides, and designated places to stop for breaks that provided repair kits (see figure 2). The ANWB initiated many of these system-building resources. By 1904, for instance, the tourist organization had placed 1,000 road signs throughout

37 Ebert, "When Cycling Gets Political," 120.

the country.[38] Through a system of local representatives, the organization was well aware of local conditions and worked with local elites, while also effectively lobbying at a national level, as section 1.3 shows.

Such private initiatives to create cycling resources were an international phenomenon. In Germany, cyclists financed most cycling infrastructures until at least the 1920s.[39] Cycling clubs were also behind America's Good Roads Movement in the 1880s. Evan Friss shows that similar to the ANWB, the League of American Wheelmen lobbied for establishing national road agencies and supported the government with expertise and funding.[40] Many local clubs also constructed cycling paths based on a user-fee model.[41] According to historian James Longhurst, the Good Roads Movement was "urban, middle class, efficiency minded, and nationally organized."[42] This approach to road improvement was an indictment of the US government's failure to acknowledge cyclists' demands. In the long run, however, American clubs failed to secure a solid policy basis for cycling within public institutions. The initiatives – patchwork and incidental – were vulnerable. Longhurst contends that "because cyclists could not successfully argue that the proposed side-paths were a public good that would benefit all society, they could not make a claim on public funding, and the financing was limited to charitable contributions and a user-fee model."[43] After 1900, the American framing of cycling, and subsequent policies, were limited to recreation.[44] The Dutch government did not recognize recreational cycling infrastructure as a national public good either, arguing that the provinces were responsible (see 2.5). Before the 1950s, citizen organizations built two to three thousand kilometers of recreational cycling paths. These worked successfully until mid-century, when the provinces took over this task.

Like its international counterparts, the ANWB's attempts to persuade the state to create cycling paths around 1900 initially had limited success. Especially where road surfaces were poor or bumpy, like the cobbled roads in the province Noord-Brabant, cycling tourists required paths. Occasionally the ANWB financed paths, but it did not have enough funds for larger networks. Instead, the organization lobbied local government to construct these paths,

38 Veraart, "Geschiedenis van de fiets in Nederland," 67.

39 Briese, *Besondere Wege für Radfahrer*, 2.

40 Friss, *The Cycling City*, 82-99.

41 Ibid., 103.

42 Longhurst, *Bike Battles*, 63.

43 Ibid., 55.

44 Evan Friss discusses New York's recreational network initiated by famous planner Robert Moses in the 1930s. Friss, *On Bicycles*, 75-100.

Figure 2 The Dutch tourist organization ANWB was an omnipresent stakeholder in tourism and mobility in the early phases of cycling and driving. State engineers valued working closely with this non-governmental organization thanks to its unparalleled knowledge of local (cycling) conditions. This 1928 map was for cycling tourists exploring the central Utrecht region of the Netherlands. On the red cycling paths, cyclists encountered the ANWB's road signs. [Source: Collectie Het Utrechts Archief, cartographer A.W. Stork, catalogue no. 818513]

pointing to cycling's increasing recreational and emerging utilitarian value. In the early 1900s, the bicycle's use was growing among occupations like postmen and doctors.[45] When welfare state policies increased leisure time, initially for middle-class and later for working-class citizens, many independent organizations initiated (cycling) tourism and recreation in nature.

45 In the 1890s, Dutch cities experimented with cycling telegraph messengers or postmen and military cyclists. Veraart, "Geschiedenis van de fiets in Nederland," 68-69.

To open up the Dutch countryside to recreational cyclists, the ANWB worked with local elites to create (non-governmental) cycling path organizations. Regional in scope, funded through membership fees, donations, and local governments subsidies, the local clubs sought to build cycling paths, often independently and separate from the existing roads in nature reserves. Most local organizations were created in the 1910s and 1920s: Het Wielerpad (Meppel 1906);[46] Gooi en Eemland (1914); Het Drentsche Rijwielpad (1916); EMO (Eindhoven 1916); Twente (1917); Noord-Veluwe (1917); UMO (Utrecht 1918); Noord-Kennemerland (1920); Walcheren (1926); West-Veluwe (1926).[47] They thrived but experienced growing difficulties during the economically difficult 1930s and were "provincialized" in the late 1940s and early 1950s.[48]

From a governance perspective, these were non-governmental organizations. Their leaders, however, had connections with (local) political elites, which helped members secure subsidies and permits for building local cycle paths. Historian Buiter describes how ANWB officials worked closely with prominent local politicians and industrialists to form these cycling path organizations. A key figure in establishing such organizations was Alexander Emil Redelé, a former military engineer who directed the ANWB's roads committee. He worked with local ANWB representatives who were familiar with the political and infrastructural landscape.[49] For instance, Buiter explains that Redelé invited forty prominent local figures to an inaugural meeting in Eindhoven. Local mayors and industrialists like Anton Philips, founder of the electronics factory, attended. Redelé also approached local dignitaries (schoolteachers, notaries, etc.) to join the organizations' boards.[50] After establishing these organizations, the ANWB provided technical advice on construction. The cycling path organizations demonstrated an informal and personal network of local dignitaries, state officials, and the ANWB, committed to recreational cycling.

To construct their cycling paths independently, the organizations all had a small board of volunteers, hired their own construction workers, and relied on subscription fees and subsidies. The provinces were most

46 Ibid., 89-90.

47 The original organization was founded in 1911 in the town Barneveld.

48 The Gooi en Eemland organization was an exception: it continued and still exists formally: https://www.rijwielpadengooieneemland.nl/

49 A leading ANWB figure in the UMO area advised Redelé to bring in J.C. Gerlings (ANWB representative in De Bilt) and baron Taets van Amerongen, who was passionate about cycling and had a large network in Utrecht. ANWB archive, inv. no. 1197, letter Pos (ANWB) to Redelé (ANWB), September 6, 1918.

50 Buiter, "Als een spin in het web," 79.

receptive to the clubs' requests, while municipalities provided small sums, as did local Chambers of Commerce and tourist organizations. In practice, the ANWB advanced the money to speed up the process.[51] Funding did not buy the government any influence on location or design: the cycling path organizations operated separately from the authorities and constructed their own network.[52] At the time, such governance initiatives were not unique. Conservationist clubs like the Society for the Preservation of Natural Monuments privately purchased scenic heathlands for instance.[53] The state and civil society shared the responsibility for these tasks.

Between 1900 and 1920, the citizens clubs focused on creating a leisure infrastructure. After World War I, they began to lobby public authorities. To convince policymakers that their cycling paths were worthy of public funding, organizations presented their paths as useful not just for weekend cycling trips in the countryside, but also for commuters, and persuaded local authorities to award larger subsidies. The argument that cycling paths served both recreational and utilitarian functions strengthened their case as a public good. Later in the twentieth century, cycling advocates still used this strategy. For instance, in 1923, the club UMO (Utrecht and surrounding areas, *Utrecht Met Omstreken*) targeted three groups: "city people," who could now enjoy "the beauty and quiet of the surroundings," "farmers" benefiting from "the short and good connections to the village, church, or market," while the doctor, veterinarian or postman could now "easily reach their patients or destinations."[54] In his 1935 regional zoning design for Utrecht, well-known social geographer and urban planner Theodore Karel van Lohuizen stressed that the paths' dual purpose enabled cyclists to enjoy nature undisturbed by motorized traffic, and "greatly matter for the mutual contact of the many small settlements, both purely practically as for the great freedom of movement they allow residents."[55] Van Lohuizen's emphasis on cycling infrastructure serving recreational and commuter purposes was the same argument the organizations and ANWB used. As a major figure in the surveys for Amsterdam's 1930s

51 Ibid., 81.

52 As these paths did not follow existing roads but attractive natural areas, obtaining permission from landowners to construct paths over their ground was a time-consuming task. The well-connected board members made this process easier.

53 Frank Veraart, "Agriculture and Nutrition: The End of Hunger," in *Well-being, Sustainability and Social Development: The Netherlands 1850-2050*, eds. Harry Lintsen, Frank Veraart, Jan-Pieter Smits and John Grin (Cham: Springer International Publishing, 2018), 265-66.

54 ANWB archive, inv. no. 1186, booklet "UMO 1 mei 1918-1 mei 1923," 4.

55 Th. K. van Lohuizen, "Het schetsontwerp voor het streekplan 'Utrecht-Oost'," *Tijdschrift voor Volkshuivesting en Stedebouw* 16, no. 4 (1935): 69.

Extension Plan and detailed traffic counts, Van Lohuizen realized how highly city-dwellers valued the bicycle for both work and recreation.[56] While we do not know to what extent people used the paths, historian Buiter agrees with these contemporaries who saw that the widespread cycling paths provided new and shorter connections between towns and villages.[57] The club EMO, active around Eindhoven, and the UMO southeast of Utrecht, worked in densely populated areas, where people probably used the paths not just on Sundays or after work. Twente cycling path organization, near the textile city Enschede, also had mixed-use paths, as Chapter 3 discusses.

Some organizations even considered constructing separate cycling paths alongside roads that the state authorities built. Given the government's mandate in building roads for cars, this raised the question of who was responsible for the construction of the separate cycling infrastructure: citizens or the state? The UMO built its first cycling path in 1919 parallel to a national road going east from Utrecht.[58] In the 1920s, planners started to consider this cycling infrastructure was the state's responsibility. Increasingly the issue was the demarcation between public and private responsibility. ANWB official Redelé tried to convince provincial engineers to support the regional cycling path associations' work. He lobbied engineer Linden van den Heuvell at the eastern province of Overijssel's Public Works department for subsidies, arguing that farm and factory workers living far from their workplace needed cycling paths: a state agency like the province should provide these.[59] Redelé convinced another province – Noord-Brabant – to change its statutes to allow funding for its main city Eindhoven's cycling path organization.[60] These initiatives fit into the ANWB's strategy in the 1910s and 1920s of framing cycling broadly to involve the government in cycling governance (see 1.3). In the long-term, the national government adopted the ANWB tactic of not involving the national or local, but targeting provincial government.

The cycling path organizations made considerable headway from the 1910s to the 1930s. After a decade, the Gooi en Eemland club (north of Utrecht), founded in 1914, had 2,000 members and constructed 100 km of cycling paths.[61] In 1928, ten years after it was founded, club UMO had the same

56 Arnold van der Valk, *Het levenswerk van Th. K. van Lohuizen 1890-1956: De eenheid van het stedebouwkundige werk* (Delft: Delftse Universitaire Pers, 1990).

57 Buiter, "Als een spin in het web," 81-85.

58 Ibid., 81.

59 ANWB archive, inv. no. 1219, letter Redelé (ANWB) to Linden van den Heuvell (PWS), December 12, 1918.

60 Buiter, "Als een spin in het web," 81.

61 Ibid., 77.

amount of paths around Utrecht and 3,000 members.[62] By far the most successful was the eastern area Twente, where the textile barons dominated, with about 1,000 km of cycling paths constructed between 1917 and 1937.[63] From 1914 to 1940, local organizations together built 2,500 km of cycling paths like those in figure 3, equivalent to what the Dutch government achieved in this period for cycling paths alongside roads.[64]

This private governance model had its limitations, however. The informal arrangements between the national ANWB, local government, and the cycling path organizations, could lead to friction at times. Some clubs, such as the Gooi en Eemland cycling path association, did not always get along well with the ANWB. Local clubs found the national organization bureaucratic and paternalistic in requiring detailed insight into their organization and finances before considering subsidies.[65] Government Public Works' increasing role in roadbuilding since the late 1920s also led to conflicts about the division of responsibilities – by creating paths for traffic alongside roads, the cycling path organizations were potentially encroaching on Public Works terrain. The Gooi en Eemland club, for instance, tried to build a separate cycling path along a busy commuter road between Baarn and Hilversum in the center of the country. By the 1920s, the national government started to see the construction of cycling paths along major arteries as a public task which should not be left to non-governmental organizations.[66] To present a united front, the organizations created an umbrella lobby in 1933, the Federation of Dutch Cycling Path Organizations (*Federatie van Nederlandse Rijwielpadverenigingen*, FNRV). Unlike the national tourist organization ANWB, the local clubs shared their expertise with each other. The federation, however, did not nurture a relationship with the government's Public Works. The federation did not even invite national or provincial Public Works' representatives to its annual meetings.[67] It goes to show that the

62 Ibid., 83.

63 Ibid., 85. There were also less successful organizations. If they had no subsidies, difficulties obtaining permission to use land, or if path construction was expensive, the organizations led little more than a paper existence, as in Walcheren and Kennemerland. Walcheren did build a popular cycling path through the dunes.

64 Buiter and Staal, *Het avontuur van de ANWB*, 75.

65 ANWB archive, inv. no. 1108, letter ANWB to Gooi en Eemland, April 13, 1915, and multiple letters in inv. no. 1107.

66 ANWB archive, inv. no. 1111, Gooi en Eemland 1932 Annual Report, 3; also inv. no. 1117, letter De Bruyn (ANWB) to Gooi en Eemland, October 31, 1926, when the organization tried to extend a cycle lane along a national road, the ANWB advised against it since this was the government's responsibility.

67 ANWB archive, inv. no. 1233, letter Louwerse (FNRV) to De Bruyn (ANWB), May 23, 1939.

Figure 3 This cycling path, likely constructed by UMO (Utrecht met Omstreken) cycling path organization in the woodland near Bilthoven, just east of Utrecht, is an example of the citizens' initiatives that constructed 2,500 km of cycling paths. The signpost in the centre is the iconic ANWB "mushroom." [Source: Collectie Het Utrechts Archief, no. 8544]

various actors involved in cycling path construction did not necessarily communicate with each other and that parallel governance coalitions existed. The national ANWB representatives' extensive participation created a partial overlap. It did not unify cycling governance structures.

The economic crisis of the 1930s changed the dynamic. Social organizations found it harder to acquire local state subsidies: they could not upscale their activities when local communities had less funding than the national government. Nor was the ANWB willing to step in and fund cycle paths. At its general meeting in 1935, the board argued that – while supportive of promoting recreational cycling – it deemed financial support problematic because it feared municipalities would be unwilling to subsidize, assuming the ANWB would keep the cycling path organizations afloat.[68] Instead, the board preferred a form of public-private partnership, with (local) government providing more generous subsidies.

The ANWB's position forced cycling path organizations to seek closer collaboration with municipalities. Often in vain. Despite supporting letters from ANWB arguing the public would stand to benefit from

68 "Notulen van de vergadering van het Algemeen Bestuur … op Zondag 22 december 1935," *De Kampioen* 53, no. 6 (1936): 133.

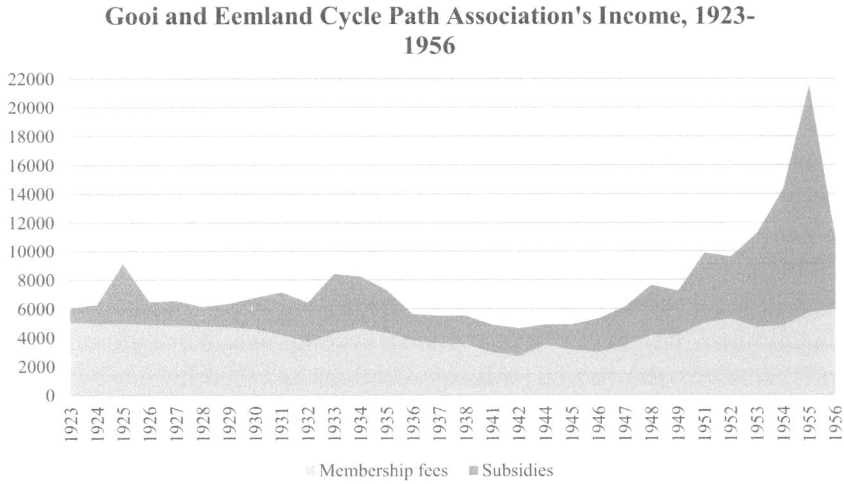

Figure 4 Income (in guilders) was modest but stable throughout the 1920s to 1940s. Higher subsidies in the 1950s indicate growing governmental acknowledgement of recreational cycling infrastructure as a public good. Source: ANWB archive, inv. nos. 1111 and 1112 (data for 1939, 1940, 1943 and 1950 missing).

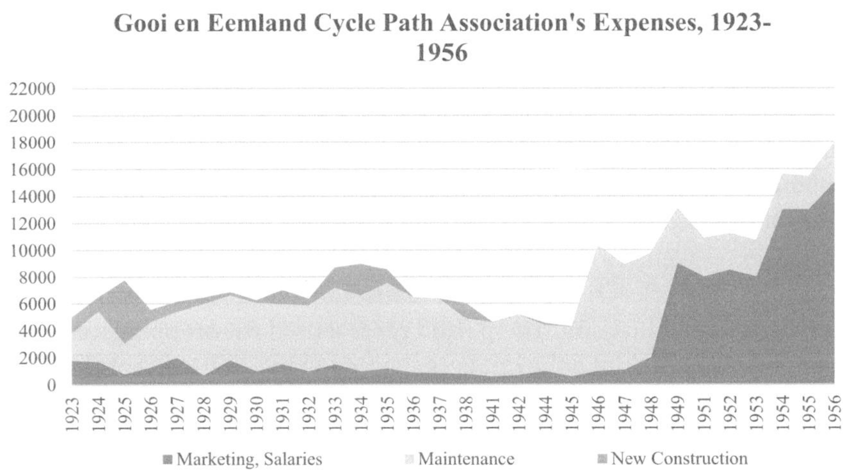

Figure 5 While subsidies rose, higher (labour) costs meant that the citizen organizations could not translate this higher income into a larger network. Source: ANWB archive, inv. nos. 1111 and 1112 (data for 1939, 1940, 1943 and 1950 missing).

subsidizing privately initiated cycle paths, local authorities were reluctant, fearing that without control over these private associations, granting public money would mean fully relegating control over the recreational network. If local authorities raised funds for their cycling networks, why not carry out the work under their direct control? The idea of public responsibility gained more ground in the late 1930s and 1940s – and in the long run rendered cycling path organizations obsolete. In figures 4 and 5, we see the income and expenses of the Gooi en Eemland organization from the 1920s to the 1950s. These figures show how subsidies and membership fees stagnated during the economic crisis of the 1930s. All income went to maintenance instead of expanding the cycle network. After the war in the late 1940s, subsidies increased sharply, but as the data shows, costs also rose. By then, the citizen organizations reached their limits and their role in recreational cycling governance diminished. In other words, the cycling path organizations had their heyday between 1920s and 1950s – as their 2,500 km of paths demonstrate. Had local government agreed to funding, they could have continued their work. Ultimately, the private regional organizations represent the path not chosen.

The local clubs did not disappear. Instead, the cycling path organizations were upscaled to a higher governance level in the provinces. Over time, the argument that the government, at some level, should be responsible for providing recreational cycling infrastructure gained traction among Dutch provincial and municipal policymakers – in large part because the nationally oriented ANWB succeeded in convincing provincial policymakers to cooperate with the cycling path organizations and create new cycling paths as a public good, serving tourism, recreation, and (rural) commuters. As a result, several non-governmental provincial organizations were subsumed later in the 1950s and 1960s into regional governmental recreational agencies (*recreatieschappen*). These built on the physical and personal networks created decades earlier by recognizing recreational cycling as a public good. The period represented the steppingstone for later, when provincial policymakers succeeded in ensuring that the civil society efforts continued.

1.3 Advocating Cycling as Part of Car-Centric Planning

In 1923, a sharp attack on the ANWB appeared in numerous newspapers. Signed simply "R.," the letter's author was Pierre van de Rivière, chairman of the Dutch tourist organization VVV's branch in the rural, northernmost province Groningen and editor-in-chief of the *Provinciale Groninger Courant*. According to Van de Rivière, the government showed more interest in cars than in bicycles, possibly because the authorities "as a general rule have become drivers, and no longer ride bicycles."[69] The ANWB had similarly turned its back on cyclists. Its support of cycling path construction was a ruse, "not favoring cyclists, but drivers who want an empty road for their high speeds."[70] New separated cycling paths were often narrow, forcing cyclists to ride single file to allow overtaking, hardly conducive to the leisure rides Van de Rivière promoted as chairman of a tourist organization. This proved that both the ANWB and the government went to much greater lengths to create a pleasant experience for motorists than for cyclists. Van de Rivière criticized the hugely expensive attempt to increase drivers' speed as politically misguided. By backing this scheme, the ANWB revealed its true colors as a "degenerate cyclists' union," and so he called upon "the meek cyclist" to resist this mistreatment and use their majority to create a "real cyclists' organization."[71] The critique shows the claim of the ANWB to represent (all) cyclists was contested.

In the 1920s – just as again decades later in the 1970s – the ANWB indignantly rejected such criticism. Still wanting to be considered a cycling organization, the ANWB was evidently alarmed at the potential reputational damage Van de Rivière's letter could cause.[72] Equally, it wanted to be seen as the preeminent car organization in the Netherlands. A 1938 survey revealed that many tourists still regarded the ANWB primarily as a cyclists' club. Its leadership deplored the impression, calling it "outrageous" that an organization with tens of thousands of driver members and that published a widely read car journal (*Autokampioen*), was still seen as purely for cycling.[73] The ANWB was convinced it could promote cycling and driving equally. In the spirit of typically Dutch political compromise, an agreement was feasible, even if cyclists had to give up a little more than motorists, the organization seemed to suggest. The resulting principle of traffic separation applied to major roads may have favored drivers (financially). In the long run, as I argue, it also created better conditions for cyclists and drivers' co-existence to succeed than elsewhere. The next chapter explains how this compromise came about. Combining cycling and driving as a lobby organization in the 1920s and

69 R. [P.F.J.H. van de Rivière], "De fiets in het nauw," *Provinciale Geldersche en Nijmeegsche Courant*, September 22, 1923.

70 Ibid.

71 Ibid.

72 ANWB archive, inv. no. 1665.

73 "Misvattingen over het werk van den A.N.W.B.," *De Kampioen* 55, no. 23 (1938): 413.

1930s, while also trying to retain its function as a tourist organization, was a true balancing act.

Early on, the ANWB's cycling advocacy defended the value of cycling for the whole country. Framing cycling infrastructure as a public good involved emphasizing the bicycle's utilitarian potential – rather than just its predominantly recreational use. This distinction between recreational and utilitarian use is crucial to understanding every aspect of Dutch cycling. In practice, the distinction may not have been very sharp – after all, cycling infrastructure often served dual purposes. State engineers and decision-makers, however, created separate governance structures and funding streams for both. And uniquely, the ANWB leadership managed to position the organization as the only stakeholder in both the private governance coalition for recreational cycling as well as the public policy coalition for commuter cycling infrastructure. The ANWB technocratically considered its own organization the only common factor in cycling policy free from the ephemeral state party politics, and therefore a more efficient, reliable stakeholder.

The monopoly is quite remarkable. The ANWB carved out its key mediating role in Dutch mobility governance thanks to its board members' elite status and its unrivalled knowledge of local road conditions. While civil society organizations can have a role in political decision-making, interest groups still need to prove themselves reliable and useful governance partners. By successfully presenting itself as representing the interests of tourists (from pedestrians to car drivers) and an exceptional source of expertise, the ANWB succeeded in becoming a key presence in many government bodies and on relevant committees, like the Committee for Consultation on Roads (see 2.2). As state engineers came to regard roadbuilding (for the new car, and to a lesser extent the bicycle) as their task, they required new road engineering expertise for which they had not really been trained. The ANWB organized conferences, published trade journals, and disseminated new knowledge of roadbuilding and mobility.

Until the 1920s, the tourist organization developed more expertise in road engineering and (access to) user knowledge than Dutch state engineers. This also allowed it to become a powerful lobby group after 1920. To obtain this specialist "road" knowledge, the organization appointed regional representatives, called consuls, who informed the central board of local developments and lobbied at a local level.[74] Since 1898, this knowledge

74 Linders-Rooijendijk, "Gebaande wegen I," 138-44, 225-29; Buiter and Staal, *Het avontuur van de ANWB*, 20-22, 50. Germany did not have such a local network: Ebert, *Radelnde Nationen*, 166.

came together in the Roads Committee that the organization formed with the rival Royal Automobile Club (KNAC).[75] According to Mom and Filarski, Dutch Public Works historically focused on rail- and waterways instead of roads. The state agency found it difficult to introduce new organizational forms and procedures to address the roads issue.[76] The non-governmental ANWB stepped in as an indispensable partner for making infrastructure policy – a role it gladly claimed and fiercely defended.[77] State engineers found it important to participate in its network.[78] A close relationship emerged between the state and the non-governmental organization. In 1917, several national and local engineers joined the organization's Roads Committee to benefit from its expertise.[79] Key members of the ANWB studied engineering at the Military Academy in Breda, like Alexander Redelé and Hermanus Willem Otto de Bruyn, both former engineers in the Royal Netherlands East Indies Army. After World War II, the ANWB's traffic department employed University of Delft-trained civil engineers.[80]

The ANWB, rather than state officials, became the go-to advisor for all questions relating to cycling and roadbuilding.[81] This pattern also emerges from the ANWB archives: (state) organizations with a question or a complaint about cycling contacted the ANWB because it most visibly represented cyclists. Whether it was for advice on the best bicycle parking system, help with lobbying, or paying for a local cycle path, citizens from all over the Netherlands turned to the ANWB.[82] Most people would not know how to go about making a complaint or a request to the government about cycling. The ANWB often put petitioners in contact with the relevant authorities.

75 Despite working with this car club, the ANWB saw it as a rival and tried to stop it becoming too powerful by also embracing car advocacy: Buiter and Staal, *Het avontuur van de ANWB*, 46.

76 Mom and Filarski, *De mobiliteitsexplosie*, 177. On Public Works administration before the 1920s, Eric Berkers, *Technocraten en bureaucraten: Ontwikkeling van organisatie en personeel van de Rijkswaterstaat, 1848-1930* (Zaltbommel: Europese Bibliotheek, 2002); A. Bosch et al., *Twee eeuwen Rijkswaterstaat: 1789-2015*, 2nd revised and expanded edition (Asten: Nieuwe Uitgevers, 2015).

77 This also happened with the French Automobile-Club du Rhône in Lyon. Gardon, "Gouverner la circulation urbaine: Des villes françaises face à l'automobile (années 1910 – années 1960)," 2.3.

78 Mom and Filarski, *De mobiliteitsexplosie*, 93-94.

79 Linders-Rooijendijk, "Gebaande wegen I," 142.

80 These included A. Boost (trained at Breda), L. Lorié, K. Tusenius, H. Puister, C. Kuysten, G. de Regt and J. Barkhof, see Chapter 4. The Technical University of Delft had university status since 1905 but was not officially named a university until 1986, operating as a *Technische Hogeschool*.

81 Mom and Filarski, *De mobiliteitsexplosie*, 178; Marcus Popkema, "Tussen techniek en planning: de opkomst van het vak Verkeerskunde in Nederland, 1950-1975" (PhD diss., TU Eindhoven, 2014), 74.

82 The ANWB archive has numerous requests for advice on bicycle parking systems.

Over the course of the twentieth century, a complex governance network emerged around cycling, ranging from state agencies and industry to user groups. Yet the ANWB succeeded in acquiring a near monopoly over cycling governance before 1920. Its legitimacy as an expert foreshadows the user expertise that Dutch cycling activists would claim to possess in the 1970s.

In turn, the ANWB managed to place its members in key positions within new institutions and forums that channeled knowledge about traffic and road engineering. Through all its ventures, it had a strong impact not just on policy but also guiding (local) traffic experts. From 1925, the non-governmental organization issued the leading roadbuilding journal *Roads* (*Wegen*) and organized annual roadbuilding conferences through its affiliated organization The Dutch Road Congress (*Het Nederlandsche Wegencongres*).[83] A leading organizer and *Roads* editor-in-chief was engineer De Bruyn. During army leave, he had become involved in the ANWB, heading the Roads and Traffic section. After retiring from active service in 1922, De Bruyn became a "spider in the web of the roadbuilding lobby."[84] He often served as the ANWB's representative on governmental consultation committees (see 2.3).

To retain its power, the ANWB was forced to take up utilitarian transport and traffic when these became important topics around 1920. An emerging car lobby made the case for urgent road improvements based on economic arguments.[85] The ANWB initiated the first Dutch Road Congress in 1920 in cooperation with the Royal Institute of Engineers (KIvI), not with the government Public Works Agency.[86] This culminated in the 1925 "Grand Meeting" where some 200 stakeholders gathered to discuss road improvement and funding.[87] Thanks to the ANWB's advocacy, the debate led to a compromise between motorists and cyclists, as we will see in the next chapter. The ANWB's positioning involved a subtle shifting in political emphasis between the 1890s and 1920s.

Since the 1890s, the question of determining cycling's status as a public good had centered on whether cycling should be considered simply a leisure

83 Linders-Rooijendijk, "Gebaande wegen I," 226.

84 Mom and Filarski, *De mobiliteitsexplosie*, 178.

85 M.L. ten Horn-van Nispen, "Het Rijkswegenplan 1927. De aanpak van het verkeersvraagstuk in de jaren twintig," *Jaarboek voor de geschiedenis van bedrijf en techniek* 9 (1992): 185. In 1915, Transport Minister Cornelis Lely had already proposed in vain to improve some roads: Dirk Ligtermoet, *Beleid en planning in de wegenbouw: De relatie tussen beleidsvorming en planning in de geschiedenis van de aanleg en verbetering van rijkswegen* Rijkswaterstaat-serie, vol. 51 (Den Haag: Rijkswaterstaat, 1990), 15-16.

86 Mom and Filarski, *De mobiliteitsexplosie*, 174-80; Linders-Rooijendijk, "Gebaande wegen I," 226.

87 Mom and Filarski, *De mobiliteitsexplosie*, 179-80.

activity for the well-to-do or whether it should also be a utilitarian mobility for commuting and going to school. The ANWB framed cycling in both ways to claim its larger role in state policy: cycling was not just a luxury for the happy few, but useful for the entire nation – and the government should invest in promoting it. Back in 1895, when the province Noord-Brabant was considering a bicycle tax, the tourist organization wrote to the provincial authorities, opposing their argument that cycling was a luxury good or activity. The ANWB countered: "the bicycle is no longer a recreational activity for the well-to-do, but has surely become the man in the street's and the manual laborer's horse."[88] This claim was an exaggeration, but it suited the ANWB's political strategy. The ANWB objected furthermore to the prohibitive provincial bicycle tax, arguing that cycling brought the rural population closer to urban centers. Finally, echoing modern arguments, the organization claimed that the bicycle was "of great importance for the health of persons who lead a busy but sedentary lifestyle."[89]

These arguments had entered the national stage when the ANWB fought a proposed national bicycle tax in 1895. At the time, conservative liberal MP Jan Willem Hendrik Rutgers van Rozenburg (1830-1902) introduced the idea when parliament discussed a general tax reform.[90] MPs debated which new consumer goods and services – for example, domestic servants, luxury furniture, and horses – should be included in the "personal tax" (*personele belasting*). According to Rutgers van Rozenburg, since horses were taxed as a luxury good, the bicycle as their successor ("the iron horse" as bicycles were called) should be as well.[91] He claimed that "enthusiasts use at least 90 percent of the bicycles in this country, only a few are used by a pharmacist, a baker, or a butcher."[92] This view was not shared by many at the time.

The Dutch did not invent bicycle taxation: across Europe, cycling was taxed as a luxury activity in the 1890s. By 1895, France already had a bicycle tax.[93] Individual provinces in Belgium introduced a tax between 1893 and

88 NA 2.04.57 (Ministry of Internal Affairs), inv. no. 2280, ANWB petition.

89 ANWB archive, inv. no. 1657, petition ANWB against bicycle tax Noord-Brabant.

90 Background political information from www.parlement.com, a site run by and based on research at the University of Leiden's Centre for Parliamentary Documentation.

91 *Handelingen II* 1895/96, 17 (November 19, 1895), 266, where Rutgers van Rozenburg defended the comparison with the horse, pointing out that its utilitarian use on farms exceeded all utilitarian cycling's share.

92 *Handelingen II* 1895/96, 16 (November 15, 1895), 248.

93 Part of France's tax laws in 1893. The initial rate was 10 francs, reduced to 6 (1898), then 3 (1907), raised again to 6 (1924), then 12 (1926), and 25 (1942), before it was abolished in 1958. As Dutch cyclists already had a metal disc as proof of payment, this idea was obviously not pioneered in the Netherlands (adopted in 1924).

1895, centralizing this in 1900, but did not allocate the revenues to cycling paths or road improvement.[94] Only around the capital Brussels was bicycle tax money reserved for cycling infrastructure. Similarly, some cities and provinces in the Netherlands considered a bicycle tax in the mid-1890s, but the national tax scheme prevented this.[95] According to Albert de la Bruhèze and Oldenziel, around 1900, various countries either used the bicycle tax for building cycling lanes or justified it by promising them. Internationally, cycling was still considered an upper-class leisure activity and hence taxable as a luxury.[96]

Dutch national politicians contested the bicycle's luxury status already in the 1895 debate. In contrast to his conservative liberal colleague Rutgers van Rozenburg, Finance minister Sprenger van Eyck opposed a luxury tax "since most bicycles are being used for people's occupation or business."[97] Rutgers van Rozenburg, commenting on a draft of the bill in 1898, asserted that he had always meant to exempt utilitarian cycling from the tax and missed an "exemption for farm laborers and other workmen, for schoolchildren and for everybody who uses the vehicle in their livelihood; I also unfortunately miss the semi-exemption, at the time proposed by us, for so-called second-hand bicycles and bicycles for mixed use."[98] Commenting in 1924, social democratic MP Arie IJzerman found this a striking development: Rutgers van Rozenburg had argued for these exemptions even though he "could not be suspected of democratic leanings" and had "wholehearted contempt" for the working class.[99] If a politician of this persuasion could regard utilitarian cycling as a national public good in the 1890s, there would be an even stronger reason for more left-leaning politicians.

In the 1920s, the ANWB would change its position. In the 1890s, it still objected to a bicycle tax. To do so, the organization – perhaps opportunistically – claimed cycling was as much a commuter activity as a recreational one. The bicycle was simultaneously a useful vehicle for workers and a

94 Donald Weber, *De blijde intrede van de automobiel in België 1895-1940* (Gent: Academia Press, 2010), 23.

95 Cities such as Enschede (1895, 6 guilders) and Brielle (1895, 4 guilders) considered a bicycle tax. The province Noord-Brabant proposed a 5-guilder tax in 1896 but was asked to suspend this in anticipation of a national law. After lobbying, it could levy the tax, which meant a double tax burden for cyclists until the 1930s, (N.A. Lunsingh Tonckens, *Bekostiging van wegen* (Hengelo: H.L. Smit & Zn., 1929), 89-93.) Attempts by social democrat MP Drop to change this failed: *Handelingen II* 1929/30, 28 (December 3, 1929), 838.

96 Albert de la Bruhèze and Oldenziel, "Who Pays, Who Benefits?," 75-77.

97 *Handelingen I* 1895/96, 16 (November 15, 1895), 248.

98 *Handelingen I* 1897/98, 27 (July 12, 1898), 385.

99 *Handelingen II* 1923/24, 69 (May 13, 1924), 2007.

recreational toy for the upper classes. In a petition, it argued that a bicycle tax should acknowledge the different types as well as the widely diverging costs of (second-hand) bicycles.[100] As a young and new organization, the ANWB was still building up its credibility and legitimacy. Rutgers van Rozenburg dismissed the ANWB's involvement by accusing its relatively young members of "youthful rashness" and "tilting at windmills" in their petition.[101] The liberal MP especially resented what he considered the club's disingenuous emotional appeal regarding the plight of poor farmers and their bicycles, questioning the frequent references to working-class cyclists and the "fairy tales that there are numerous workers who to ride to their work, buy a bicycle with 25 guilders taken out of the savings bank."[102] In other words, Rutgers van Rozenburg, staking his defense of a bicycle tax entirely on framing cycling as recreational, had to dismiss the claim that cycling was also, and increasingly, a utilitarian activity.

While exaggerated, there was something to the ANWB's claim that the bicycle had a utilitarian use around 1900. The organization's chairman Bergsma attached a list to its 1898 petition against the bicycle tax. He had begun his long public career as mayor of the Frisian village Het Bildt (1892-1896) and listed all his villagers who needed a bicycle for their work. In a community "connected neither to the tramway nor to the railroad ... a cyclist purely for pleasure can hardly be found."[103] Cyclists in Het Bildt included manufacturers, merchants, the doctor, veterinarian, teacher, reporter, sales representatives, blacksmiths, housepainters, carpenters, and farm workers. The petition argued that those residents had "not purchased their bicycle out of luxury." Even if they also used their bicycle for recreation, this did not justify an "increase of the taxes, already oppressive" which would "weigh heavily" on the average cyclist.[104] The argument may have been self-serving, but when the tax was introduced, the number of bicycles in the cheapest tax category (the working class) increased rapidly, showing that the bicycle had become affordable for many workers.

The ANWB's 1898 petition stipulated that if the government introduced the tax, cyclists should be compensated with proper cycling infrastructure – to be paid in part from the bicycle tax revenues. The income should at least partly be used for "the improvement of roads or the construction of cycling

100 *Handelingen II* 1895/96, 17 (November 19, 1895), 273.

101 *Handelingen II* 1895/96, 18 (November 20, 1895), 284.

102 Ibid.

103 NA 2.21.026.08, inv. no. 49 (personal archive Mr. H.J. Smid, bicycle tax collection 1895), ANWB petition.

104 Ibid.

paths."[105] This point is crucial: it is one of the first instances when bicycle taxation was linked to public cycling path construction. The argument would weigh heavily for justifying a new bicycle tax twenty-five years later. It demonstrates the ANWB's willingness to compromise and its belief that cyclists' interests were best served by working with the government. From the start, the ANWB was willing to consider a bicycle tax, if done in a fair and balanced way, using a progressive tax rate and/or a generous system of exemptions for working-class cyclists using the bicycle for work.

By the early 1920s, the ANWB came to support the bicycle tax. By then, the organization's mission had fundamentally changed. It had been founded in the 1880s as a single-issue cycling club. By the 1920s the ANWB had evolved into a complex organization. It defended middle class recreational cycling and claimed to defend utilitarian cycling, while balancing the interests of cyclists against those of car drivers. As an all-purpose tourist and lobby organization trying to prevent rival traffic organizations from becoming too powerful, the cycling club quickly converted to a car lobby group. The tourist organization competed with the car organization KNAC for members. Despite the ANWB claiming to represent all cyclists, when the number of cyclists grew, an ever-smaller percentage joined the ANWB.[106] It nevertheless continued to present itself as the cyclists' representative, avoiding specification of the term "cyclist" while in practice doing more for middle-class recreational cyclists than urban working-class cyclists, for instance. Over the course of the twentieth century, the ANWB was increasingly a lobby organization for motorists, yet continued its cycling advocacy. Reconciling cycling and driving, the demand for a large road network, and support for nature reserves was a delicate balancing act. Ultimately, various parts of this "heterogenous, multi-layered" organization pulled in different directions.[107]

ANWB chairman Bergsma, who served from 1884 to 1937, best represents the balancing act. A strong supporter of cycling from the start, he decades later considered the car as the bicycle's natural successor. In 1936, he wrote that the car performed on a larger scale what the bicycle did on a local level – ending the countryside's isolation and bringing more jobs and recreational facilities within people's reach. When the bicycle was introduced in the 1880s the ANWB promoted its usefulness and safety. In the 1930s, the

105 *Handelingen II* 1897/98, 59 (June 9, 1989), 1007. The ANWB also demanded a better legal status for cyclists by creating uniform national traffic rules, which did not yet exist.
106 Exact numbers are hard to give, as car drivers also became ANWB members, not least because of the roadside assistance service: Linders-Rooijendijk, "Gebaande wegen I," 355.
107 Mom and Filarski, *De mobiliteitsexplosie*, 173.

organization did the same for the car. It argued the public had to accept the car as a useful transport mode of the future. At the same time, drivers had to learn how to drive responsibly and in a civilized manner.[108] To this end, the ANWB published articles targeted at drivers, for example, explaining how overtaking cyclists at speed created a gust of wind that threatened their balance and safety.[109]

After the 1920s, alongside its continued support for cycling, the ANWB believed that a high level of car ownership was a sign of modernity. The organization considered the slightly declining number of cars on Dutch roads during the 1930s economic crisis "extraordinarily grave" and the high car tax "an attack against the car."[110] Compared to other European countries, the Dutch "cut a poor figure as far as car density is concerned."[111] According to historian Vincent van der Vinne, the fiscal pressure on drivers was indeed higher in the Netherlands than elsewhere in the 1930s.[112] The Danish Bicycle Federation (*Dansk Cyklist Forbund*, DCF), founded in 1905, created a similar alliance with automobility – although the relationship was contested and took time to materialize.[113] Contrary to the Dutch case in the 1920s, the DCF even proposed that "the luxury taxes on cars should be used for building bicycle paths."[114] Danish drivers retaliated by proposing a bicycle tax, which was never introduced in Denmark.

While becoming increasingly pro-car, in the 1930s, the ANWB leadership still argued that the Dutch government should not forget cycling infrastructure as a matter of public good. At the ANWB's 1939 annual meeting, chairman Bloemers thought it "incomprehensible" that the government did so little for the four million cyclists in the Netherlands and accused it of "seeing traffic only from the viewpoint of fast traffic."[115] Bloemers gave numerous examples of situations where engineers failed to allot enough space to cyclists. Often separate cycling paths were lacking or suddenly ended at bridges over major rivers. This was an insult to cyclists, who paid a bicycle tax, he argued. If anything, this tax should ensure cyclists had

108 E. Bergsma, "Middelen om verkeersonveiligheid te bestrijden," *De Kampioen* 53, no. 3 (1936): 49-50.

109 "Wanneer een auto een fietser inhaalt," *De Kampioen* 53, no. 4 (1936): 83.

110 "Het 53ste jaarverslag (Vervolg)," *De Kampioen* 53, no. 16 (1936): 346.

111 [H.J. Peppink], "Er zijn 43.297.587 automobielen op de wereld," *De Kampioen* 56, no. 17 (1939): 319.

112 Vincent van der Vinne, *De trage verbreiding van de auto in Nederland 1896-1939: de invloed van ondernemers, gebruikers en overheid* (Amsterdam: De Bataafsche Leeuw, 2007), 411.

113 Carstensen and Ebert, "Cycling Cultures in Northern Europe," 35-36.

114 Ibid.

115 "Notulen van de jaarlijksche Algemeene Vergadering," *De Kampioen* 56, no. 18 (1939): 346.

the same status as drivers. Not only should there be a network of separated cycling paths alongside main roads, but he also pled for an extensive network of recreational cycling paths.[116] How the state took up this call to facilitate cycling alongside driving is the topic of the next chapter.

1.4 Conclusion

Non-governmental organizations played a foundational role in Dutch cycling governance. The tourist and lobby organization ANWB in particular earned a long-term position in Dutch cycling governance, unlike clubs elsewhere. This chapter explored their attempts at putting cycling on the political agenda. This advocacy's success was predicated on an inclusive discourse presenting cycling infrastructure as a public good. While cycling's advocates were middle class and well-connected politically, they did anticipate and promote a broader working-class use of the bicycle – even if some found the arguments opportunistic. According to Mom, 1930s car lobbyists found it advantageous to spread myths about the car as a utilitarian vehicle, even though most drivers used cars for pleasure, adventure, and as status symbol.[117] In a similar vein, by connecting different types of cycling to the pressing social issues of the day, claims for government to support cycling and its infrastructure became more powerful.

Depending on political circumstances, the organizations foregrounded either non-urban or urban cycling, recreational, or utilitarian cycling. Even in the first fifty years of Dutch cycling (1880s-1930s), when they predominantly promoted recreational activities and only gradually became more affordable, the non-governmental organizations also presented the utilitarian framing of cycling as a political tactic to get funding. Cycling advocates within the ANWB and Parliament linked concerns over the unhealthy living conditions in cities with the bicycle's potential for recreation in the countryside. Conversely, some also argued the bicycle could also solve the issue of people living in the countryside being isolated (from work, school, and the rest of the nation). Many of these arguments returned later in the twentieth century, as cycling advocates kept having to make the case for cycling infrastructure as a public good.

116 An additional argument for separate cycling infrastructure was the claim that accidents on mixed traffic roads were 32 percent higher than on roads with separated cycling paths.
117 Gijs Mom, *Atlantic Automobilism: Emergence and Persistence of the Car, 1895-1940* (New York: Berghahn Books, 2015), 343-46.

This chapter, "Citizen Power," identifies why a civil-society stakeholder like the ANWB managed to gain a large governance role early on. First, the openness of the Dutch political system to representatives of civil society was significant. The political rules of the game enabled the ANWB to become key to cycling's policy coalition. The organization made full use of this. Its board members were at home in policymaking and administration circles. Second, the organization's leadership was adroit at preventing rival tourist or mobility organizations from gaining prominence. The organization built up expertise between the 1890s and 1920s that also benefited state engineers, while lobbying engineers at different state levels to give cyclists a place alongside the future car infrastructure. When the topic of utilitarian infrastructure for cars and cyclists became a political issue in the 1920s, the ANWB protected cyclists' interests when policymakers embraced the economic value of cars and trucks, leaving cyclists out of the equation.[118] In contrast to its sister organizations in other countries, the leadership believed it could support both sides and continued to lobby, in the form of countless requests to local authorities for constructing and funding cycling facilities, and managed to represent cycling in the policymaking process.

For all its wide-ranging lobbying work, the ANWB had little to say about cycling in cities, even though many local cycling organizations were started up in cities by well-to-do urbanites as recreational clubs to help citizens escape and explore the countryside. To get funding or when lobbying, the ANWB discourse also frequently invoked workers who used their bicycle to commute. Its lobby focused on cycling infrastructure outside cities, along the busiest roads, and in nature reserves. Given the low levels of automobility in the Netherlands until the late 1950s – especially from an international perspective – there was perhaps less political pressure for dedicated infrastructure to separate cyclists from motorists in cities. Still, the disparity between cycling facilities constructed in the countryside and those in cities characterized the period. It would take another half century before the needs of the urban cyclist were addressed. The ANWB and cycling path organizations' extensive network of recreational infrastructure helped integrate the Netherlands materially, even though it barely involved the government. When the state's role in this policy domain of leisure grew, the private governance model clashed with a potential public role.

118 Popkema, "Tussen techniek en planning," 50. Attempts to calculate the economic benefits of cycling: O.J. Boot and Jan Ploeger, *De economische waarde van het fietsverkeer in het bijzonder met motief onderwijs: Eindrapportage* (Den Haag: Adviesbureau Van Roon, 1987).

2 A Contested Compromise: National Government Supports Commuter Cycling

In 1930, the accountant Duyts had to build a cycling path in his road because his town authorities refused to take on this task. Citizens like Duyts across the country organized to construct their cycling paths because they could not count on the government. Yet something was changing. In the late 1920s, unbeknown to Duyts, engineers and interest groups at the highest level of Dutch policymaking were debating the principles of a new program for roads and cycling paths. In the summer of 1927, leading national Public Works engineers paid a work visit to the busy road between the manufacturing town Delft and the harbor city Rotterdam along the river Schie. Local provincial engineer Arnold Kempees gave them a walking tour so that they could observe with their own eyes the traffic situation and road conditions. The engineers expected traffic to increase rapidly, and concluded "this narrow road with busy car traffic is especially dangerous for cyclists."[1] Gerrit Jan van den Broek, a leading Public Works engineers thought, "at least something should be done for the safety of cyclists."[2] Engineer Ludolf Reinier Wentholt suggested constructing a cycling path on the other side of the Schie, but his colleague George Rooseboom shot down this plan: without a bridge to cross the river, cyclists would waste time waiting for the ferry, and prefer to cycle on the road – it would simply be faster, so who could blame them? The public works' administration August 1926 traffic counts had shown that at least 800 cyclists used the road daily. The engineers decided that the situation was urgent enough to justify a separate cycling path, a solution benefiting both cyclists and drivers.

In 1929, Public Works inspector-general and engineer Rooseboom found the cyclists' requests for infrastructure "reasonable demands," arguing that engineers took "insufficient account of the interests of the cyclists … One should not forget that the cyclists produce a large share of the Road Fund."[3] This sounded promising. Why was cycling infrastructure now seen

1 NA 2.16.59, inv. no. 43, committee meeting June 8, 1927.

2 Ibid.

3 NA 2.16.59, inv. no. 44, meeting March 7, 1929.

as a public good, a task for state engineers? And why was the "Road Fund" important to Rooseboom?

Struggling to fund their recreational infrastructure, cycling organizations were in no position to also initiate commuter paths. The state became involved in the 1920s for two reasons. Some engineers and politicians – echoing their international counterparts – believed that the Netherlands needed an ambitious highway system for cars to prepare for the modernist future. This early lobby for automobility is somewhat surprising: as historians have pointed out, the Netherlands was one of the first countries to undertake such a program in Europe, yet its car levels were the lowest. To facilitate such a network, the many cyclists obstructing the flow of drivers on planned major car routes had to be moved aside, one of the motivations in the 1927 Delft-Rotterdam road debate. This development inadvertently coincided with the government's new bicycle tax in 1924. Originally introduced to reduce budget deficits, the tax was earmarked for road improvement in 1926 after lobbying efforts from car interests and national engineers. What arguments were employed to justify this? And what were the (long-term) consequences for Dutch cycling (infrastructure)?

Historians Adri Albert de la Bruhèze and Ruth Oldenziel as well as Anne-Katrin Ebert have studied the inequalities and profound consequences of the bicycle tax for the future of Dutch cycling. Ebert evaluates German and Dutch cycling clubs' cultural policies to explain why the Dutch ANWB was more successful in framing cycling as a national activity. According to Ebert Dutch cyclists used the bicycle tax to demand (utilitarian) cycling paths: "The bicycle tax put cyclists on the political map and helped to create a tradition of traffic engineering devoted to cycling paths and regulation. This would form an important basis for the 'survival' of the Netherlands as a cycling nation in the second half of the twentieth century."[4] Compared to German clubs, the Dutch ANWB could make a stronger claim. In Ebert's view, this explains the stronger embedding of cycling in Dutch (middle-class) culture and politics. In contrast, Albert de la Bruhèze and Oldenziel emphasize the unjust distribution of bicycle tax revenue and its highly contested political nature. Not only did tax revenues prioritize the few drivers over the much larger group of cyclists, it also benefited non-urban cyclists more than urban working-class cyclists. Over a ten-year period, the tax forced a political compromise between cyclists and car drivers, providing the overwhelmingly urban and taxpaying cyclists with roads that included cycling paths outside the city. The numerous cyclists generated

4 Carstensen and Ebert, "Cycling Cultures in Northern Europe," 37-38.

a higher revenue than the small group of drivers, yet most proceeds from the tax went to building highways. Working-class urban cyclists did not use these paths and did not benefit from them, despite paying tax.[5] Using newly sourced material, this chapter, "Contested Compromise," adds depth, detail, and corrections to these claims. By studying a longer timeframe than Ebert, Albert de la Bruhèze, and Oldenziel, the following chapters trace the long-term impact of the bicycle tax.

The justification for taxing cyclists and drivers in relation to road and cycling path building was the subject of lengthy parliamentary debates. State engineers deliberated with social representatives about roadbuilding norms. Records of these debates shed light on the behind-the-scenes negotiations. This rich material reveals how the political compromise for the bicycle tax led Dutch state engineers to regard cycling commuting infrastructure as a public good. As national and provincial engineers had almost no influence on urban infrastructure, however, these policymakers largely ignored urban cycling (and commuters) in the 1920s and 1930s. Only in the 1970s would urban cycling appear on the national agenda. I show that social democrat MPs ensured the debate included the value of utility cycling in the countryside. As Ebert did not focus on government sources, her claim that engineers constructed Dutch cycling paths on account of the bicycle tax needs further examination. Public Works archives reveal that cycling paths were indeed considered a political compromise for the controversial bicycle tax. I also show that cycling paths were already on the agenda (and considered justified) before the bicycle tax. Cycling's legitimacy did not rest exclusively on the existence of a bicycle tax. I examine how the translation of principles into government agency road engineering norms had a significant long-term effect on cycling path construction. This chapter takes up the political debate on where cycling belonged in governance terms. In the spirit of compromise between interest groups, Dutch policymakers found a way to facilitate driving combined with cycling.

2.1 Justifying Road Funding and the Bicycle Tax

Traditionally, Dutch engineering had been focused on building water- and railways before it invested heavily in roadbuilding as well. In 1913, Public Works Minister Cornelis Lely, a famous engineer responsible for some of the Netherlands' major water works, initiated the government's plans for

5 Albert de la Bruhèze and Oldenziel, "Who Pays, Who Benefits?," 73-100.

roadbuilding. He ordered Public Works engineers to design a new (and expensive) national roads network: their 1915 plan projected reserving a 10-meter-wide space for roads so that cycling paths could be included if necessary. A state role in cycling infrastructure was at least on the agenda. Financial constraints during World War I dashed any hope of realizing the expensive plan.[6] Early on, Dutch engineers and social stakeholders envisaged the future motor age and the roads network required. Since most Public Works engineers were railway specialists, lacking knowledge of roadbuilding, the tourist non-governmental organization ANWB managed to play an outsized role in providing advice.[7] Projecting a rapid increase in motorized traffic, politicians and engineers claimed that the Dutch road network was not ready for – what they assumed would be – a car-centered future. They felt that the existing roads were too narrow and could not withstand heavy vehicle loads.[8] Previously, lobbyists for better roads presented tourism as their main argument. After 1920, they adopted utilitarian terms like the economy, industry, and agriculture. Although the Netherlands was the premier cycling country, its leadership also embraced cars early on. The country built exclusive automotive highways in the 1930s – within Europe only Italy and Germany did this earlier, and the UK waited until long after World War II.[9] Those boosting an automotive future realized it would be extremely expensive and required new financial instruments. Consequently, taxation considerations in the mid-1920s proposed earmarking the 1924 bicycle tax (an emergency measure to balance the budget) for road construction. This decision had major implications for the construction of suburban and rural cycling paths alongside major roads.

Funding road improvements meant answering the question: should cyclists be included or excluded in road taxation? To jumpstart the discussion on financing roadbuilding, stakeholders in the new car industry organized the Dutch Road Congress and decided in the early 1920s to draft a road tax bill. The driving force behind this congress, and opponent of the toll road model, was the ANWB. It set up a committee with representatives of its Roads Committee, automobile club KNAC, commercial car owners (BBN),

6 Mom and Filarski, *De mobiliteitsexplosie*, 100-103.

7 Ibid., 91, 100, 177; Willem van der Ham, *Heersen en Beheersen: Rijkswaterstaat in de twintigste eeuw*, PhD diss. (Zaltbommel: Europese Bibliotheek, 1999), 30-31, 47-48; Auke van der Woud, *Het landschap, de mensen. Nederland 1850-1940* (Amsterdam: Prometheus, 2020).

8 Especially in the West of the Netherlands, where soil conditions made subsidence a major risk.

9 Gijs Mom, "Roads without Rails: European Highway Network-Building and the Desire for Long-Range Motorized Mobility," *Technology and Culture* 46, no. 4 (2005): 745-772, here 747.

government officials from the provinces Groningen and Noord-Brabant (where a road tax was already in place), and state agency legal and financial experts.[10] Once more, interest groups rather than the Dutch government initiated this discussion – although they did cooperate closely with the state. One issue was whether or not to include cyclists in the road taxation.

In their 1924 report, the road and car coalition stakeholders considered taxing cyclists, but decided against it. The main rationale behind their tax proposal was to pay for road maintenance, based on a vehicle's weight and number of wheels. This engineering logic of load differed from taxing road *use*: traditionally the toll model had generated funds for national roads until 1899, although tariffs varied depending on vehicle type and weight.[11] If the main problem was that roads could not withstand the weight of heavy cars and trucks, it would be justifiable to tax drivers.[12] Bicycles were so light that they caused little road damage or necessary maintenance.

Organizations did consider – but rejected – the possibility of taxing cyclists to fund separate cycling paths. According to the 1924 report, cyclists preferred separate cycling paths to improved roads. Free from heavy motorized traffic, the paths would have low maintenance costs. If the state constructed a network of separate cycling paths, cyclists would not benefit from road improvements, making it unjustifiable to incorporate a bicycle tax in the Road Fund. A separate bicycle tax could pay for cycling paths. Because these paths were cheaper to construct than roads – and because there were many cyclists – a tax would have to be modest. However, the plan was unworkable because the sheer cost of the bureaucracy to raise the tax would undermine its actual purpose. For this practical reason, most of the Dutch Road Congress committee decided cyclists should not be taxed under the new bill.[13]

For opportunistic reasons, the more car-oriented stakeholders (industrialists for instance) did push for cyclists paying tax: implementing road improvements with the modest tax revenues from the small number of

10 The committee report: Vereeniging Het Nederlandsche Wegen-Congres, *Ontwerp voor eene wegbelastingwet* ([Den Haag]: Haagsche Drukkerij, 1924).

11 Local toll roads existed well into the twentieth century. The decentralized structure of road governance, allowing local government significant autonomy, meant national government had to buy out local government, a drawn-out and expensive process.

12 Wegen-Congres, *Ontwerp voor eene wegbelastingwet*, 17-20. The issue of funding road use through a pricing system returned in the 1960s and is ongoing: Maarten Smaal, *Politieke strijd om de prijs van automobiliteit: de geschiedenis van een langlopend discours, 1895-2010* (Delft: Eburon, 2012), PhD diss., University of Tilburg.

13 Wegen-Congres, *Ontwerp voor eene wegbelastingwet*, 21.

drivers would delay construction.[14] By contrast, the millions of cyclists on the roads – surprisingly more than the government ever realized before starting to count them – provided a tempting source of income. The committee's report included its dissenting members' opinions. The tourist organization ANWB, representing cyclists, automobile club KNAC, and motorcycle club KNMV, claimed the proposal unjustly singled out motorized vehicles and that both horse-drawn vehicles and bicycles should contribute to road improvements.[15] Committee member Andries Johannes ten Hope, a Rotterdam industrialist representing the transport lobby BBN, penned a strongly worded letter signed by eighteen factory owners.[16] Alarmed by the committee proposal, the entrepreneurs claimed their businesses depended on road transport, and echoing the ANWB, KNAC, and KNMV car lobby comments, argued that bicycles, not cars, caused costly road improvements: cyclists demanded expensive, smooth road surfaces and extraordinarily wide roads "as a consequence of the very large number of cyclists in our country."[17] Cyclists should pay their fair share for road improvements.[18] The road-width argument is problematic. Cyclists only demanded wider roads if car drivers wanted an unobstructed flow without reducing their speed. The real motivation was the remark that cyclists were an easy source of income: because of the vast number of cyclists, even a small levy would generate substantial funds.[19]

Most of the actors involved saw through this transparent defense of a bicycle tax. The committee chose not to comment on this tax. Instead, it objected to the BBN's far-fetched argument for absolving truck drivers and motorized traffic of all responsibility for funding new and better roads. According to the committee, it was undeniable that the need for better roads – which had sufficed until then – resulted from the increase in heavy motorized traffic, especially trucks. The road tax, moreover, would help in constructing a long-distance network of national roads, mainly benefiting industrial and commercial traffic. Taxing local cyclists or even horse-drawn

14 Albert de la Bruhèze and Oldenziel, "Who Pays, Who Benefits?," 80-83.

15 Wegen-Congres, *Ontwerp voor eene wegbelastingwet*, 50.

16 They launched a marketing campaign with posters 'Keep the road clear' and dramatic images of chained factories and collisions between trucks and trains: S.A. Reitsma, *Laat den weg vrij! Een ernstig woord tot het denkende deel der natie* (Den Haag Moorman's Periodieke Pers, 1933).

17 Wegen-Congres, *Ontwerp voor eene wegbelastingwet*, 21.

18 Ibid., 58.

19 Ibid., 21. Noord-Brabant cyclists paid 2.50 guilders tax; in the fiscal year 1921-22, this yielded 287,000 guilders from a total tax income of 744,000, of which only 150,000 was from motorized vehicles.

vehicles was indefensible. Business owners favoring motorized transport should pay for their own roads.[20]

The committee's proposal neither included nor defended a bicycle tax, but offered many arguments in favor. Published in 1924, the report coincided with the introduction of the government's bicycle tax. When the governmental committee, working in parallel on how to fund road improvements, received the report, it proposed including the 1924 bicycle tax revenues in the road fund.[21] In other words, the car lobby's report may not have introduced the bicycle tax, but it did provide arguments to transform a temporary measure into a permanent tax earmarked for road improvement.

In Parliament, however, the bicycle taxation led to extensive debates about cycling infrastructure as a public good. Initially, budget deficits rather than roadbuilding ambitions had motivated the government to reintroduce a bicycle tax in 1924. It was not Public Works but Finance Minister Hendrik Colijn who proposed this. A prominent member of the Protestant ARP party, he became Prime Minister in 1925. Invoking a financial crisis, Colijn argued that the government urgently needed new sources of revenue. Cyclists had great fundraising potential. Colijn stated bluntly: "If the tax levies a moderate amount, it cannot be deemed to weigh too heavily upon the owners of bicycles, whereas it will obtain a quite substantial benefit for the treasury."[22] He estimated one million Dutch cyclists, so a 3-guilder tax would bring in 3 million. To the government's surprise, there were several more million cyclists, so the tax would generate at least double that amount: between the late 1920s and 1940, the number of cyclists rose from almost two to four million.[23] While Colijn might have been inspired by the ongoing debate about road funding, he did not use any of the old arguments for taxing bicycles as luxury goods. Contested in the 1890s, this argument was even harder to make in the 1920s because bicycle ownership and use had spread throughout society. There was simply an urgent national financial need and here was some low hanging fruit, ripe for the picking. Because Colijn had the backing of a majority coalition, parliament passed the proposal, albeit with amendments. Cyclists now had to display a disc shown in figure 6. His major concession was a clause stipulating the tax would expire in 1930.

20 Ibid., 65-66.

21 Mom and Filarski, *De mobiliteitsexplosie*, 179.

22 *Kamerstukken II* 1923/24, 268, 3, 2. These records are digitized and accessible at https://zoek.officielebekendmakingen.nl/uitgebreidzoeken/historisch.

23 Albert de la Bruhèze and Veraart, *Fietsverkeer in praktijk en beleid*, 44-47.

Figure 6 Under Dutch bicycle tax regulations (1924-1941), cyclists always had to have this disc fixed to their bicycle (although to prevent theft, many kept it on their person). Valid for one year (here 1939-1940) and costing most people 3 guilders, the disc was personalized with the bike owner's name and address. The country's millions of cyclists generated large revenues that the state used to construct highways for cars. Working-class people exempt from the tax got a perforated disc for free. To avoid this stigma, some scraped together the money for a disc. [Source: Belasting & Douane Museum Rotterdam, no. 58776]

The temptation of easy revenues proved too big. Within two years, the tax became permanent. This time, it was specifically allocated to road improvement. While the average cyclist would not have been aware of this – they just had to pay the fees – the fundamental debate marked a key development in Dutch road policy. The Ministry of Public Works proposed a Road Fund (*Wegenfonds*), into which new taxes on cars and fuel would flow. In addition, the bicycle tax would become an earmarked tax (*bestemmingsbelasting*) also deposited in the Road Fund. This Road Fund would be used for national roads, but partly distributed over provinces to reconstruct the provincial road network. The austerity measures of 1923 and 1924 had increased government revenues significantly and improved the country's economic situation after 1925. The bicycle tax – approved by Parliament only as a temporary crisis measure – was no longer needed to balance the budget.

To make the bicycle tax permanent, a more fundamental Parliamentary debate about cyclists' political rights and duties ensued in October 1926. Supporters of a bicycle tax presented traffic separation as the ideal solution for (un)mixing drivers and cyclists. This was a Dutch engineering compromise for solving political conflict: without choosing sides, give each party something. Giving cyclists separate cycling paths was possibly the only justification for including the bicycle tax in the Road Fund. Christian democrat Finance Minister De Geer pointed out that under the new law, cycling paths could be financed from the Road Fund if they served a traffic function – in other words, if they "unburdened" the main roads that would be overcrowded if cyclists and drivers intermingled.[24] This line of reasoning gives us insight into what type of cyclists the politicians were considering: those who used the roads to commute between cities and suburbs or went cycling for recreation on weekends.

Politicians across the spectrum were well aware that Dutch cyclists were too numerous for policymakers to ignore. Catholic MP – and future Royal Commissioner of the province Noord-Brabant – Augustinus Bernardus Gijsbertus Maria van Rijckevorsel thought the unique number of cyclists in the Netherlands required traffic separation because "their life is constantly endangered ... [and] depends on the competence and the sense of responsibility of car drivers." Consequently, "bicycle traffic has to be separated from car traffic; separated cycling paths have to be constructed."[25] Van Rijckevorsel later served on the Roads Committee, advising the Minister of Public Works, where he also tried to give cyclists some (protected) road space. Other politicians agreed that the appeal for cyclists' safety seemed little more than an excuse to get them out of the way and ensure motorists an uninterrupted flow. Conservative liberal Walrave Boissevain, an Amsterdam shipping merchant, argued that "the extraordinarily large number of bicycles in the Netherlands has created a situation which requires special measures, because they hinder fast traffic on normal roads to such an extent that fast traffic and bicycle traffic should no longer be allowed together from a safety viewpoint."[26] A car advocate, he nevertheless believed that the Dutch state should be involved with cycling governance, if only to get bikes out of the way. The ANWB's cycling advocacy and the growing popularity of bicycles for everyday use made cyclists an inescapable political problem that could not be easily ignored.

To separate cyclists and drivers, a bicycle tax for building both roads and cycling paths seemed a good if contested compromise. According to Albert de la Bruhèze and Oldenziel, however, most cyclists who lived in cities and tended to be working class, were not part of this political compromise.[27] While urban cyclists had to pay the tax, the Road Fund was not used to

24 *Handelingen II* 1926/27, 10 (October 27, 1926), 212.

25 *Handelingen II* 1926/27, 11 (October 28, 1926), 230.

26 Ibid., 231.

27 Albert de la Bruhèze and Oldenziel, "Who Pays, Who Benefits?."

reconstruct urban infrastructure. For one, national engineers had no jurisdiction over this area as municipal autonomy was a key doctrine of Dutch administration, while tax revenues flowed into the national and provincial coffers.[28] In Parliament, the main advocate for working-class urban cyclists' needs was social democrat Florinus Martinus Wibaut, a powerful Amsterdam alderman responsible for public housing construction. He defended workers' rights to leisure: the average worker had very little free time anyway and "for a major part of the population [the bicycle] is the only way to get away from the oppressive environment of large cities."[29] Rather than opposing the bicycle tax outright, Wibaut focused on how urban workers used the bicycle to enjoy nature. Taxing this vehicle made the bicycle tax the "worst and certainly most foolish tax the Netherlands has ever known."[30] This framing of cycling as not just utilitarian but also recreational had enduring appeal in social democratic circles: in the 1950s, provincial policymakers – members of his own social democrat party – would later use the argument to defend investing in recreational cycling facilities for the highly urbanized Zuid-Holland region.

Other social democratic opponents of the bicycle tax, however, spoke more about the plight of rural than urban cyclists. MPs like Arie IJzerman argued that rural cycling should be a crucial public good: "much more than in cities, the bicycle has become totally indispensable," he said, for attending school, and for both children and adults' livelihood in the countryside.[31] IJzerman's argument represents the wide support rural cyclists could count on in Parliament from MPs who had a rural background. While there were also numerous MPs with an urban background, urban cyclists only became the focus of attention in the 1970s. Many MPs felt that due to expensive or infrequent public transit and low wages, the bicycle was the only means available for rural people and agricultural workers to extend their action radius.[32] This required proper and safe cycling infrastructure. Instead, the proposed bicycle tax would go to a highway network considered economically necessary for the country. Most people could not afford a car, nor did they need one because the bicycle fulfilled all their mobility

28 Mom and Filarski, *De mobiliteitsexplosie*, 199.

29 *Handelingen I* 1923/24, 38 (June 19, 1924), 707; Harry Lintsen et al., *Well-being, Sustainability and Social Development*, 306-8.

30 *Handelingen I* 1927/28, 17 (December 29, 1927), 183.

31 *Handelingen II* 1923/24, 69 (May 13, 1924), 2008, an argument reiterated by Arend Braat of the Plattelandersbond (2010) and Hiemstra (2021).

32 The Swedish government had a very similar argument: Rautio and Östlund, "'Starvation Strings' and the Public Good," 42-63.

needs – a valid argument even for the ruling coalition parties. According to Chris van den Heuvel of the conservative Protestant party (ARP) dominant in rural areas, "to levy a bicycle tax in the future on our mine workers or farm laborers who would then pay for these modern roads on which they will never drive and which they will not even see from a distance" was unfair.[33] Like the social democrats, Van den Heuvel realized that the proposed network of national main roads would only benefit elite business travelers and entrepreneurs.[34]

MPs from a rural or working-class background could see how important cycling was to people with the same roots. Social democrat Piet Hiemstra, a former farm laborer and dairy factory worker from the northern province of Friesland, spearheaded the agricultural laborers' fight against the bicycle tax throughout the 1920s. His "Hiemstra Amendment" in 1924 proposed to exempt laborers who paid no income tax and could provide proof from their local authority that they needed a bicycle for work.[35] Catholic MPs Toon Loerakker, a laborer at a flower bulb grower in the Haarlemmermeer, and Cornelis Knigge, farmer and dairy farm director in the province of Utrecht, proposed similar amendments. They were both from rural areas and thought the bicycle was important there. As Loerakker testified, the provincial bicycle tax in Noord-Brabant had been a heavy burden on the poorest classes: "I have repeatedly witnessed how collections had to be organized to help people pay the tax and thereby enable them to get to work."[36] In 1926, social democrat Jan van den Tempel, a former house painter's assistant who came from a large family and later in life obtained a doctorate in economics, insisted that large households with several cheap and rickety bicycles would be severely limited in their mobility if they had to pay 3 guilders for every (second-hand) bicycle they owned. Cartoonist Johan Braakensiek satirized this in 1924 (see figure 7).[37] Even according to progressive liberal Jan Berent Westerdijk, a wealthy farmer from the province Groningen, "living in the countryside" meant he witnessed the large cohorts of men and women cycling to their work on farms. In the countryside, he

33 *Handelingen II* 1926/27, 9 (October 26, 1926), 180. Van den Heuvel had conflicting feelings: he thought Parliament was too hostile to car drivers. And the proposed Road Fund was unfair not just to cyclists but also car drivers in remote areas; both would pay but not benefit from the road construction.

34 Former railway official and public commentator Reitsma also objected: S.A. Reitsma, *De rijwielbelasting als bestemmingsheffing voor het motorsnelverkeer* (Assen: Van Gorcum, 1938).

35 *Handelingen II* 1923/24, 69 (May 13, 1924), 2020-22.

36 Ibid., 2021.

37 *Handelingen II* 1926/27, 11 (October 28, 1926), 231-32.

Figure 7 Social democrat politicians opposed the bicycle tax because it put a disproportionate burden on working-class families, especially if they owned multiple bicycles. This 1924 cartoon by Johan Braakensiek, a well-known cartoonist for the magazine De Amsterdammer, *shows a man who, because he cannot afford three taxes, has to transport his whole family on one bike. The caption reads: "The consequences of the bicycle tax" [Source: Belasting & Douanemuseum Rotterdam, no. 04922]*

believed, "most bicycles are used in this way."[38] In other words, knowing how important the bicycle was for workers, these politicians were convinced that cycling should be considered a crucial public good. They felt they had to pressure the government to take cycling seriously. In later decades, cycling advocates in parliament would often refer to their personal experience of cycling. In this mundane but important way, the everyday practice of cycling in the Netherlands entered the country's national arena, influencing cycling advocacy and governance.

Another objection to the bicycle tax targeted the logic of the road tax and improvement system. The new tax on motorized vehicles was based on weight and tried to link a vehicle's weight and road use in relation to road construction and maintenance. The heavier the vehicle, the more damage it inflicted and the more its owner had to pay towards maintaining the road. This engineering logic could not be applied to the bicycle, however,

38 *Handelingen I* 1923/24, 38 (June 19, 1924), 707. Westerdijk stressed that many working women relied on bicycles. Exempting male heads of household or children would not help them.

because its light weight caused hardly any damage.[39] At an international level, the roadbuilding association PIARC discussed these issues at length.[40] University of Delft-trained engineer and social democrat MP Theo van der Waerden argued that as far as financing road improvements was concerned, in principle "each mode of transport mainly has to pay for the construction and maintenance of its own lane."[41] The introduction of the car led to ideas about roads exclusively for its use, a revolutionary idea with serious implications for other road users. If regulations forbade cyclists to use the highway and they had to use their own cycling paths, it was fundamentally unjust to have them contribute to highway construction, so his argument went.

The compromise – that cyclists would get cycling paths in return for tax money – was contested. Were cycling paths needed? And should car drivers not pay for them? According to progressive liberal Fekko Ebels (VDB), a farmer from rural Groningen, only a few roads were busy enough to require such separated cycling paths. This would definitely not exceed the 6 to 7 million guilders raised through the bicycle tax.[42] Social democrat Van den Tempel went further, stating that car drivers should fund separate cycling paths: after all, they were responsible for endangering cyclists in the first place. He found it absurd that "cyclists have to pay for limiting the danger that is created by the other side."[43] Backing this argument was Steven Anna Reitsma, editor of a tram journal and an outspoken commentator on Dutch traffic policy.[44]

Others did not raise principal objections. They saw opportunities. According to Protestant MP Hendrik Leenstra, cyclists would be well-advised

39 *Handelingen II* 1926/27, 11 (October 28, 1926), 228.

40 Gijs Mom, "Building an Infrastructure for the Automobile System: PIARC and Road Safety (1908-1938)," in *Proceedings 23th World Road Congress 17-21 September 2007* (Paris: PIARC, 2007), 2004-30; Massimo Moraglio, "Transferring Technology, Shaping Society: Traffic Engineering in PIARC Agenda, in the early 1930s," *Technikgeschichte* 80, no. 1 (2013): 13-32; Ruth Oldenziel, "Accounting Tricks: How Pedestrians and Cyclists were Thrown under the Bus, 1910s-1940s," (Unpublished paper, 2016).

41 *Handelingen II* 1926/27, 9 (October 26, 1926), 175.

42 *Handelingen II* 1926/27, 11 (October 28, 1926), 228.

43 Ibid., 232.

44 Reitsma wrote many critical pamphlets in the 1920s and 1930s: S.A. Reitsma, *Open brief aan de leden der Tweede Kamer van de Staten-Generaal n.a.v. den Nederlandschen verkeerschaos* (Utrecht: Drukkerij J. van Broekhoven, 1932); S.A. Reitsma, *De weg naar coördinatie van het verkeer in Nederland* (Den Haag: Moorman's Periodieke Pers, 1933); S.A. Reitsma, *Laat den weg vrij!* F. de Vries and S.A. Reitsma, *Twee lezingen over het verkeerswezen* (Arnhem: G.W. van der Wiel & Co., 1933); Reitsma, *De rijwielbelasting als bestemmingsheffing voor het motorsnelverkeer*; S.A. Reitsma, *Herwaardeering van verkeerseconomische waarden* (Den Haag: Moorman's Periodieke Pers, 1942).

to support the bicycle tax since it gave them a stronger bargaining chip with which to demand state-sponsored cycling infrastructure. If attempts to abolish the bicycle tax succeeded, their negotiating position would become very weak, since "the cyclist does not have a right any more to ask for the roads to be in good condition, but they can if the bicycle tax is deposited in the Road Fund."[45] His reasoning implies many considered cycling paths not a public good to the same extent as roads for cars. The Road Fund not only drew money from car, fuel, and bicycle taxes, but also considerable funds from the general treasury: parliament considered automobility important enough to receive public funding, even if most Dutch citizens did not benefit directly. Leenstra portrayed cyclists as citizens only because they paid taxes – not because cycling should be considered a public good. This rather limited concept of citizenship was, however, a key argument: engineers, civil society actors, and business groups employed it often in this period.

A parliamentary majority ultimately voted in favor of this bicycle tax. Still, the working-class cyclists' defense did have political consequences. In 1924, heads of households who paid no income tax and needed the bicycle to commute were exempted from the tax.[46] In 1934, at the deepest point of the Great Depression, the government created an exemption for the unemployed, greatly increasing the number of free bicycle tax discs distributed.[47] Between 1930 and 1931, the government handed out about 82,000 free discs. By the late 1930s, the number rose to more than 400,000.[48] Despite the exemptions, most cyclists had to pay the 3-guilder tax. The millions in revenue largely went to roads for car drivers, who at this time formed only some 3 percent of road users.[49]

The political compromise did not end the controversy. When Germany occupied the Netherlands, the German authorities abolished the bicycle tax in 1941, sensing a populist opportunity to curry favor with workers. Leader of the populist Dutch national socialist NSB party, Anton Mussert, announced during a speech in the southern mining town of Heerlen in February 1941 that the NSB would strive to abolish the bicycle tax, so that the "fellow citizen with the dog tag" would no longer be seen in the streets.[50]

45 *Handelingen II* 1926/27, 11 (October 28, 1926), 230.

46 Grapperhaus, *Over de loden last*, 25-26.

47 Ibid., 60-62. Cyclists were fined for not displaying this small metal disc. Free tax discs had a hole in them, stigmatizing the cyclist as poor or unemployed.

48 "De rijwielbelasting verdwijnt," *Haagsche Courant*, April 24, 1941, 1.

49 Albert de la Bruhèze and Oldenziel, "Who Pays, Who Benefits?," 80-81.

50 "Gemengd nieuws. N.S.B. contra rijwielbelasting," *Nieuwsblad van Friesland*, February 24, 1941, 2. Albert de la Bruhèze and Oldenziel, "Who Pays, Who Benefits?

The next month, in an important speech outlining his policy, Reich Commissioner Arthur Seyss-Inquart argued that the Dutch tax system was a liberal, capitalist, and socially unfair system and announced his intention to abolish the bicycle tax.[51] Then, on 24 April 1941, the newspapers reported that in a few days (starting on May 1) the Germans would abolish the tax, as they did elsewhere in Europe, to win round the local working class.[52] In the official communiqué, Seyss-Inquart justified the end of the bicycle tax by referring back to the parliamentary debates, arguing that the tax had not distinguished between rich and poor cyclists, nor cyclists who used the bicycle for work versus those who only used it for leisure, and that this would rectify the injustice.[53] Dutch newspapers dutifully reprinted this message.[54]

What does this debate tell us about political views of cycling in the mid-1920s? Discussions about cyclists took up a substantial amount of politicians' time during the Road Fund debate. The debate was inspired by an imagined future of high automobility and the need for road infrastructure investments. Cyclists entered this debate as an afterthought once they clearly became a valuable source of income. In addition, making way for cars meant separating traffic so that drivers would have the road to themselves and force cyclists to one side. Unintentionally, cycling rallied defenders in parliament as well: many members – including but not limited to the social democrats – testified to its usefulness, in cities and the countryside, for work and pleasure. Car drivers were the financial priority. In the process, national politicians articulated the status of cycling infrastructure as a public good.

2.2 A Polder Model for Cycling Governance

The bicycle tax and Road Fund debate of the mid-1920s forced Public Works engineers responsible for the nation's roadbuilding program to think about cyclists and cycling infrastructure. How did the engineers and policymakers in executive government departments translate these political debates into cycling policy? What were these engineers' views on cycling and its place on the road? To what extent did they cooperate with civil society representatives in this process? And what engineering norms did they create

51 "Belangrijke rede van den Rijkscommissaris," *Nieuwsblad van Friesland*, March 14, 1941, 1.

52 Albert de la Bruhèze and Oldenziel, "Who Pays, Who Benefits?," 84.

53 Ebert, *Radelnde Nationen*, 408.

54 "De rijwielbelasting verdwijnt," *Haagsche Courant*, April 24, 1941, 1; "De rijwielbelasting verdwijnt," *De Tijd*, April 24, 1941, 1.

that would be applied for decades to come, creating a network of suburban and rural cycling paths?

When the Dutch parliament and Ministry of Public Works took on the role of providing more and better roads and prioritizing cars, their decision did not sideline civil society actors. In 1915, Public Works Minister Lely had created an advisory National Roads Committee (*Rijkswegencommissie*) with his Ministry's engineers. In 1926, parliament discussed appointing a road construction advisory body, also to engage with major social groups. The envisaged road tax and Road Fund would generate large sums of money, which MPs could to some extent control. If the distribution of these funds was left entirely to the discretion of Public Works engineers, it would be difficult for Parliament to control the process. That is why Protestant MP Jan Krijger (CHU), a former civil servant in Public Works argued that "tax-paying stakeholders can hardly be denied a moral right to some influence on the way their tax money is spent."[55] The government and conservative MPs opposed Lely's proposal, arguing that an advisory council would slow down road improvements.[56] At stake was the issue of how technocratic the roadbuilding development would be – that is, to what extent experts rather than parliament would govern the process. A parliamentary majority strongly favored a more indirect form of social representation, a decision-making process that represented social rather than technical interests.

Non-governmental actors already played a very large role in mobility. Government agencies regularly sought advice informally from the ANWB. The proposal for a Road Fund formalized this consultation process between governmental and non-governmental actors. Catholic MP Leonardus Gerardus Kortenhorst recommended H.W.O. de Bruyn, secretary of the ANWB and KNAC's joint road committee, the key expert outside the official government bureaucracy, as valuable member of a state committee.[57]

As a result, a Public Works engineers advisory committee and a non-governmental committee operated in tandem since the 1920s. Through regular meetings, both aimed to create a consensus-driven road policy amid a politically charged environment.[58] Cooperation proved difficult, however. The civil society committee complained about the engineers' lack

55 *Handelingen II* 1927/28, 68 (May 9, 1928) 2010.

56 *Handelingen II* 1926/27, 9 (October 26, 1926) 186-87.

57 *Handelingen II* 1927/28, 67 (May 8, 1928) 1989.

58 The engineers were on the National Roads Committee (*Rijkswegencommissie*, 1915-1928), changed to National Committee for Roads (*Rijkscommissie voor de wegen*, 1928-1932) in 1928. The civil society committee was known as National Committee for Consultation on Road Improvement (*Rijkscommissie van overleg voor de wegenverbetering*, 1928-1932).

Table 1 Civil society representatives in the Roads Consultation Committee (COW), 1920s-1930s

Person	Organization	Interest/Expertise
Edo Bergsma	Dutch Tourist Organization (ANWB)	Tourists, Cyclists, Car drivers
Hermanus de Bruyn	Dutch Road Congress (Het Nederlandsch Wegencongres)	Roadbuilding industry
Emile Schiff	Royal Dutch Automobile Club (KNAC)	Car drivers
Andries ten Hope	Dutch Federation of Company Vehicle Owners (BBN)	Transport Sector/Industry
Cornelis Sormani	Dutch Union of Drivers and other Automobile Personnel (NUCA; Nederlandsche Unie van Chauffeurs en overig Automobielpersoneel)	Car drivers
J. van der Molen	Dutch Highways Organization (NEVAS)	Highway construction industry
H. Molhuysen	Royal Dutch Agriculture Committee (KNLC)	Agriculture/Rural
Henri Bloemers	Dutch Institute for Public Housing and Urban Development (NIVS)	Public Housing, Urban Planning
Pieter van Tienhoven	Dutch Society for Nature Conservation (Natuurmonumenten)	Nature Conservation
Piet Bakker Schut	The Hague Urban Planning Department	Public Housing, Urban Planning

Source: COW archives, NA 2.16.11 and 2.16.59

of cooperation.[59] At the instigation of the government, the two committees merged into one Roads Consultation Committee (*Commissie van Overleg voor de Wegen*, hereafter COW) in 1932. Through the consultation committee, key members of roadbuilding lobby and interest groups discussed the principles of Dutch road planning with leading engineers.

Besides the ANWB, the civil society committee comprised a cross-section of interest groups (see table 1) but was dominated by car lobbyists. According to historian Hans Buiter, the committee enabled the car lobby to access government Public Works.[60] Dutch industrialists – closely aligned with the car lobby for the transportation of goods – were well represented, as was the Dutch highway lobby (NEVAS).[61] Several road user and industry associations tried, as Mom and Filarski explain, to present their particular interest as the

59 The engineers committee admitted in 1928 that it had been difficult "to stay in touch with those who regard themselves as representatives of road users." NA 2.19.46, inv. no. 43, committee meeting April 18, 1928.

60 Hans Buiter, "'Hoogviadukten in het polderland?' De introductie van de autosnelweg in Nederland," *NEHA-jaarboek voor economische, bedrijfs- en techniekgeschiedenis* 60 (1997): 303.

61 Ibid., 285-306 for more on NEVAS.

public good.[62] Mayor Henri Bloemers, a biologist and future ANWB director, was a very active member, as was Pieter van Tienhoven, representing the nature conservation organization *Natuurmonumenten.* Urban and spatial planning representatives also participated, including social democrat and urban planner Pieter Bakker Schut from The Hague. Bakker Schut was part of a generation of engineers trained around 1900, with common social democratic sympathies, as historians Nil Disco and Harry Lintsen show. Influential engineers and planners at the heart of the formative period of Dutch road and cycling governance in the 1920s and 1930s, trained in this milieu.[63] By giving various sectors and factions of society a seat at the table, Public Works ministers strengthened the corporatist and consensus-driven political culture which political scientists characterize as the hallmark of Dutch political history.[64]

According to ANWB director Bergsma, it was important to consider various interests in addressing the problem of ever busier roads. In 1928, he argued that "the progress in the use of motorized power is larger than was thought." This did not, however, mean the bicycle, "that individual mode of transport that has become so truly and specifically Dutch," was disappearing. On the contrary, Bergsma believed the bicycle "is used more and more. In short, our roads are quickly being filled with modern means of transport."[65] The policymaking process to address these pressing issues should not be left to engineers. Road users and taxpayers had to have a voice. Bergsma argued that "road users, who have such an interest in the construction and improvement of roads, bridges and other connections – and who, in addition, provide a tax entirely out of their own pockets – should be involved in the

62 Mom and Filarski, *De mobiliteitsexplosie*, 173.

63 On Bakker Schut: Yvonne van Mil, "Pieter Bakker Schut (1877-1952), manager in stedenbouw," *Bulletin KNOB* 116, no. 4 (2017): 209-23. Influenced by the progressive liberal professor Pekelharing, these students formed the Sociaal-Technische Vereeniging van Democratische Ingenieurs en Architecten (Social-Technical Organisation of Democratic Engineers and Architects). Leading Dutch urban planners such as Piet Bakker Schut (later active in The Hague), Th. K. van Lohuizen (central to the 1930s Amsterdam Extension Plan) and Arie Keppler (social housing in Amsterdam) were members. Others such as Willem Albarda and Theo Van der Waerden would become MPs for the SDAP party, where they defended cyclists in debates around bicycle taxation: N. Disco and H.W. Lintsen, "De vervlechting van ingenieursberoep en industrie: 1890-1925," *Tijdschrift voor Sociale Geschiedenis* 9 (1983): 343-69; Harry Lintsen, "De Delftse Polytechnische School als bakermat van het socialisme 1900-1925," in *Het tweede jaarboek voor het democratisch socialisme*, eds. J. Bank, M. Pos, and B. Tromp (Amsterdam: Wiardi Beckman Stichting/De Arbeiderspers, 1980), 81-109. See also Nil Disco, "Made in Delft: Professional engineering in the Netherlands 1880-1940" (PhD diss., University of Amsterdam, 1990).

64 Kickert, *The History of Governance in the Netherlands.*

65 *De Kampioen* 45, no. 51 (December 21, 1928): 1429, quoted in NA 2.16.59, inv. no. 47

preparation and execution of the plans of road administrators and consulted in ongoing matters."[66] He specifically mentioned nature conservation and city planning groups, who were concerned about natural and cultural beauty and wanted to safeguard them during road improvements. Bergsma, like his successor Bloemers, saw the potentially negative effects of roadbuilding and cars for nature and recreation. Their leadership created a compromise between road space for drivers and cyclists. As Mom rightly notes, the ANWB's "roots as a bicycle club with a long tradition of enjoying touring in nature, prevented these advisers from reducing planning issues to a mere problem of road adjustment."[67]

An analysis of the financial basis for mobility policy shows that the bicycle tax unintentionally made cyclists politically visible. Examining the policy coalition around mobility (involving non-governmental actors) as well as the applicable rules of the game (balancing trade-offs), reveals that Dutch traditional political values of accommodation and compromise meant both drivers and cyclists got something from the deal – albeit clearly benefiting drivers more than cyclists financially. In the mid-to-late 1920s, a formative period in Dutch road policy, cycling advisors played a key part in shaping Dutch car and cycling infrastructure outside major cities.

2.3 Is Cycling Infrastructure a Public Good?

As a result of the parliamentary debates and the mediating role of the ANWB around the bicycle tax, national state involvement in cycling governance started in earnest after the mid-1920s – though several (lower-level) state engineers had been involved in cycling infrastructure before that. By around 1915, some already believed that national roads (*rijkswegen*) should include separated cycling paths. Speaking in 1916, provincial engineer Lambertus Tjerk van der Wal thought national roads should be built with a separated cycling path on one side and a separated pedestrian path on the other side, both measuring 2.50 meters.[68] On secondary, provincial roads (*provincial wegen*), engineers should design a 1-meter-wide cycling lane, but not necessarily separated, he thought.[69] Van der Wal had joined the ANWB in 1910

66 Ibid.

67 Mom, "The Emancipation of the Urban View," 89.

68 L.T. van der Wal, *Het wegennet en de wegenverbetering in de provincie Zuid-Holland* (Middelharnis: Flakkeesche Boek- en Handelsdrukkerij, 1916), 17.

69 Ibid.

and would later sit on its board (1941-1959).[70] Between 1928 and 1952, he headed Public Works in the rapidly suburbanizing province Zuid-Holland, allowing him to realize his vision of separate cycling infrastructure. It was not unusual for a state engineer to be a member of the tourist organization ANWB: Dutch cycling governance typically entwined state and civil society.

Engineers were aware that suburban commuters constituted the large numbers of bicycle traffic. In preparation for the road construction plan and to improve forecasting, the national Public Works agency organized regular traffic counts when in the 1920s, the state's sporadic traffic counts became more systematic.[71] Its counts in 1923, 1926, 1929, and 1932 showed that the busiest cycling routes were to and from the major Dutch cities.[72] The busiest in 1923 were between The Hague, Utrecht, Haarlem, Arnhem, and their suburbs.[73] The traffic counts showed this suburban traffic (*voorstadsverkeer* in Dutch engineers' jargon) was much heavier than inter-city or through traffic.[74] To ensure all commuters were counted, the national 1926 count included the 6-8AM and 6-8PM rush hours when workers cycled to factories and construction sites.[75] By 1932, the counts showed how at most points in the country, bicycle traffic had increased compared to three years earlier. On some routes, cycling dropped, according to the reports because of the economic downturn between 1929 and 1932, reducing "the suburban traffic of factory and office workers" near the major industrial hubs.[76] Through their statistical data, national engineers discovered the many blue- and white-collar workers commuting by bicycle to cities during rush hour.

It led national engineers to start building separate cycling infrastructure for commuters along all the national roads they noted in their traffic count datasets. Take for instance a discussion in 1924 – before the bicycle tax

70 "Ir. L.T. van der Wal †," *De Kampioen* 79, no. 10 (1964): 616.

71 In 1915, engineers considered improving the national road at Duivendrechtse Bridge in the industrial district near Amsterdam. To make an informed decision, they asked local Public Works officials to organize a two-week traffic count at the bridge in October 1915. The daily average was 1,000 cyclists and 600 pedestrians, but only 50 cars and 30 motorcycles. There were more horse-drawn vehicles (200). Source: NA 2.16.59 (Commissie van Overleg voor de Wegen), inv. no. 1, dossier 15, letter November 3, 1915.

72 *Verslag betreffende de verkeerswaarnemingen in 1932. I. Rijkswegen*, ('s-Gravenhage: Algemeene Landsdrukkerij, 1934), 10; for the map Bijlage IIIc. Gemiddeld rijwielverkeer per dag van 14 uur. In aantallen.

73 *Verslag betreffende verkeerswaarnemingen in 1923 op de rijkswegen in Nederland*, ('s-Gravenhage: Algemeene Landsdrukkerij, 1925), 8.

74 Successive reports note that due to rising automobility, inter-city traffic grew in importance.

75 *Verslag betreffende de verkeerswaarnemingen in 1926. I. Rijkswegen*, ('s-Gravenhage: Algemeene Landsdrukkerij, 1928), 5.

76 *Verslag betreffende de verkeerswaarnemingen in 1932. I. Rijkswegen*, 11.

had been linked to roadbuilding – concerning the need for a cycling path alongside the national road from Amsterdam to Germany at Naarden, one of the first Dutch commuter towns. Engineer Wentholt found the option of an on-road and unseparated cycling lane "too dangerous for cycling traffic, because regular drivers will see the path as part of the regular road surface and not recognize it as a cycling path."[77] His colleague Van Heijst proposed to skip the cycling lane altogether. The influential G.J. van den Broek, head of Public Works' national roads department disagreed, however. He pointed out that 900 to 1,000 cyclists used this road daily, far outnumbering cars, and justified a separated cycling path.[78] The committee's engineers agreed that "suppressing a cycling path on this road with very busy cycling traffic cannot be recommended."[79] The engineers preferred separate infrastructure because drivers tended to consider on-road cycling lanes part of their space.[80] The engineers' conviction that separate cycling infrastructure would reduce accident rates was another reason for promoting this engineering solution. The ANWB cited traffic statistics seemingly showing that cycling accidents were less likely on roads with separate infrastructure.[81] In short, just before the bicycle tax took effect, the basic assumption was that on major roads, cars and cyclists should not mix, but have their own, clearly recognizable place.

For national engineers and policymakers, the introduction of the bicycle tax was a powerful additional argument for separate cycling paths. According to many Roads Committee (COW) members, the bicycle tax morally obliged policymakers to "give something back" to cyclists. In 1932, engineer Van der Steur stated plainly that "cyclists bring in a lot of tax and they should be taken care of."[82] Nature conservationist Van Tienhoven pointed out that cyclists brought in 7.5 million in taxes, giving them a right to decent

77 NA 2.16.59, inv. no. 43, committee meeting July 9, 1924.

78 *Verslag betreffende verkeerswaarnemingen in 1923 op de rijkswegen in Nederland*, 17 + Bijlage IIIC.

79 NA 2.16.59, inv. no. 43, committee meeting July 9, 1924.

80 NA 2.16.59, inv. no. 43, committee meetings August 15, 1924 and February 19, 1926. Engineer Van Heijst said: "If a paved cycling path or road shoulder is adjacent to the road, drivers will soon see it as a three-lane road and drive on the cycling lane more often, which endangers cyclists." Different paving was the compromise solution: less expensive than separate paths but more visible for drivers. A meeting on December 12, 1924 emphasized the same point for a bridge: cycling lane paving should be a different colour, or with a band of lightly coloured stone to highlight the separation. See also NA 2.16.59, inv. no. 44, committee meeting December 21, 1928, that found this solution more practical (and cheaper) than buying land for a separated path.

81 "Vrijliggende rijwielpaden zijn noodzakelijk," *De Kampioen* 56, no. 11 (1939): 184.

82 NA 2.16.11 (Raad van de Waterstaat en zijn Taakvoorgangers), inv. no. 371, meeting December 29, 1932.

cycling facilities.[83] Committee chairman and MP Van Rijckevorsel remarked: "given that cyclists bring in a lot of taxes it is not right that this category of road users is fobbed off with a partial solution."[84] According to Cornelis Gerrit Timo Sormani, chairman of the automobile transport workers union, "the cyclists are being treated in a 'step-motherly' fashion."[85] Even though separate cycling paths were more costly and complicated to administer due to expropriation procedures, cyclists had a basic right to them.[86] Committee members from very diverse interest groups agreed on this point. The bicycle tax decision clearly changed the perceptions of cyclists' public status and increased support for a state role in cycling governance.[87]

Most key stakeholders approved of cycling paths. A few engineers opposed constructing them. At one committee meeting in the mid-1930s, engineer Ludolf Reinier Wentholt raised a series of far-fetched counterarguments: for one, on the new highways, cars would go so fast that elderly cyclists from the countryside (*bejaarde buitenlieden*) would not be able to properly estimate the speed of the cars, which might cause accidents at intersections. Wentholt made it clear that people who could literally not keep up with the speed of progress would not receive any concessions but would be forced to stay in their own quiet zones.[88] During another meeting, Wentholt tried to play down the need for cycling infrastructure, calling cycling predominantly a leisure activity. In the clash over resources, Wentholt framed cycling as recreational and motorized traffic as utilitarian and asked: "Should utility not transcend tourist interests?"[89] He meant that investments in roads for cars and trucks should get priority over cycling infrastructure, ignoring the utilitarian function of bicycles. This recreational framing of cycling was a minority position in 1936 when so many traffic counts had shown the opposite. Few committee members supported his views. Many fellow engineers considered cycling more a public responsibility than Wentholt.

83 NA 2.16.11, inv. no. 371, meeting October 17, 1935.

84 NA 2.16.11, inv. no. 372, meeting July 13, 1936.

85 Ibid.

86 NA 2.16.59, inv. no. 47, meeting December 13, 1929.

87 Proof that cycling's status as public good depended on class connotations and tax-paying status, is a case in 1916 at Soest, a village in the province Utrecht. A headmaster requested more connections between his school and where most workers lived. Reporting to the ANWB, cycling path association Gooi en Eemland wrote that the alderman in Soest did not want to consider a cycling path. The workers' low incomes meant "the municipality is not willing to contribute sufficiently in the costs considering the very small tax contribution of the inhabitants of this borough." ANWB archive, inv. no. 1108, letter Gooi en Eemland to ANWB, October 14, 1916.

88 NA 2.16.11, inv. no. 371, meeting October 17, 1935.

89 NA 2.16.11, inv. no. 372, meeting February 11, 1936.

Figure 8 National engineers' ideal was to separate cyclists and car drivers completely. At this provincial road between Soestdijk and Bilthoven, Utrecht province, the engineers achieved that (c. 1930) by turning the former road into a cycling path while constructing a new road to the right of the row of trees. [Source: Collectie Het Utrechts Archief, Photographer unknown, no. 220773]

Still, some within the (national) engineering community took a more outspoken stance in favor of the car.

These national engineers had little to say about urban cycling infrastructure.[90] After all, they had no jurisdiction over urban space, where traffic separation was not the norm. And if their schemes impacted cities, they tended not to consult urban planners.[91] On road plans and traffic counts, national planners literally drew circles around cities and left the inside blank.[92] Provincial road plans also ended at city boundaries, as did the cycling paths constructed on national or provincial roads. Few cities constructed downtown cycling paths and kept street space shared. Urban cyclists not only needed their own paths, but also a place to park their bicycles. In these instances, the ANWB acted on behalf of urban cyclists, complaining that not enough was done for cyclists living in the city: urban planners and architects failed to provide enough parking facilities around housing.[93] Urban cyclists also experienced problems

90 Oldenziel and Albert de la Bruhèze, "Contested Spaces," 29-49.

91 Mom, "The Emancipation of the Urban View," 83-85.

92 Visible on the 1923 map of cycling traffic: *Verslag betreffende verkeerswaarnemingen in 1923 op de rijkswegen in Nederland*, Bijlage IIId. http://publicaties.minienm.nl/documenten/verslag-betreffende-de-verkeerswaarnemingen-in-1926-1932

93 "De verschoppeling der steden," *De Kampioen* 61, no. 5 (1946): 3.

like long waiting times at railway crossings. They had to fight a long battle for underpasses that would reduce these delays.[94] The Netherlands' relatively late automobilization probably prevented negative effects on everyday cycling in cities in the prewar era. Well into the second half of the twentieth century, national policymakers ignored urban cyclists, while urban policymakers and engineers reiterated the argument that separate cycling infrastructure was either unnecessary or impossible in narrow city streets.

2.4 Making Cycling Infrastructure the Default Norm

The roadbuilding and taxation discussions led to national engineers creating norms for national roads that included separate cycling paths as the standard. On the subject of road profiles, engineers often referred to the bicycle tax as the reason for cycling infrastructure, defining more generous norms for national roads, at least on paper, than the earlier cycling paths. In 1928, engineers created different standard road profiles, with 2.50-meter-wide cycling paths for the busiest roads, as in figures 8 and 9.[95] Cycling paths were the default on national roads. Engineers could only deviate from the guidelines under specific conditions.[96] In practice, these norms were flexible. It was common to have a single broader path on one side of the road for two-way cycling traffic. This was the case with the cycling path along a new, highly visible landmark, the Afsluitdijk, a 32-kilometer causeway and major dam between the North Sea and the Zuiderzee, connecting the provinces

94 Frank Veraart, Martin Emanuel, and Ruth Oldenziel, "Eindhoven: Engineering a Path for Bikes?," in *Cycling Cities: The European Experience. A Hundred Years of Policy and Practice*, ed. Ruth Oldenziel, et al. (Eindhoven: Foundation for the History of Technology, 2016), 52-54; Henk-Jan Dekker, "Jeremiades over de Jeremiebrug: Fietsfiles bij spoorwegovergangen in Utrecht, 1915-1940," *Oud-Utrecht* 91, no. 2 (2018): 32-38.

95 NA 2.16.59, inv. no. 43, committee meeting May 9, 1928. On road profiles for urban roads: A. van den Bogaard, "De geplande stad, 1914-1945," in *Techniek in Nederland in de twintigste eeuw. Deel 6. Stad, bouw, industriële productie*, eds. J. Schot, et al. (Eindhoven: Stichting Historie der Techniek, 2003), 66-67.

96 NA 2.16.59, inv. no. 44, committee meeting December 21, 1928, *Leidraad voor het aanleggen van rijwielpaden langs wegen, voorkomende op het rijkswegenplan*, attachment to letter to Ministry of Public Works, Dec 28,1928. According to the guidelines, in built-up areas it was left to the engineers' discretion; they could omit cyclists if space for cycling paths was lacking, a criterium open to interpretation. Outside the built-up area, cycling paths could only be omitted if (a) constructing separate cycling paths would leave less than 5.5 meters road space for cars; (b) if the road was wider than 6 meters and less than 500 tons/day total traffic intensity; or (c) if the new road was a car highway.

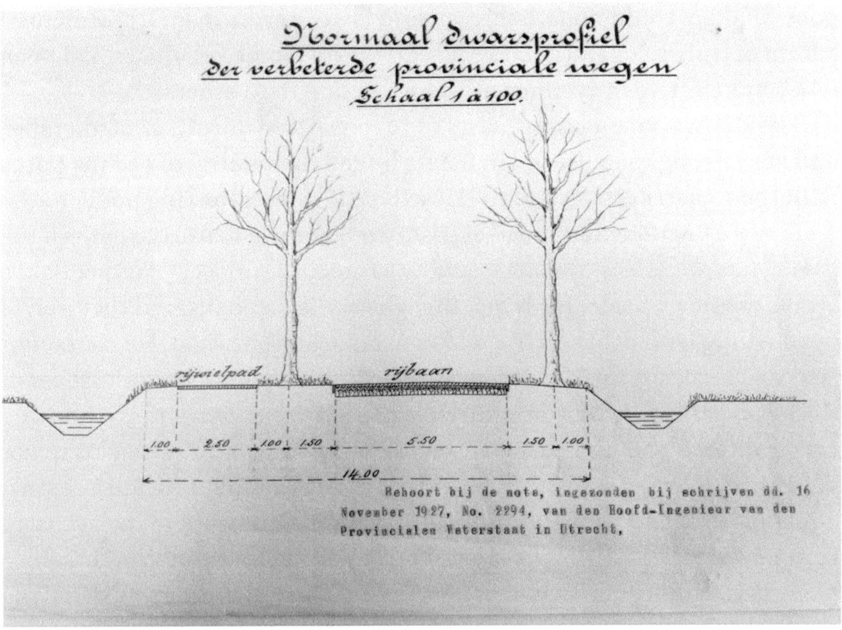

Figure 9 Around 1927, the standard profile that engineers in the Utrecht area applied for a provincial road (rijbaan) included a 2.5-meter-wide cycling path (rijwielpad). [Source: Het Utrechts Archief 1201, inv. no. 5316]

Noord-Holland and Friesland.[97] ANWB chairman Bloemers found a single wide path "very pleasant for cyclists" and had the additional advantage of accommodating large streams of cyclists during rush hours or "special occasions like football matches."[98]

Given the respect for decentralized autonomy in the Netherlands, provincial engineers had considerable leeway and could not be legally compelled to comply with these national norms. In practice, national Public Works could influence lower government's road and cycling policies through its control of funding. The provinces had few funds of their own and relied on national contributions from the Road Fund; the percentage they received was linked to compliance with official road profiles. With its preference for separate cycling paths over traffic lanes on roads, the government adjusted the subsidies so that it was financially more attractive for provinces to reconstruct roads with separate cycling paths.[99] Through these "soft" norms, national

97 NA 2.16.59, inv. no. 43, committee meeting May 20, 1927.

98 NA 2.16.11, inv. no. 372, committee meeting July 13, 1936.

99 HUA 1201, inv. no. 5300, correspondence Utrecht province with Minister of Public Works, 1929; inv. no. 5316, "Nota betreffende het normaal dwarsprofiel van de verbeterde provinciale

policymakers could nonetheless ensure lower government implemented a form of traffic separation. By the early 1940s, some provinces had been much more active in cycling path construction than others.[100]

Because (working-class) cyclists were considered unruly, unpredictable, and headstrong road users, engineers linked the quality of cycling paths with their user: cyclists had to be "seduced" into preferring paths to the main road. Engineers assumed cyclists would refuse to use the paths if the paving was not at least the same quality as the road surface.[101] Former Public Works engineer, Taeke Huitema, thought cyclists endangered themselves by mingling with other traffic and not staying on the right. He characterized this behavior as "dissoluteness, a desire to annoy, thoughtlessness or thick-headedness, which only punitive measures can prevent."[102] However, even Huitema opposed obliging the use of cycling paths if they were not of sufficient design standards. So did car enthusiast Hendrik Jan Peppink, editor of ANWB journal *Autokampioen*, who felt "the right of the cyclist ... is respected all too little" because "the condition and construction of so many cycling paths is such, that it would be a 'moral murder' to 'drive' the cyclists onto them."[103] Again though, "the interests of fast traffic are also served through the construction of many cycling paths."[104] Using bricks for surfacing alongside asphalt roads was a recipe for disaster: the cycling path would remain unused while drivers and cyclists would have to mingle. In 1929, Frederik Kanstein, an engineer from the city of Arnhem, wrote in the roadbuilding journal *Wegen* that cyclists were lazy by nature, "choosing the line of least resistance, like electricity."[105] The rise of asphalt roads equally required high-quality cycling paths since cyclists much preferred them to the bumpy road surfaces: figure 10 provides a good example. Most cycle paths alongside national roads had a concrete (paved) surface, however.[106] For car advocates like Peppink, Kanstein, and Huitema, calling for cycling infrastructure was perhaps informed more by the needs of drivers to cruise

wegen" (attachment to letter from provincial head engineer Mussert to Provincial Executive of Utrecht, November 16, 1927).

100 "Neerlands rijwielpaden en hun bewijzering," *De Kampioen* 59, no. 9 (1942): 65-66.

101 For the debate on road surfaces: Gijs Mom, "Inter-Artifactual Technology Transfer: Road Building Technology in the Netherlands and the Competition Between Bricks, Macadam, Asphalt and Concrete," *History and Technology* 20, no. 1 (2004): 75-96.

102 T. Huitema, "Rijwielpaden," *Wegen* 5, no. 11 (1929): 270.

103 H.J. Peppink, "Aparte rijwielpaden ... òf: wielrijders op den hoofdweg?," *Wegen* 5, no. 9 (1929): 220.

104 Ibid., 221.

105 F. Kanstein, "Rijwielpaden," *Wegen* 5, no. 7 (1929): 161.

106 Mom and Filarski, *De mobiliteitsexplosie*, 311.

Figure 10 Traffic separation was not possible everywhere in the Dutch landscape but getting cyclists off the main road remained a priority. This provincial sand road between De Meern and Oudewater in Utrecht province was paved somewhere between 1924 and 1934. The adjacent asphalt lanes were intended to "seduce" cyclists into not riding in the middle of the road. [Source: Collectie Het Utrechts Archief, Photographer: Jochmann Disco, no. 220755]

along without interruption than those of cyclists. Still, they went out of their way to call for improvements in the quality of cycling paths.

This broad support for cycling infrastructure alongside Dutch roads led to norms which looked impressive on paper and were applied in practice. The network of cycling paths was already growing before the days of the bicycle tax. The 1923 traffic count report includes a map of cycling traffic on roads *with* cycling paths.[107] While certain urbanized areas (notably Rotterdam and the South-East) lacked cycling paths, other regions (The Hague-Amsterdam-Utrecht) already had intensively used cycling infrastructure. By the late 1930s, the National Roads Plan counted over 2,300 km of roads. Of these, almost 1,800 km had a cycling path (3,050 km, counting roads with cycling paths on both sides). A further 1,350 km of (secondary) provincial roads also had cycling paths. Engineers and policymakers considered these paths primarily for utilitarian purposes. A separate recreational network of cycling paths, not alongside roads and often in nature, was twice as long: 3,900 km by 1942.[108] This more extended recreational network had been some forty years

107 *Verslag betreffende verkeerswaarnemingen in 1923*, attachment IIId.
108 Mom and Filarski, *De mobiliteitsexplosie*, 197.

in the making. The utilitarian interurban and rural network materialized in the decade after the introduction of the Road Fund in 1927.

Judged separately, these results are impressive. However, comparing the investments in roadbuilding to cycling infrastructure shows a glaring disparity as the ANWB argued. In 1932, the government spent 18 million guilders on national road construction for 80,000 drivers, and reserved 1.5 million for separate cycling paths for 2.8 million cyclists. By 1939, the roadbuilding budget had risen to 34 million, whereas the cycling budget had not kept pace at 1.4 million. This suggested that space for cars was growing at a much faster rate than for cyclists. Even considering the increased spending on the new highways, which the ANWB did not oppose in principle, the disparity was too great. With the exception of new car-only highways, the ANWB argued, every road on the 1938 National Road Plan should have at least one separate cycling path.[109] Cyclists got paths in return for their investments, but also paid for the roads the car lobby demanded.

2.5 Governing Cycling Publicly or Privately?

Once the bicycle tax led to agreement on cycling paths, the question remained who was responsible for constructing and coordinating them. Citizen organizations could receive subsidies, as was the case with the recreational cycling clubs, or the various government levels could take on this task. Over centuries, the growing network of roads and waterways built and governed by various authorities had led to a patchwork of responsibilities. The national, provincial, and municipal government were responsible for certain roads: the national, regional, and local network.[110] By the 1920s, each governmental layer had to provide its roads with cycling infrastructure. In contrast to car infrastructure, it was unclear who was responsible for (coordinating) the countrywide cycling network or represented cyclists in general. The question of who governed cycling in the Netherlands hinged on whether cycling was defined as a predominantly recreational or utilitarian activity, and would shape policy for the next hundred years.

In the 1930s, state policymakers disagreed on the national cycling governance role beyond constructing cycling paths on national roads. In 1936, COW consultation committee chairman Van Rijckevorsel proposed designing "a national cycling-connections plan, in which the matter can be viewed from

109 "Vrijliggende rijwielpaden zijn noodzakelijk," 184.

110 A few roads were controlled by other governmental agencies such as the water authorities.

a larger perspective," although he qualified this by saying "the interests of cyclists are more of a local nature."[111] The question created a clear analogy with the national government's initiative of building a road network for cars. According to committee member Jenze Jan Talsma, mayor of the small town Renkum, cycling governance was the national government's responsibility. He believed "cycling connections for through traffic [*doorgaande rijwielverbindingen*], ... [was] a task for the national government." Acknowledging "the felicitous local initiative that is being shown by the cycling path associations," he nevertheless believed the national government should intervene.[112] Some national policymakers agreed that the national government should provide cycling infrastructure.

The cycling governance issue became more urgent when the possibility arose of delegating road coordination to non-governmental actors. In 1936, when the umbrella organization Federation of Dutch Cycling Path Organizations applied for an annual subsidy to support their burgeoning cycle path network and to cope with the rising costs and stagnating income, the non-governmental organization claimed to be a valuable governance partner by pointing to the 2,500 km of paths they had already constructed. Multiple arguments supported their request. Not only was it in the interest of recreational cyclists to have paths and access to nature areas but the shorter connections also served government employees like postmen, telegraph messengers, and police officers who used bicycles to get around. Because the network had become so big, the associations were financially overstretched: "there is now a danger, that many of these cycling paths cannot be properly maintained anymore and will consequently have to be closed for cycling traffic, while expanding this network of cycling paths will have to be limited to a minimum, due to the decreasing contributions from state institutions."[113] In need of funds, the non-governmental organizations argued for the public status of their cycling paths.

According to the Federation, the subsidies also created employment opportunities in times of economic crisis. The Federation proposed depositing a small part of the bicycle tax (e.g., 10 cents per bicycle) in a special fund dedicated to the construction of cycling paths under the associations' control: a public-private partnership, with control over spending state funds delegated to private associations. The proposal sparked a debate that sheds light on how major policymakers viewed recreational cycling's political

111 NA 2.16.11, inv. no. 372, meeting February 11, 1936.

112 Ibid., meeting July 13, 1936.

113 NA 2.16.11, inv. no. 436, letter July 1936.

status. A COW subcommittee consisting of state engineers Van Heyst and Wentholt, plus non-state representatives Bloemers (ANWB chairman), De Bruyn (ANWB engineer) and Van Tienhoven (conservation group), ruled against awarding the subsidy.[114]

During this debate, the policymakers minimized the national government's potential role in coordinating and funding (recreational) cycling infrastructure. They did not consider the cycling paths these private associations had constructed important enough: the paths served "primarily a provincial interest, because the connections are inter-local and the associations are more or less provincially oriented."[115] A few years later, engineer Pieter Johannes van Voorst Vader also doubted "whether the national government should involve itself with the construction of these cycle paths" because leisure and tourism were only of "regional interest."[116] The national responsibility for cycling, in Van Voorst Vader's opinion, ended with creating cycling paths that freed main roads of cyclists. That was not what the average association paths did, as they were primarily for tourism. Some committee members disputed this restrictive stance, however. Talsma pointed out that cycling tourists used the road just like utilitarian cyclists if there were no cycling paths. Van Rijckevorsel added, "one should not look too much to the goal of the cycling path. It only matters whether there is a traffic need."[117] What precisely a "traffic need" meant was open to interpretation. For some, this meant drivers' needs, while others recognized cyclists' needs. The debate gave only minimal attention to cycling, mostly on account of the interest in car drivers, without much representation from cities or workers.

The national policymakers on the COW committee preferred to assign provinces the leading role. Despite "an interest in these connections," most cycle path organizations could not finance construction because many were "near financially less well-to-do municipalities."[118] The national government felt it could not be responsible for these paths because cycling was a local mode of transport, yet local communities, while given the mandate, could not bear the financial burden. The only remaining options were the provinces. National policymakers decided to "involve the provinces, while the financing could be arranged with the help of state subsidies, the province, municipalities, unemployment relief, and bodies like the ANWB and the cycling path

114 NA 2.16.11, inv. no. 436.

115 Ibid., memorandum January 21, 1938.

116 NA 2.16.11, inv. no. 372, meeting October 26, 1939.

117 Ibid.

118 NA 2.16.11, inv. no. 436, memorandum January 21, 1938.

associations."[119] In short, the provinces should have administrative control. The money had to come from different sources, *including* the national government, as well as local bodies and non-governmental partners.

ANWB leaders Bloemers and De Bruyn lobbied through the COW committee for national investments as well as provincial coordination. The committee agreed. However, the distinction between utilitarian and recreational cycling complicated matters. Legislation prohibited the national Traffic Fund (for all transport infrastructure investments) from subsidizing recreational infrastructure. The fund could only subsidize cycling paths if these served "to unburden the traffic along main connections for through traffic ... irrespective of whether these paths are part of the roads."[120] Again, recreational cyclists were framed as completely different than commuting cyclists. According to Bloemers and De Bruyn, communities wanting to attract tourists by building recreational paths lacked the financial means to do so and while "we can definitely speak of a task for the Provincial Government here," at the same time "the large cultural value for the entire population of our country of a judiciously constructed network of cycling paths separated from the large roads can hardly be underestimated."[121] A subsidy was entirely justified. The ANWB kept framing cycling as an activity for the entire nation, requiring state involvement. Funding and coordination demanded a national governing body: "This committee could then also be charged with the highly important task of coordinating the different provincial plans and promoting the execution of them, among others by encouraging the cooperation of the different relevant institutions."[122]

The ANWB lobby was not entirely successful either, as Public Works Ministry engineers were committed to a narrower definition of cycling as a (national) public good that fell to local governments. The state engineers were willing to consider a modest subsidy, but no more. And there was another dilemma: what types of cycling deserved support and which governmental bodies should receive subsidies? Van Voorst Vader and Bloemers argued that more money had to go to the provinces with a strong interest in tourism and the most attractive nature. According to Van Voorst Vader "in a region with natural beauty, there will be a larger need for cycling paths than elsewhere."[123] Politician Van Rijckevorsel opposed this blatant neglect

119 Ibid.

120 Ibid., memorandum "Nota naar aanleiding van het Adres van de Federatie van Nederlandsche Rijwielpadvereenigingen."

121 Ibid.

122 Ibid

123 NA 2.16.11, inv. no. 373, meeting February 15, 1940.

of other cyclists' needs: "one should not in the first place construct cycling paths for tourists, but take into account the opening up and establishment of industry in the countryside."[124] Bloemers admitted that cycling paths should also serve this goal, like for industrial workers around the textile cities Hengelo and Enschede.[125] While the committee members got bogged down in long legal discussions, however, the outbreak of World War II halted the debate temporarily.[126] After the war, the provinces took on the governance of (recreational) cycling together with individual organizations (see Chapter 3).

By the late 1930s, the consensus was that the national government would only build interurban commuter cycling paths alongside major roads – to create unobstructed roads for cars and compensate for cyclists' tax money. Other cycling infrastructure, whether recreational or, as the cycling clubs argued, mixed-use, was the responsibility of a coalition consisting of provincial and local government and the tourist organization ANWB.

Determining cycling infrastructure's status as a public good depended on two key factors. By invoking the scale and range of cycling, policymakers framed the bicycle as local (in stark contrast to the car as national). Cycling was lower government's responsibility. By stressing the distinction between utilitarian and recreational cycling, Public Works officials limited their involvement in cycling governance to providing separate infrastructure along major roads. Recreational cycling infrastructure was not seen as a national public good, but a local task, if that. In other words, the cycling governance debate was about the proper level as well as sector of government involvement.

There was no permanent solution to the responsibility issue. Key civil society actors like the ANWB and smaller tourist organizations continued to contest the national government's limited role in cycling governance, particularly as they thought it contrasted sharply with the government's far more generous car policies that did not make this distinction. The distinction between utilitarian and recreational cycling that divided committee members in the 1930s played a huge role in postwar discussions. The questions they debated – what was a cycling path? and for whom? – were not straightforward and had major implications.

124 Ibid.

125 Traffic counts in the 1920s and 1930s show the Eindhoven region's very high cycling share. Oldenziel et al., *Cycling Cities: The European Experience*, 42.

126 After 1942, the COW committee did not meet till after the war.

2.6 Conclusion

By the mid-twentieth century, politicians and engineers across Europe and the United States redefined the road as an exclusive space for automobiles.[127] In the decades-long process, cyclists were disregarded and marginalized. In this international process, a car lobby was also active in the Netherlands, but not with the same negative repercussions for cycling practices as elsewhere. This has everything to do with the introduction of a bicycle tax linked to national roadbuilding that put Dutch cycling policy on a different track. As a result, and in line with Dutch accommodationist tradition, policymakers did not emphatically choose either travel mode. It became a key principle for Dutch traffic planners to cater for both. Many wanted to remove cyclists from the road to facilitate drivers – but unlike elsewhere, they built an extensive network of separate cycling paths, not simply accepting a "survival of the fittest" system without protection for cyclists. The national and provincial government started (re)constructing roads for cars, simultaneously providing separate cycling paths. The state resisted taking on a larger role in cycling governance (regarding recreational cycling), but independent actors stepped in. The cycling infrastructure served suburban commuters. Firmly embedded in engineering norms, and with significant support from engineers and civil society representatives, this type of infrastructure continued thereafter. In short, the origins of the unique Dutch trajectory lie in this period.

Ebert and Albert de la Bruhèze and Oldenziel discuss this development in broad brushstrokes. Ebert explains that the bicycle tax gave Dutch cycling advocates a unique political weapon that cyclists in other countries lacked. The bicycle tax also meant that, unlike cyclists elsewhere, they had a legitimate claim on the Dutch government to see cycling as a public good. This chapter confirms Ebert's argument using previously unexplored sources – parliament and committee records – to paint a richer and more nuanced picture of how not only the ANWB, but also engineers and politicians saw cycling. The bicycle tax debates in parliament established the connection – and the compromise – between the tax and cycling paths, albeit contested. Using the advisory committee's reports to the Dutch minister of public works, I show how engineers, working with civil society stakeholders, did indeed understand that the bicycle tax's mandate and political compromise allowed cyclists some claim to public funds.

The bicycle tax did not, however, convince Dutch policymakers to consider cycling as a public good: motorized traffic had a higher priority.

127 Norton, *Fighting Traffic*.

For national policymakers, cycling was only a public good when it came to getting (commuting) cyclists off future interurban car connections. While recreational cycling advocacy was strong and eventually successful, the bicycle tax argument did not procure state support for recreational cycle paths. One of Albert de la Bruhèze and Oldenziel's central arguments is that urban (working class) cyclists paid the tax but got nothing in return. Indeed, in engineering circles and strikingly in parliament, urban cyclists remained largely invisible. They could use suburban infrastructure on recreational trips, but cycling's status as a public good remained limited to the middle classes.

Ebert summarizes the problem facing German cyclists: "Because cyclists paid nothing, they could not ask for anything."[128] Albert de la Bruhèze and Oldenziel rightly qualify her conclusion, stating: "the paths were also designed to discipline rather than facilitate cyclists." In short, mobility policy was less benign than Ebert argues.[129] It is fair to say that the bicycle tax, especially its third and final instalment, was both unjust and disproportional. Yet it also set the Netherlands on a path-dependent course of cycle path construction, creating an institutional culture of engineering standards with long-lasting effects on the viability of Dutch cycling. Once traffic separation became a norm, and cycling remained popular, mostly because of this infrastructure, it was very difficult to stop constructing these commuter paths. The bicycle tax certainly prioritized drivers' needs. Yet it had unintended, positive long-term consequences – a kind of path dependency – for cyclists. It indirectly contributed to the survival of a cycling culture and cycling expertise among engineers, creating favorable conditions for introducing cycling infrastructure in Dutch cities after 1970 when cycling reemerged as part of the political agenda: by that time there were still large groups of cyclists who legitimated activist demands. Moreover, the expertise and experience of mainly provincial engineers, some national engineers, and the ANWB's engineers, enabled a quicker implementation of pro-cycling measures than elsewhere, as we will see.

This exploration of Dutch policymakers' role in cycling governance reveals a large involvement for the period 1920-1950, from MPs to engineers and interest groups. Because of the bicycle tax, long and heated discussions about cycling took place in parliament. The social democrats fiercely tried to defend working-class cyclists against the bicycle tax. When only very few people owned a car (1 percent of the population), car-minded politicians from

128 Ebert, "When Cycling Gets Political," 236.
129 Albert de la Bruhèze and Oldenziel, "Who Pays, Who Benefits?," 85.

different parties had to reckon with cycling's popularity as everyday mobility mode in the Netherlands. The bicycle tax and debate led to a compromise and an engineering standard of traffic separation. National engineers preparing the way for automobility could not either ignore the cycling majorities that became visible in their agencies' traffic counts. Engineers, crucially working with civil society representatives, spent a significant amount of time and effort reconciling drivers and cyclists' interests. A division of tasks emerged: The national Public Works department's involvement in cycling was limited to providing separate cycling paths alongside major roads – a pattern still visible in later decades. Provincial Public Works departments provided regional roads with cycling paths, while any coordination of the cycling infrastructure network as a whole was informally left to the ANWB. The interaction between stakeholders was what we might call accommodating. The relationship between the influential tourist organization ANWB and engineers was close and cooperative. Urged on by politicians whose electorates increasingly depended on cycling, cyclists had institutional backing. More so than in other countries, commuter cycling infrastructure became a public good.

Conclusion Part I

Cycling as a practice underwent major changes across the world between 1880 and 1950. Starting as an upper-class leisure activity, it gradually became more affordable and more popular for commuting and transport. In the process, the bicycle acquired increasingly working-class connotations. Meanwhile, the elite adopted the new symbol of modernity, the automobile, and shifted their lobbying efforts from better roads for cyclists, to better roads for cars – preferably with no space for bicycles. The declining status of cycling and the declining role of both government and lobby groups reinforced each other, making cycling an increasingly dangerous and unattractive practice. By the mid-twentieth century, large groups of people all over world were still cycling as they could not afford or lacked access to public transit or car ownership. The conditions for cyclists, however, were deteriorating. In the Netherlands this played out differently. Cycling put down deeper roots in governance structures and engineering norms. The role of the bicycle tax compromise can hardly be overestimated. Nor should we overlook the regional cycling path organizations working on recreational infrastructure. Both laid the groundwork for a material infrastructure supporting the practice of cycling to a greater extent than elsewhere.

The political compromise of the bicycle tax formed a crucial though contingent starting point which set the Netherlands on a track of traffic separation. The late adoption of cars combined with an early and ambitious commitment to highway building in a high-cycling context resulted in a unique situation when politicians and engineers faced the dilemma of funding these highways. A coalition of non-governmental cycling advocates, politicians, and national and provincial engineers agreed upon a political compromise in which national policymakers exploited cycling's popularity as a source of revenue to finance a road network for future motorization. This system still shapes Dutch cycling today. Giving cyclists a separate path while prioritizing cars was a way of engineering and designing traffic space with important path-dependent implications. The persistence of cycling in the Netherlands during the 1950s and 1960s, much more so than in other countries, has its roots in policies put in motion in the 1920s. In engineering norms and culture, engineers adopted road and cycling path norms that acquired their own path-dependent logic. Once traffic separation was established as desirable, engineers everywhere adopted this measure, a process that was increasingly hard to reverse. The annual reports of all the public works departments listed the construction of new stretches of cycling paths as a matter of course. For young engineers entering the Public Works departments, including cycling facilities in new or reconstructed roads was standard procedure. Outside the Netherlands, cycle path construction never became such a standard practice. In the struggle for resources and legitimization, cycling advocates in other countries lost out. In my view, because cycling paths became official policy and standard practice, Dutch Public Works continued to construct cycling infrastructure even during the heydays of car-centered planning post-World War II, as we will see.

Alongside this system, a second coalition of recreational cycling governance existed, as discussed in the first chapter. This separation of utilitarian and recreational cycling policy coalitions would also shape Dutch cycling in the long run. Cycling paths not following the road network were ostensibly recreational and did not involve the national government or national policymakers at all. From the 1910s onwards, a coalition of the ANWB, here in its role as a tourist organization, together with local elites, set up several cycling path organizations. They managed to build a network of several thousand kilometers of cycling paths in rural areas. The citizen initiative role in Dutch cycling governance was large and predated state involvement. As a result, these citizen groups built up expertise which gave them legitimacy and usefulness as governance partner. Coupled with the Dutch polder model, where social representatives were routinely included in consultation procedures, this explains why the tourist and (since 1920) lobby organization ANWB played such a prominent role in Dutch cycling governance.

The political appreciation of cycling was highly uneven. For one, the ANWB mostly ignored cycling within cities, so did most MPs. Moreover, national investments in car infrastructure far exceeded those in cycling. And as we have seen, not all cyclists were equal. The distinction that cycling advocates and politicians made from the start was between recreational and utilitarian infrastructure, investing in the latter while leaving the former to citizen organizations. This was linked to the idea that they required different funding streams and different governance coalitions. Throughout the twentieth century, these two separate streams would remain. Urban cycling also remained out of focus for national and provincial policymakers. The policies in place supported suburban commuters and recreational (including urban) cyclists, but fell short for utilitarian cycling within the city. The late adoption of the car in the Netherlands meant that cyclists in cities, sharing the road with cars, had some safety in numbers, as figure 11 illustrates. This would change in the decades to come.

Figure 11 Cyclists dominated city streets, like here in Utrecht in 1931. A traffic policeman, standing in his chequered box behind the tram pole, has stopped the throng of cycling traffic from the right to allow another stream of cyclists to cross the road. Urban engineers did not apply traffic separation within cities, giving cyclists ample space to use the entire width of the road. This was necessary too, given the droves of cyclists during rush hours. This picture appeared in the illustrated magazine Utrecht in Woord en Beeld *(Utrecht in Text and Pictures). [Source: Collection Het Utrechts Archief, no. 83925]*

Between 1880 and 1950, Dutch cycling governance expanded from private recreational cycling to a public-private partnership supporting a wider range of cycling practices. The ANWB, working with regional cycling path organizations, initially dominated private initiatives. By the mid-1930s, cycling governance had a larger if contested place in national governmental policy. An informal coalition of engineers and non-profit organizations governed cycling at multiple levels. Responsibility for utilitarian (commuter) cycling had landed with Public Works departments, whose design norms for roads now included separate cycling paths. Responsibility for recreational cycling on the other hand gradually shifted to the provinces, but still largely controlled by non-government organizations. As a knowledge actor and intermediary, the ANWB was the stakeholder keeping cycling on as many policymakers' agendas as possible.

Rising income levels in the 1950s and 1960s enabled more people to drive a car, often with disastrous consequences for cycling levels and policies. How did utilitarian and recreational cycling fare in the Netherlands? Did the governance coalitions established before the 1950s survive, or were they dismantled, leaving cyclists to fend for themselves, just like urban cyclists up to that point?

Part II

Divergence: How Dutch Cycling Policy and Practice Persevered, 1950s-1970s

The scarce cycling scholarship on the postwar period tends to focus on the disappearance of urban cycling as a result of its increasing marginalization, both on the road and in policymaking. Adri Albert de la Bruhèze and Frank Veraart's work is exceptional in this regard. Building on their research, mobility historians Gijs Mom and Ruud Filarski acknowledge the presence of cyclists, moped riders, and extensive cycling infrastructure in their overview of Dutch mobility in this period, but highlight the rise of the car and the extension of the road network.[1] In cycling historiography, Ruth Oldenziel, Martin Emanuel, Veraart, and Albert de la Bruhèze frame the period 1950-1970 in a similar way, as we see from their subsection titles "Planning a Car-friendly Future 1945-1970" for the city of Utrecht or "Catering to Cars in a City of Cyclists 1950-1970" on Enschede.[2] Most cycling historiography focuses on how policymakers and engineers facilitated cars in the city and the negative impact on urban cycling practice. This general direction of the historiography is well documented and beyond doubt. Policymakers and engineers did prioritize cars. Cycling everywhere suffered as a result – an injustice that urban cyclists would fight to correct after the 1970s, as discussed in Part III.

Rather than concentrating on how car-centered policies marginalized cycling, Part II, entitled "Divergence," analyzes cycling's contested but enduring status in policy circles as a Dutch public good from the late 1940s to the 1960s. This approach does not contradict but, rather, enriches the current scholarship. My focus on national and regional rather than urban politics sheds a different light on cycling governance during these neglectful years. It helps explain the divergent trajectory of Dutch cycling in comparison with elsewhere. The following three chapters argue that while urban cyclists were still largely left to their own devices, older policies aimed at commuter cycling outside cities survived in governmental circles. In addition, new groups of cyclists became more central: in the 1950s, (provincial) policymakers developed an interest in recreational cyclists. Previously governed by independent organizations, the promotion of cycling tourism for public health or economic benefits triggered new governance initiatives. In the 1960s, the safety of children cycling to school gradually emerged as an additional concern. Social actors, planners, policymakers, and car boosters may have expressed negative judgements about (commuter) cycling, considering it outdated, yet a large group of Dutch

1 Mom and Filarski, *De mobiliteitsexplosie*, 246. Mom, "The West and the Rest," 143-99. The same applies to Nil Disco, "De bewogen stad, 1945-1980," in *Techniek in Nederland in de twintigste eeuw. Deel 6. Stad, bouw, industriële productie*, eds. Johan Schot et al. (Eindhoven: Stichting Historie der Techniek, 2003), 74-99.

2 Oldenziel et al., *Cycling Cities: The European Experience*, 32, 43.

engineers and politicians remained committed to publicly supporting bicycle commuting or developed new support for cycling tourism.

The argument spans three chapters. Chapter 3 takes up the governance of recreational cycling where it left of in Chapter 1. It analyzes how responsibility shifted from citizen organizations to provincial policymakers. Chapter 4 focuses on what happened to the utilitarian cycling policies described in Chapter 2 of Part I once the bicycle tax ended in 1941. It examines the extent to which older engineering norms for cycling infrastructure still applied in the 1950s and 1960s. Finally, Chapter 5 shows how the new moped boosted cycling infrastructure in unintended ways as its undecided status as a vehicle (neither a bicycle nor a car) challenged policymakers. Chapters 4 and 5 stress policymakers' ambivalence towards utilitarian cycling in this period. Together, these chapters show why cycling persisted in the Netherlands as policy and practice, unlike in many surrounding countries. While Dutch politicians and engineers still considered many forms of cycling a public good at a time when policymakers in other parts of the world found cycling outdated, urban cycling became increasingly dangerous when policymakers started accommodating cars in cities – the topic of Part III. This chapter constitutes a crucial pivot linking the cycling tradition established before 1950, to the cycling revival of the 1970s. Tracing this trajectory is essential for understanding each of those periods: to discover not just the long-term effects of policies set in motion in the 1920s, but also to fully understand the conditions cycling activists and policymakers faced in the 1970s.

To understand the context of the late 1940s to 1960s, we need to place the development of cycling in wider mobility developments. As a practice, cycling remained crucial to Dutch mobility in the 1950s for commuting, shopping, and touring. Many European cities saw their cycling levels drop sharply around 1950.[3] In the UK, for instance, the total distance cycled per capita declined after 1952.[4] In the Netherlands, as the *Cycling Cities* research group has shown, the bicycle's relative decline – its share of total trips vs. automobility and public transit – was already visible in the 1950s.[5] Even

3 Adri A. Albert de la Bruhèze and Ruth Oldenziel, *Cycling Cities: The Munich Experience*, 8. Volker Briese notes it was common to transform existing cycling lanes into car (parking) lanes in Germany, but that hardly ever happened in the Netherlands and Denmark: Briese, *Besondere Wege für Radfahrer*, 20.

4 Malcolm J. Wardlaw, "History, Risk, Infrastructure: Perspectives on Bicycling in the Netherlands and the UK," *Journal of Transport & Health* 1, no. 4 (2014): 243-250, here 245.

5 Ton Welleman, *The Dutch Bicycle Master Plan: Description and Evaluation in an Historical context* (The Hague: Ministry of Transport, Public Works and Water Management, Directorate-General for Passenger Transport, 1999), 28-29.

so, the bicycle's relative share remained significant; the growth in total distances traveled confirms that cycling remained popular for the Dutch in the 1950s and 1960s. The high cycling levels dropped sharply in the 1960s, reaching rock bottom around 1970, but even at this relative low-point in Dutch urban cycling, the bicycle still had a major modal share of roughly 30 percent in many cities.[6] In Denmark's capital, Copenhagen, where cycling levels also remained significant before plummeting between 1955 and 1970, substantial numbers of commuters (20 percent) and schoolchildren (30 percent) continued cycling in the 1960s.[7] In many cities across Europe, cycling's decline was much more dramatic than in the Netherlands.

Dutch cycling policies and practices turned out to be more resilient than in other countries in the 1950s and 1960s, when governmental support for automobility made cycling increasingly hard across the Western world. This partly has to do with the deep roots cycling governance put down in the first half of the twentieth century, but also with new developments. Cycling practice remained very strong in the Netherlands, even while policymakers planned – and built – car infrastructures and kept investing in public transit.[8] Rather than seeing the bicycle and car as successive stages, it is important to realize that new technologies often undergo a significant period of overlap with older ones. In the 1950s and 1960s, cycling co-existed with driving in the Netherlands. While engineers and planners around the world might have framed cycling as obsolete, the reality was different.

Indeed, as scholars like David Edgerton and Elizabeth Shove stress, technologies do not simply disappear overnight, but experience a gradual process of decline. Certain elements of the technological system remain in place.[9] People do not discard technologies as quickly as policymaking elites and engineers do. Many elements of technical systems that innovators, policymakers, or opinion makers consider outdated can survive long beyond the actual use of an artefact. Shove calls these elements "pockets of persistence" and argues that they can serve the revival or reintroduction

6 Oldenziel et al., *Cycling Cities: The European Experience*, 13.

7 Emanuel, "Making a Bicycle City," 507.

8 As Jan Ploeger at Eindhoven University of Technology argues in his forthcoming book on the bicycle-train combination, there was early opposition to car-oriented planning among Dutch planners, which fostered the emergence of the bicycle-train combination as an alternative.

9 David Edgerton, *The Shock of the Old: Technology and Global History since 1900* (London: Profile Books, 2006). Already questioning the easy periodization of cycling and automobility: Monika Burri, *Das Fahrrad: Wegbereiter oder überrolltes Leitbild? Eine Fussnote zur Technikgeschichte des Automobils* (Zürich: Eidgenössische Technische Hochschule, Institut für Geschichte, Technikgeschichte, 1998).

of these technologies at a later date.[10] Taking cycling history as an example, she writes: "Regimes of cycling declined across Europe during the 1960s but different forms of vanishing resulted in the patchy and partial endurance of selected elements."[11] Indeed, my argument in the following chapters is that the decline of cycling in the Netherlands was much less radical than elsewhere, and that important elements of the cycling system and network persisted throughout the 1950s and 1960s. Cycling engineering expertise, governance relations, and practices survived to such an extent that they partly explain why the campaign for creating cycling cities after 1970 was so successful. Dutch cyclists still formed a large group, engineers still had knowledge about cycling infrastructure, and policymakers and engineers still formed cycling governance coalitions.

To sum up, understanding cycling in the 1950s and 1960s is crucial to the overall project. This is the period when European countries wrestled with the car. The approaches they chose might have been only marginally different, but in the long run, these minor initial choices created major differences, as path dependency theorists also argue. Explaining the revival of urban cycling since the 1970s, something which interests both historians and cycling practitioners, too often fails to consider the circumstances when it started around 1970. While these do not explain everything, they are crucial to understanding the conditions that supported campaigns to limit the car's place in the city and redistribute space for cyclists and pedestrians. Cycling activists needed a base to recruit members and gain legitimacy with policymakers; engineers tasked with building cycling infrastructure needed the expertise to do so. Many countries lacked this base and knowledge due to cycling's dramatic decline after mid-century. We urgently need to examine why cycling persisted in the Netherlands after the 1950s. The *Cycling Cities* project is addressing this diversity in urban cycling. In the following chapters, I fill a gap in the scholarship by taking a first look at other cycling policies and practices at a national and provincial level. Older policies lived on, now as routine procedures, leaving fewer traces in the public debate, but have led to material infrastructure. New policies regarding mopeds and recreational cycling had major impacts on cycling and mobility in the Netherlands. It is easy to lose sight of this: the automobilization in these decades is one of the most fundamental infrastructural transformations of the twentieth century. We need to study this in order to understand why Dutch cycling survived alongside the car.

10 Elizabeth Shove, "The Shadowy Side of Innovation: Unmasking and Sustainability," *Technology Analysis & Strategic Management* 24, no. 4 (2012): 363-75.

11 Ibid., 372.

3 A Right to Recreation: Provincial Policymakers Design Cycling Networks

"Actually, the construction and maintenance of cycling paths is a government task for municipalities," a journalist wrote in 1949. "It is fortunate that this Twente organization, founded on January 6, 1917, has taken this governmental work upon itself because otherwise, the condition of many cycling paths in Twente would almost certainly be a lot worse than it is now."[1] He praised the Twente cycling path organization for its valuable service to society – but also implicitly critiqued local government for ignoring a public task. Constructing 1,000 km of cycling paths in the eastern region of Twente had not been easy: local municipalities invested very little, and funding had to come from elsewhere. Both tourists and commuters were served: Twente had a significant textile industry, so to provide mobility options for its workers, major industrialists sponsored path construction through the Chamber of Commerce. The region also received small sums from the tourist organization (ANWB), the tourist board (VVV), the district water board, the farmers union, the agricultural loan bank, and dairy factories. The plethora of contributors reveals the broad support for cycling infrastructure in this prominent textile region. By the late 1940s, a growing chorus of voices called for local authorities to take up a larger role in cycling governance and not leave it all to the energy and initiative of others. Finally, the 1930 call from the Loosdrecht accountant with which our story started seems to resonate: building cycling path should be the government's task.

Some of the earliest policymakers to realize this were in the province of Drenthe. Wanting to take up the construction of cycling paths as a state task, policymakers decided to establish a cycling path organization in 1940. At their founding meeting, a key member of the successful sister organization in neighboring Twente (in the province Overijssel), L. Vincken, advised Drenthe "to concentrate not only on paths with a recreational interest, but also on utilitarian paths [*utiliteitspaden*] to obtain financial support from residents."[2] Vincken's underlying strategy was opportunistic: get funding for recreational cycling paths by stressing cycling paths' economic uses. Hendrik

1 "De Rijwielpadvereniging Twente: Haar werk en haar vele moeilijkheden," *Twentsch dagblad Tubantia,* April 30, 1949, 2.

2 DA 0196, inv. no. 544, minutes foundation meeting Drenthe cycling path organization, May 7, 1940.

Wytema, mayor of one of Drenthe's municipalities, agreed, pointing to the 1930s crisis funding that deployed jobless workers to construct cycling paths. Those plans were also for "utilitarian paths" and had a broader political support than the purely recreational paths, he counseled.[3] In the context of the 1930s economic crisis, the liberal clubs found tourism was less of a priority. They needed other arguments to convince social democrat and confessional parties to invest in cycling. Vincken and Wytema perhaps realized that utilitarian cyclists formed a larger group supported by labor parties than purely recreational ones reputed to benefit the well-to-do. Their claims for state involvement were more likely to succeed if both groups would benefit from cycling paths.

What arguments did the organizations use to convince local government to become more active in cycling governance? And why were provincial and municipal politicians so eager to build cycling paths at a time when around the world, policy attention on cycling was reaching a low point? Framing cyclists as either predominantly utilitarian or recreational was a strategy that non-governmental organizations applied opportunistically to secure government subsidies. Stressing the bicycle's utilitarian potential turned out to be a good strategy. And it proved to be successful.

While scholarship has paid attention to recreational cycling in its early bourgeois culture, it has ignored its crucial history in mid-century as missing link to the 1970s expansion of utilitarian cycling. Moreover, for the 1950s, the scant scholarship focuses on urban commuters, but not on how recreational cycling helped sustain and expand cycling infrastructure. This chapter, entitled "A Right to Recreation: Provincial Policymakers Design Cycling Networks," uses untapped archival material to reveal how provincial politicians and public works departments initiated new cycling policies for regional recreational cyclists in the 1950s and 1960s. My argument provides a counterweight to the dominant narrative of cycling's decline in this period. A coalition of state and non-state actors believed that recreational cycling could solve concerns over public health and boost tourism – as two case studies illustrate – and invested in cycling accordingly. Policymakers in the densely populated province of Zuid-Holland introduced policies to improve its residents' lives, while the poor agricultural province of Drenthe saw the economic potential of attracting middle-class cycling tourism. Interested in social uplift, social democrat politicians pushed to create recreational opportunities for the city's working and middle classes. Only in the 1960s would the national government start to follow suit and support provincial initiatives.

3 Ibid.

3.1 Pioneering Recreational Cycling Governance in the 1940s

When non-governmental cycling path organizations had sought Dutch state funding, they had made a case for their cycling infrastructure by underlining that it was a public good, serving both recreational and utilitarian purposes. However, their attempts to acquire national funding in the 1930s had failed. National policymakers argued that (recreational) cycling was a local concern, to be handled by local government. At the same time, as we have seen, public works engineers at national and provincial levels remained committed to building separate cycling paths alongside roads, which they considered utilitarian. In contrast, the cycling paths discussed in this chapter had their own trajectory. These did not follow the road network, but were recreational, meandering through the landscape. Over time, policymakers began to realize that this division was too simplistic: separate cycling paths that were not alongside main roads also served commuters and were mixed-use. In 1940, this problematic distinction between utilitarian and recreational cycling infrastructure was what had prompted Drenthe's policymakers to establish a cycling path organization.

The effects of the war proved important. Around mid-century, local policymakers, increasingly recognizing the public service that the cycling path organizations provided, for example, financed much-needed repairs to the paths damaged in the war.[4] State reconstruction agencies offered the organizations interest-free advances for the repair work. The tourist organization ANWB mediated between these agencies and the local cycling clubs in complex and time-consuming administrative procedures.[5] The state did accept the war damage claims. It shows the government started to recognize the private organizations' work as a public service. In 1950, for instance, Utrecht's provincial government strongly recommended that municipalities raised subsidies to repair war damage to the local cycling path organization's network (UMO). Municipalities had to "award and/or

4 Two ANWB officials, engineers Lorié and Tusenius, visited cycling path organizations throughout the country and together with local officials made an inventory of the war damage: ANWB archive, inv. no. 1225 and for specific organizations ANWB Archive, inv. nos. 1113, 1115 (Gooi en Eemland), 1184 (Noord-Kennemerland), 1191, 1193 (UMO), 1209 (Walcheren), 1220 (Twente), and 1231, 1243 for an FNRV overview. The ANWB urged organizations to take advantage of the financial claims as some boards had become rather inactive over the years.

5 The ANWB was an intermediary in these contacts with the Ministry of Finance's Reconstruction Agency (*Bureau Financiering Wederopbouw Publiekrechtelijke Lichamen*): ANWB archive, inv. no. 1191, letter UMO to Bureau, May 15, 1946 and letter ANWB to Bureau, June 15, 1946. UMO submitted a total of forty-four damage claims. After approval by the provincial state engineer, the ANWB forwarded requests to the Ministry's policymakers.

increase a reasonable yearly contribution" to UMO "which in effect takes care of the government's tasks."[6] The ANWB and the UMO lobby succeeded in getting state recognition.

In the immediate postwar period, provincial authorities were increasingly involved in recreational cycling. Due to the Netherlands' decentralized governance structure, each province had slightly different plans and funding procedures, as a 1938 ANWB report demonstrates.[7] Zuid-Holland was working on its newly designed cycling path plan; as did other provinces like Drenthe, Noord-Brabant, and Limburg. Some provinces contributed by subsidizing a fixed amount or percentage per kilometer of constructed cycling path.[8] These provincial subsidies often stipulated that the cycling paths had to serve a public need: in Utrecht subsidies were only awarded to "paths which were considered of sufficient general interest."[9] The province Gelderland's chief engineer emphasized that provincial subsidies were only meant for "bicycle tourism ... because bicycle traffic indicates an entirely different concept."[10] The funding streams for cycling paths alongside roads (seen as utilitarian) and recreational cycling paths were separate, as were the governance coalitions.

In 1946, the Federation of Dutch Cycling Path Organizations resubmitted its prewar request for national subsidies. The federation stressed the bicycle's all-purpose postwar use, because due to shortages, "in the coming years, the bicycle will (have to) be used a lot."[11] The organizations employed their preferred strategy of stressing cycling paths' mixed use to argue that tourism was not the primary goal. The paths served "numerous daily bicycle commutes for agricultural and factory-workers" and diverting cyclists from main roads was an additional reason to invest heavily in cycling infrastructure.[12] The

6 HUA 1007-3, inv. no. 30791, letter Utrecht province to municipalities, September 12, 1950.

7 Note again how hand in glove the Public Works department worked with the ANWB; the distinction between state and non-state actors was almost meaningless.

8 Overijssel provided 7.50 guilders per year/km of cycling paths. In Gelderland, the provincial executive chose paths it would finance (25 guilders per year/km to a maximum of 180 km). Shortly after World War II, in a renewed attempt to boost tourism, this subsidy rose to 75 percent of maintenance costs, with a maximum of 75 guilders per km: Gelders Archief (hereafter: GA), 0039 (Gedeputeerde Staten 1814-1950), inv. no. 9482, subsidies to cycling path organizations 1927-1949; GA 3204 (GS 1950-1979), inv. no. 7173, letter GS to provincial roads committee, November 26, 1947.

9 NA 2.16.11, inv. no. 436, memorandum 'Nota naar aanleiding van het Adres van de Federatie van Nederlandsche Rijwielpadvereenigingen.'

10 GA 0039, inv. no. 9482, letter HID Kempees to cycling path organization Oostelijk Gelderland (BROG), February 11, 1932.

11 NA 2.16.11, inv. no. 436.

12 Ibid. Letter May 14, 1946.

ANWB agreed this was in the public interest: "a network of well-constructed utilitarian and recreational cycling paths would fulfil ... a pressing need."[13] The Minister of Public Works' national advisors also deemed "a cycling path network covering the entire Netherlands of such great importance for the Dutch people ... there is every reason to systematically design such a plan."[14] The Ministry, however, decided that coordination and construction fell to new provincial bodies rather than the national government.

In the 1940s, policymakers systematized the planning of land use, including recreation, under a new agency coordinating the efforts.[15] During the war, the German occupiers of the Netherlands introduced a national spatial planning agency (*Rijksdienst voor het Nationale Plan*) in 1941.[16] Each province had its own planning agency (*Provinciale Planologische Dienst*), also overseeing recreation. After the war, the government established the new Ministry of Social Work in 1952, becoming the Ministry of Culture, Recreation, and Social Work in 1965 that would subsidize recreational cycling infrastructure (see 5.3).[17] The provinces' recreational cycling path design reflected growing government intervention in leisure and recreation policy, as well as designs for spatial planning and nature preservation. Since the 1940s, this process tended to sideline non-governmental actors, diminishing the role for private initiative. At the same time, it also marked cycling governance's increasing professionalization. The policy's main implementors were the provinces and municipalities working together regionally.

In the late 1940s and 1950s, some local policymakers created organizations to design regional (recreational) cycling networks.[18] Others transformed existing private organizations into state agencies, taking over their privately built networks. Many municipalities, especially in the touristic Veluwe region, took up this initiative energetically in the late 1940s, seeing opportunities for developing tourism. For example, the founder of Zuid-Veluwerand, the cycling path organization around the nation's key nature reserve area in 1949, was the mayor of the town Barneveld. He invited local mayors, tourist organizations

13 Ibid.

14 Ibid. Concept advice 1946.

15 Faludi and Van der Valk, *Rule and Order*, 64, 74, 97; Mom, "The West and the Rest"; Van der Cammen and De Klerk, *Ruimtelijke ordening*, 152-65; Mom and Filarski, *De mobiliteitsexplosie*, 255-56.

16 Its first director was Frits Bakker Schut, son of prominent social democratic The Hague city planner Piet Bakker Schut.

17 Ido de Haan, Jan Willem Duyvendak, and Maarten van Bottenburg, *In het hart van de verzorgingsstaat: Het ministerie van Maatschappelijk Werk en zijn opvolgers (CRM, WVC, VWS), 1952-2002* (Zutphen: Walburg Pers, 2002).

18 E.g. the organization Zuid-Veluwerand (founded in 1949, ANWB archive, inv. no. 1199).

(VVV), along with the province's chief engineer J.G. Kruimel and ANWB engineer C.A. Kuysten to take part.[19] Kruimel clarified provincial subsidy opportunities, while Kuysten explained the wider construction context and procedures elsewhere. The governance network consisted exclusively of regional policymakers (mayors), provincial policymakers (engineers), and the (national) non-governmental tourist organizations VVV and ANWB, functioning as advisors. The national government barely played a role.[20]

3.2 Pioneering Provincial Cycling Governance in Drenthe and Zuid-Holland

Stimulating cycling in Drenthe was a means to an end. Compared to densely urbanized Zuid-Holland, which supported cycling for public health reasons for its urban working classes, Drenthe was a sparsely populated, poor, and rural province. Drenthe actively promoted cycling (tourism) as a way to boost its economy. Before World War II, Drenthe had an active private cycling path organization and, uniquely, early state involvement with cycling. The local bourgeoisie in Drenthe created a non-governmental organization in 1916. By 1937, it had constructed 100 km of paths. At that point and in the midst of the economic crisis, provincial policymakers took an interest in cycling path construction, even though the Drenthe branch of the Dutch Municipalities Association's (*Vereniging Nederlandse Gemeentes*) application to the national government for more cycling paths had been recently rejected. A key figure behind the promotion of cycling paths was the Royal Commissioner for Drenthe, the conservative liberal baron Reint Hendrik de Vos van Steenwijk. A doctor of law, he belonged to a prominent family and was a long-time member of the Upper House of Parliament. His goal was to improve the economic situation of the relatively poor province of Drenthe. To this end, he wanted to attract industry and tourism, partly by improving

19 ANWB archive, inv. no. 1199, "Kort verslag van het verhandelde in de vergadering van 10 Februari 1949."

20 Other examples of new public cycling path organizations, or private organizations taken over by the state, include Noord-Veluwe (ANWB archive, inv. no. 1226) and VRIGA in the Achterhoek region (both in Gelderland province) as well as organizations in Noord-Brabant province: Brabant Centrum, De Kempen, Markiezaat Bergen op Zoom, and later the Baronie van Breda. See also "Heden en toekomst van de rijwielpaden," *De Kampioen* 53, no. 16 (1936): 266-67; "Rijwielpaden in de Geldersche Achterhoek," *De Kampioen* 62, no. 11 (1947): 329; C.A. Kuysten, "Rijwielpaden in Zuid-Holland," *De Kampioen* 62, no. 9 (1947): 274-75; "Rijwielpadennet in het centrum van Noord-Brabant," *De Kampioen* 67, no. 12 (1952): 348-49; "Fietsen op de paden van de Zuid-Veluwe," *De Kampioen* 75, no. 6 (1960): 176-77.

Drenthe's infrastructure. He took the initiative for the Development Agency for Drenthe (*Centrale Vereeniging voor den Opbouw van Drenthe*).[21] Cycling path construction was one public works administration project that fitted in with these goals and would also address the massive youth unemployment in Drenthe during the 1930s economic crisis.[22] He appointed some mayors and the Development Agency's director Jaap Cramer to form a committee for this purpose. Cramer was a social democrat who would succeed De Vos van Steenwijk in 1951 as Royal Commissioner for Drenthe. While of different political affiliations, both were elite figures with good connections, in a position to realize their vision for Drenthe. Authorities in Drenthe pioneered the funding and building of the regional cycling path network in the late 1930s as a public works project to combat unemployment and create prosperity through a program of infrastructure construction.[23]

Drenthe also pioneered cycling governance. It was unclear which agency should be responsible for cycling paths. While the discussions in the late 1930s had shown the state's reluctance to do anything more than construct commuter cycling paths alongside roads, Drenthe decided to take the lead. The inspector for youth unemployment argued that Drenthe's municipalities were too poor to contribute: municipal plans would only cover parts of the province. A provincial initiative would be a faster and more streamlined process. The inspector also advised to start quickly on a pilot, not to wait until the completion of a comprehensive plan. Building one path immediately would "rouse interest from other parts of the province."[24] Head of the Development Agency Cramer sought funding higher up, asking the ANWB, and significantly not the national government, to "function as a funnel for national subsidy."[25] Cramer obviously hoped that the ANWB's good contacts with national government would guarantee funding, but he was mistaken. ANWB engineer De Bruyn, a member of the national Roads Consultation Committee (see Chapter 2) responded that neither the state nor the ANWB was likely to provide the funding.[26] Still, the very well-connected De Vos

21 DA 0196, inv. no. 544, letter Royal Commissioner Drenthe to nominated members subcommittee cycling paths, February 19, 1937.

22 On using unemployed people in public works: Van der Woud, *Het landschap, de mensen. Nederland 1850-1940*.

23 W.J. Simons, *Daar fietst men toch zo heerlijk heen: vijftig jaar Provinciale Drentse Rijwielpadvereeniging/Stichting Het Drentse Fietspad* (Assen: Stichting Het Drentse Fietspad, 1990).

24 DA 0196, inv. no. 544, meeting sub-committee on the construction of cycling paths by young unemployed people, March 11, 1937.

25 DA 0196, inv. no. 544, letter Cramer to ANWB (De Bruyn), February 5, 1940.

26 Ibid., letter De Bruyn to Cramer, March 7, 1940.

van Steenwijk and Cramer knew the rules of the political game – what arguments to employ and which people to approach to achieve their goals.

Provincial cycling coordination required a new agency and external expertise. When Drenthe formed its Provincial Cycling Path Organization, its founding meeting in 1940 was an interesting event from a governance perspective: it demonstrates how knowledge traveled between provinces, without the national government's involvement. Vincken attended the meeting to suggest key strategies and best practices from Overijssel and as representative of the National Federation of Cycling Path Organizations. The Twente region of Overijssel received annual subsidies from the province on two conditions: that a path's quality was approved by a provincial inspector and that it was of "provincial interest." The province apparently considered this to be the case for 300 km of the Twente organization's 900 km network. In addition, Twente received 1,000 guilders as an annual subsidy from the local Chamber of Commerce "because factory workers benefit from the paths," proving that many paths were indeed used for both recreation and commuting.[27] A broad range of actors subsidized the organization's activity. Vincken's presence and central role in the founding of Drenthe's cycling path organization demonstrate how the provincial and regional cycling path organizations networked and shared expertise and best practices at a sub-national level.

Dutch provincial authorities also shared knowledge on governing (recreational) cycling. The Cycling Path Organization Drenthe worked with the province's Public Works chief-engineer, Willem Izak Cornelis van Veelen. He quizzed colleagues in other provinces on how they went about gaining support for rural cycling path improvement and construction, in other words those paths not alongside roads.[28] Was there a private cycling path organization or a "semi-official" one consisting of municipalities, or was the province directly involved? And which organizational form did other provincial engineers prefer? Van Veelen enquired whether the other provinces already had a rural (*landelijk*) cycling path plan. If so, how was it structured. He also wanted to find out the various provincial conditions for cycling path subsidies. Van Veelen needed the information as input for Drenthe's policy to support the Provincial Cycling Path Organization, and so he urged his colleagues to reply swiftly. Within a few months, all the provinces had provided detailed responses about their plans, allowing Van Veelen to draw up a memorandum summarizing provincial cycling policy

27 DA 0196, inv. no. 544, foundation meeting May 7, 1940.

28 DA 0924, inv. no. 167, letter Public Works department Drenthe to public works in other provinces, May 16, 1951.

in 1951.[29] The provinces were aware that private organizations existed and were now seeking cooperation, having sparingly subsidized them for decades.

With the input from other provinces, the Provincial Cycling Path Organization Drenthe formulated its tourism-oriented "Central Cycling Path Plan" in 1952 – updated and expanded in 1959. A key figure ("the great promotor and animator" as one newspaper called him) behind the plan was Gerard Adriaan Maarten van den Muyzenberg, an official in Drenthe's Public Works department.[30] As the organization's secretary and treasurer from 1950 until 1974, he and his colleagues designed a 400 km cycling path network, dependent on national subsidies. Drenthe would have needed fifty years to implement the plan without such a subsidy. In its address to the national government, Drenthe Public Works administration framed a provincial cycling path network as a necessary condition for economic growth. According to the engineers, Drenthe had huge tourism potential and "the cycling path network contributes substantially in boosting a healthy provincial economic structure. A completed cycling path plan constitutes the "framework" that has to be furnished with holiday centers, catering companies, and amusement facilities."[31] In other words, recreational cycling paths were necessary to lift Drenthe from its economically disadvantaged position. Cycling tourism was the key to that end (see figure 12).

The province framed the cycling paths in a broader light, however. Chairman of the Provincial Cycling Path Organization Drenthe's board, Jan Antonie Reinders Bosma (the former mayor of Winterswijk in the province of Gelderland) refused to distinguish recreational and utilitarian cycling paths: the organization dealt with cycling paths "in general."[32] Cycling paths relieved traffic on the major roads – a remarkable argument given Drenthe's low car ownership in the 1950s.[33] Cycling paths would improve the connection between industrial and commuter areas as well as end the

29 DA 0924, inv. no. 167, "Nota betreffende het beheer, de subsidiëring, de plannen en de wijze van aanleg van landelijke rijwielpaden in de verschillende provincies van ons land," 6-7.

30 "G.A.M. v. d. Muyzenberg promotor Drents fietspad vertrekt op 13 december," *Nieuwsblad van het Noorden*, November 30, 1974, 15.

31 DA 0923, inv. no. 6258, letter Public Works Drenthe to Ministers OKW/EZ/SZ, December 4, 1959.

32 DA 0196, inv. no. 544, foundation meeting Drenthe cycling path organization, May 7, 1940.

33 Wolf, "Following America? Dutch geographical car diffusion, 1900 to 1980" (PhD diss., Eindhoven University of Technology, 2010), 75.

Figure 12 Policymakers in Drenthe felt their province could attract tourists and improve the economy through a network of appealing cycling infrastructure. Here tourists enjoy riding on a cycling path constructed near Ruinen, 1965, part of the provincial recreational cycling path plan. [Source: National Archives, Fotocollectie Nederlandse Heidemaatschappij, Photographer unknown/Heidemij., no. 161-1396]

isolation of certain rural areas.[34] These appeals were probably strategic, aimed at acquiring more funding.

The strategy targeting economic growth and utility succeeded because many government bodies funded Drenthe's cycling path plan. In 1955, the Drenthe provincial legislature approved a 500,000-guilder subsidy.[35] The local communities contributed 7.5 percent, the province added 20 percent. The remaining 72.5 percent was state funding, with the Ministry of Economic Affairs picking up 42.5 percent, and the Ministry of Social Affairs 30 percent.

34 DA 0923, inv. no. 6258, letter Public Works Drenthe to Ministers, December 4, 1959.
35 "f 500.000 gevoteerd voor rijwielpadennet," *Nieuwsblad van het Noorden*, July 12, 1955, 3.

By the 1960s, the new Ministry for Culture, Recreation and Social Work did subsidize recreational cycling. The Ministry of Public Works was not even considered for funding as it did not regard recreational cycling its responsibility. The contribution from Economic Affairs was unusual, but Drenthe's official designation as economic development area made it eligible for funding.[36]

The right people in the right place made all the difference. Drenthe's Jaap Cramer lobbied national policymakers, and what is more, he was great at playing the publicity game. When the cycling path plan's implementation stalled, Cramer's stunt was to organize a visit from the royal family, resulting in a widely disseminated picture of Queen Juliana cycling on the paths. He then used the queen's interest in the plan to put pressure on the government: there were regular meetings between the Royal Commissioners and the Queen, which helped Cramer keep up the momentum for funding the plans (see figure 13).[37] Van den Muyzenberg's longtime position in provincial Public Works made him another key spider in the web. Individuals mattered in Zuid-Holland as well.

In the 1950s, Zuid-Holland was one of the first Dutch provinces to devise a provincial "Cycling Path Plan." Revisions and updates over the following decades allow us to track the plan's changing notions of what cycling meant to provincial policymakers. Zuid-Holland forms a different case than Drenthe. The motivation behind the plan was a social democratic conviction that provincial authorities should provide healthy recreational facilities for urban working people. In densely populated Zuid-Holland, with large cities like Rotterdam and The Hague, this was a particular concern in the late 1940s and early 1950s.[38] The province commissioned a short film in 1954 to introduce the new recreational area Brielse Maas, east of Rotterdam, explicitly targeting city-dwellers. Called Free Outdoors (*Vrij Buiten*), the drama illustrates the ideology behind the plan.[39] A stressed city bank employee seeks in vain a little peace and quiet on a busy beach. Finally, he finds tranquility and relaxation by taking a cycling trip on paths through Zuid-Holland's nature, far from any noisy cars or traffic. The film represented a new insight in policy circles. According to provincial engineer Fekkes in 1956, "in a wide circle, the realization is growing that on a larger scale than hitherto, provisions have to be made" for cyclists and mopeds, particularly "by creating possibilities for

36 Simons, *Daar fietst men toch zo heerlijk heen*, 60.

37 Piter Bergstra and Martin de Bruin, "Een goede CdK opent Haagse deuren," *Nieuwsblad van Friesland*, August 27, 1994, 23.

38 Harm Kaal, "Ruimtelijke ordening," in *Behoedzaam bestuur. Twee eeuwen provincie Zuid-Holland*, eds. Roel Pots and Nico Randeraad (Leiden: Primavera Pers, 2014), 156-57.

39 The film is on archief.zuid-holland.nl under "Vrij Buiten."

Figure 13 Keen cyclist Queen Juliana of the Netherlands visited Drenthe on many occasions. In 1975, during the tenth Four Day Cycling event (Rijwielvierdaagse), she cycled over one of the province's many recreational cycling paths. [Source: Rob Mieremet, Nationaal Archief / Anefo, no. 928-0377]

recreational cyclists to travel by good and attractive cycling routes."[40] Fekkes explained that such paths would follow Zuid-Holland's many waterways or connect villages to public pools or beaches.

The Zuid-Holland cycling path policy only financed paths that were not parallel to the road network. Separate cycling paths alongside major roads did not qualify. Policymakers considered these an integral part of the road – already funded and governed by national and provincial government public works departments. Politicians and engineers tended to see these paths that lined the country's main arteries as serving a predominantly commuter function. This clashed with the plan's goal – to provide a quiet, relaxing ride in the countryside. Under the plan, paths had their own trajectories, sometimes former rail- or tramway lines.[41] Engineers had to meet high design standards for paths: 3 meters wide and asphalted.[42] While the substantial paths along roads served utilitarian cyclists, provisions for recreational traffic fell short. As this traffic needed paths with a separate trajectory, Fekkes

40 J.C. Fekkes, "Provinciaal Rijwielpadenplan voor Zuid-Holland," *Wegen* 30, no. 1 (1956): 1-2.
41 Examples from the 1950s include the tramway in the Hoeksche Waard between Rotterdam-Zuid and Oude Maas (NA 3.02.27, inv. no. 1559); between Goedereede and Melissant (NA 3.02.27, inv. no. 1339); and in the 1970s, the former tramways between Wassenaar and Leiden and in Naaldwijk (NA 3.02.27, inv. no. 2918 and 2923).
42 "Nieuwe rijwielpaden in Zuid-Holland," *Trouw*, August 13, 1959.

set "great store by opening up the province through the construction of a touristic cycling path network."[43] For provincial policymakers, recreational cycling was an important public good.

This commitment to providing recreational opportunities had a social democratic pedigree. In Zuid-Holland, Provincial Executive (*Gedeputeerde Staten*) politicians rather than provincial public works administration (PWS) engineers initiated the provincial cycling plan. Provincial politician Stien de Ruyter-de Zeeuw was the major initiator of this policy in the 1950s and 1960s.[44] With a background in radical socialist movements, in 1946 she represented the social democratic SDAP party (later the labor party PvdA).[45] When she left the Provincial Executive in 1970, she was praised for her role in creating cycling paths and recreational facilities.[46] As a workers' advocate, she believed in the right to "possibilities for recreation and nature" close to the city for "people without a car."[47]

A later successor, fellow social democrat Joop Borgman, took up the torch. He was Provincial States member from 1962 and Provincial Executive member from 1974 until 1987, the period when the provincial recreational plan was reanimated and substantially revised. His interest in (recreational) cycling is also evident from his post-retirement role as director of the National Cycling Platform (*Landelijk Fietsplatform*), an independent coordinating agency for recreational cycling, from 1991 until his death in 1999.[48] In an interview with newspaper *Trouw* in 1978, Borgman strongly supported creating and promoting a recreational cycling network. He also emphasized how much road traffic safety had improved in the past two decades thanks to widespread cycling path construction alongside most secondary and tertiary roads.[49] This reflects policymakers' growing focus on utilitarian cycling in the 1970s. In short, since the postwar decades, cycling in Zuid-Holland had strong supporters within the Provincial Executive. Implementation of the recreational

43 Fekkes, "Provinciaal Rijwielpadenplan voor Zuid-Holland" (1956), 1-2.

44 Dick Linders, "Mr. Chr. A. De Ruyter-De Zeeuw, Rotterdam 22 Aug 1907 – IJlst 5-12-1995, Onafhankelijk, sociaal-democraat," *Rotterdams Jaarboekje* Series 10, no. 4 (1996), 146-53; Kaal, "Ruimtelijke ordening," 167.

45 The SDAP merged with two other parties to form the PvdA in 1946, the major social democratic party in Dutch politics.

46 Linders, "Mr. Chr. A. De Ruyter-De Zeeuw," 152.

47 Ibid., 152.

48 For Borgman, see https://www.parlement.com/id/vg09llponlz5/j_joop_borgman, biography consulted September 7, 2018.

49 Haro Hielkema, "Zuid-Hollandse fietsers krijgen meer de ruimte: Gedeputeerde Borgman kan nog wel paar miljoen gebruiken," *Trouw*, November 10, 1978, 6.

plan may have been slow – requiring a constant battle for resources and engineers' attention – but this long-term political commitment by Zuid-Holland social democrats to recreational cycling was a crucial success factor.

Digging Up Data

Politicians in Zuid-Holland struggled to get provincial engineers to implement cycling plans. To make a proper cycling policy, a first step was an inventory of existing cycling paths in the province. No such overview existed. The lack of a coordinating agency for cycling meant that Zuid-Holland's engineers had the painstaking and time-consuming task of compiling this information, as shown in figure 14. There were paths alongside roads, governed either by national or provincial government (in green). Then there were separate paths in the provincial cycling path plan (black), or built by other organizations (blue). All these authorities had constructed paths, but many from the late 1950s only existed on paper (the paths in red). That is why this map has so many different colours.

A similar request by the Provincial Executive in the province Gelderland was also time-consuming for Public Works. It asked for the extent of existing cycling paths and the cost of further construction.

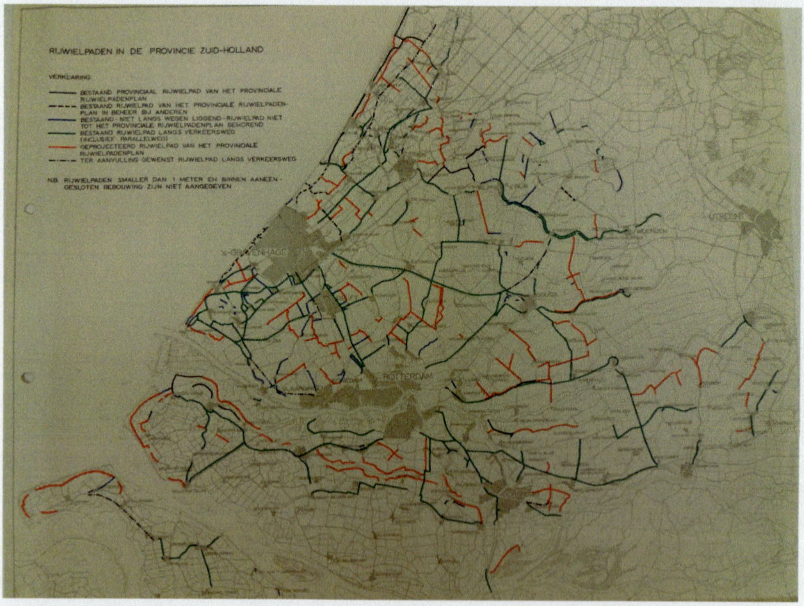

Figure 14 Map of cycling paths in Zuid-Holland province, c. 1959. [Source: NA 3.02.27, inv. no. 1559]

In February 1949, politicians reminded engineers that they had still not provided this information requested in April 1948. The Gelderland engineer eventually responded, telling them in no uncertain terms how expensive cycling path construction would be, requiring "very extensive preparation," and it could only be "implemented very gradually." Politicians were irritated by this negative response: cycling paths alongside roads were needed urgently: a temporary and cheaper construction was one option.

In the 1960s, implementation of the Zuid-Holland plan was slow. Engineers blamed this on lack of technical personnel and frankly admitted that roads were their priority. But finding out how provincial engineers spent their budgets remained an issue well into the 1970s. Politician Joop Borgman complained that the plan's progress reports gave no insight into what percentage of the budget for improving road surfaces went to cycling paths or to roads. This information was only obtained after multiple strongly worded letters and meetings with engineers. A true political decision about the distribution of funding was only possible once politicians knew how much money was actually being spent on improving conditions for drivers and cyclists. Civil servants' longevity in office fostered consistent policymaking and enabled long-term personal relationships between stakeholders, but could also lead to inertia and delay.

Sources: NA 3.02.27, inv. no. 1559, letters PWS to Gedeputeerde Staten December 14, 1959 and January 7, 1963; GA 0039, inv. no. 9488, letters GS Gelderland to HID PWS, April 22, 1948, February 22, 1949, April 21, 1949, July 1, 1949; NA 3.02.46, inv. no. 2676.

Friction arose between politicians and provincial Public Works engineers: politicians' frequent calls to action suggest engineers were slow to implement the plan (see box). Initial objectives had to be scaled back multiple times. The first proposals in the mid-1950s were for 400 km of cycling paths. For budgetary reasons, provincial engineers proposed reducing the plan to 215 km.[50] The five-year plan for realization adopted in 1959 counted 110 km, a target not even reached by 1965.[51] In 1958, provincial engineers acknowledged "the demand of your board to accomplish more in realizing the provincial cycling path plan than has happened in previous years."[52] The Provincial Public Works' chief-engineer admitted he believed that roads

50 NA 3.02.27, inv. no. 1559, letter PWS to Provincial Executive, March 26, 1959.

51 Ibid., January 7, 1963.

52 NA 3.02.27, inv. no. 1559, PWS memorandum February 6, 1958 "Overzicht van de stand van zaken met betrekking tot de voorbereiding en de aanleg van het provinciale rijwielpadenplan."

had priority over cycling paths. For the delay, he also blamed the lengthy and costly process of land expropriation.[53] This was less of a problem for paths alongside roads because they were included in road budgets and expropriation procedures; but because recreational cycling paths had their own trajectory, they required legally and financially complex procedures. The process was not just an excuse by the engineers, as Joop Borgman later claimed. If landowners refused to part with their land, there was little the authorities could do.[54] Despite fluctuating investments, the network grew steadily over the 1960s and 1970s (see figure 15).

By the 1970s, Zuid Holland's expanded version – like other provincial plans – included utilitarian as well as recreational cycling, demonstrating a changed perception of cycling over the decades. According to the provincial public works engineers in 1975, "the bicycle's transport function" (*vervoersfunctie*) was valued more.[55] This new awareness was partly due to mobility statistics research, such as a 1974 Dutch Statistics Agency report which Zuid-Holland engineers cited. This showed that 29 percent of commuters still cycled.[56] More important was the 1970s shifting political climate: cycling activism and growing concern over automobility's negative effects had changed policymakers' views on cycling, as Part III will demonstrate. Policymakers realized that the bicycle was not just for recreation, and once again recognized its actual and future role in transport. They designed the new cycling path plan with an additional 125 km of cycling paths also fit for utilitarian purposes. Likewise, the northern province Friesland introduced a recreational cycling path plan in 1962, updated in 1978 to include "cycling paths for daily use."[57] The policymakers considered their ambitious plan a product of the previous recreational plan.[58] The institutional "cost" of introducing it was low, because public works could just copy the initial plan's procedures used in the 1950s and 1960s.[59] The only major difference was launching a much

53 Ibid.

54 Hielkema, "Zuid-Hollandse fietsers krijgen meer de ruimte."

55 J.P. Bakker and R. v.d. Hout, "Herziening provinciaal fietspadenplan Zuid-Holland," *Verkeerskunde* 26, no. 7 (1975): 356.

56 Ibid., 359.

57 "Onzeker of provincie aanleg paden met geld kan steunen," *Leeuwarder Courant*, November 6, 1978.

58 "Eerstkomende 15 jaar: bijna 250 fietspaden erbij in Friesland," *Leeuwarder Courant*, December 1, 1979, 13.

59 Since funding was never sufficient to implement the envisioned provincial plans entirely, policymakers created guidelines for determining the priority of potential new paths. They also used this strategy when they saw the need for utilitarian cycling paths since the 1970s.

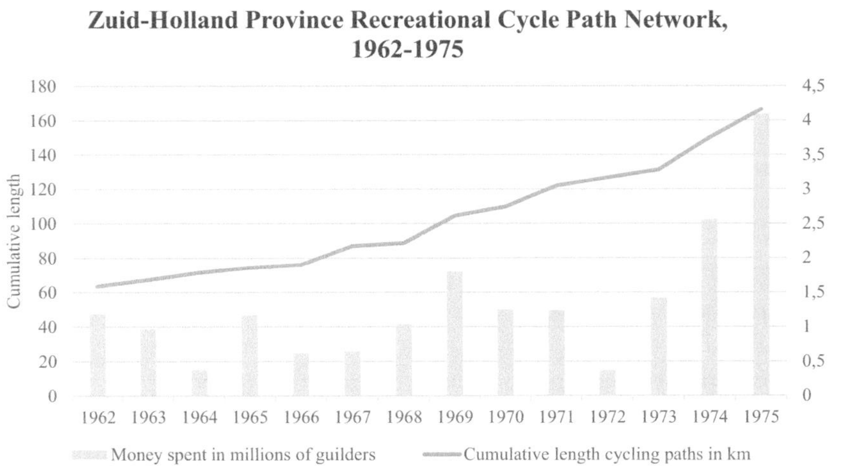

Figure 15 Funding for the Zuid-Holland cycling path plan fluctuated but supported a gradual increase of the network over the 1960s and 1970s. Source: J.P. Bakker and R. v.d. Hout, "Herziening provinciaal fietspadenplan Zuid-Holland," Verkeerskunde *26, no. 7 (1975), 356.*

more extensive participation design procedure. In a sign of the times, municipalities, recreational organizations, but also members of the public could comment on and revise the proposed plans.

Seeing cycling as a means to improve public health and wellbeing, provincial officials created a cycling path network and recreational areas for their (urban) residents. The energetic efforts of several provincial social democrat politicians who put pressure on Provincial Public Works engineers to take cycling more seriously, resulted in a progressive cycling policy for Zuid-Holland compared to many other provinces in the 1950s and 1960s.

3.3 1960s National Subsidies for Recreational Cycling

A national role in recreational cycling, briefly discussed in the 1930s, became only a reality in the 1960s. The Ministry of Culture and Recreation began subsidizing recreational facilities, including cycling paths.[60] In the 1970s, the Public Works ministry also subsidized (urban) utilitarian cycling infrastructure, as Chapter 7 explores. The national recreational subsidies were earlier, reflecting that the infrastructure for recreational cyclists received more political support in the 1950s and 1960s than for commuters, who were not entirely neglected. The national recreational subsidy

60 De Haan, Duyvendak, and Bottenburg, *In het hart van de verzorgingsstaat*, 156.

entailed a 75 percent state contribution for cycling paths separate from the road network, leading through national parks, forests, and other attractive recreation areas. Appendix 1 lists the projects funded in the 1960s, earlier than the construction of (activist inspired) urban cycling infrastructure. The 1960s paths were mainly in provinces attempting to attract tourists and recreational cyclists. Provinces differed significantly in their proactive efforts to acquire subsidies. A 1976 consultancy report on recreational cycling and the associated policy processes noted these considerably different policies from province to province: "there is not one government in particular directing the policy process concerning recreational cycling."[61] Policy coordination at a national level remained minimal.

The fact that (recreational) cycling was not declining as fast as some believed, spurred the national government to adopt a more proactive role. A 1961 Ministry of Culture, Recreation and Social Work committee meeting on outdoor recreation posed the question whether it still made sense to "invest large subsidy amounts in cycling paths in view of the advancing motorization."[62] R. van de Waal, head of the National Planning Service's recreational branch (*Rijksdienst voor het Nationale Plan*) considered this type of thinking simplistic and warned against "the great danger of "catchphrases" [*kreten*] in determining policies." He referred to one popular catchphrase: "every young person rides a moped nowadays; the regular bicycle [*trapfiets*] is disappearing." However, "results of objective research often show a much more nuanced picture."[63] He cited 1956 research showing that during weekends, a third of the population left cities for recreation, often by bicycle. Even if this share dropped to 10 percent, he argued, that would still be 150,000 cyclists leaving cities at weekends. These absolute numbers justified investing in cycling infrastructure. According to another committee member, policy could be inspirational: "The presence of cycling paths – and familiarity with them – can stimulate cycling."[64] Van de Waal pointed to recreational cycling's rising popularity in the United States as further proof that bicycles would no doubt never disappear altogether.

Cycling path construction up to this point had been mainly rural. The construction of urban cycling facilities really took off in the 1970s. As an intermediate step, policymakers increasingly focused on paths connecting

61 NA 2.27.5215, inv. no. 2167, letter consultancy agency C. Kappelhoff to A.A.H.C. van Onzenoort (Ministry of Culture and Recreation, Outdoor Recreation branch), December 20, 1976, attachment to letter, 54.

62 NA 2.27.5111, inv. no. 60, 28th meeting INCOR-committee, July 5, 1961.

63 Ibid.

64 Ibid.

the city to the countryside. In their rich vocabulary for cycling facilities, Dutch engineers coined the term "break-out path" (*uitvalspad*).[65] They were referring to paths that provided a safe, separate way to leave the city for short recreational rides or connect to nearby recreational facilities. Recreation officials increasingly saw the lack of such paths as problematic; city-dwellers who did not own a car were locked within the city's confines. According to W.G. van der Kloet, an engineer with the State Forestry Agency (*Staatsbosbeheer*) "especially during the weekend, traffic from residential areas to destinations in the vicinity can be very substantial, especially around cities. These paths fulfil an important role, especially if they can be extended to connect with the separate cycling paths being constructed – albeit not enough – in towns."[66] Such paths would not be cheap but served commuters as well as tourists, making the expense more justifiable.

Initially, the coalitions focused on recreational cycling and only involved utilitarian cycling in the late 1960s, as we will see. One typical example is the province of Limburg, where a cycling path committee (for recreational paths) created during World War II achieved very little.[67] Even a new provincial committee to help establish a Provincial Cycling Path Fund for local communities in 1956 was discontinued twelve years later.[68] This ended in 1968: according to the authorities, the municipalities did not use the fund sufficiently. A new funding opportunity for recreational *and* utilitarian cycling replaced it. Zuid-Holland province shows a similar pattern. Private organizations and provincial agencies framed cycling predominantly as a recreational activity until the late 1960s, when the focus on utilitarian cycling returned. Triggered by recreational plans and the people behind them, the attention to utilitarian cycling broadened cycling path plans and subsidies.[69]

65 NA 2.27.5111 (Ministry CRM, INCORET = Interdepartementale Coördinatiecommissie Openluchtrecreatie en Toerisme), inv. no. 60, 'Nota Rijwielpaden' by Ministry of Agriculture (*Cultuurtechnische Dienst*), attachment to letter to Ministry CRM/INCORET, August 16, 1963.

66 NA 2.27.5111 (Ministry CRM, INCORET = Interdepartementale Coördinatiecommissie Openluchtrecreatie en Toerisme), inv. no. 60, Memorandum INCOR-315, attachment to Ministry of Agriculture letter to Ministry OKW/INCOR, June 12, 1962.

67 Frank Veraart and Adri A. Albert de la Bruhèze, "Fietsen in de Nederlandse bergen. Achterblijvend fietsgebruik in het zuiden van Limburg in historisch perspectief," in *Studies over de sociaal-economische geschiedenis van Limburg/Jaarboek van het Sociaal Historisch Centrum voor Limburg*, eds. Ad Knotter and Willibrord Rutten (Maastricht: Sociaal Historisch Centrum voor Limburg, 2001), 142.

68 Ibid., 145-48.

69 When the Brabant organization De Kempen was founded in 1976, its aim was "mainly recreational paths, while now functional paths (those with a traffic function) are considered

3.4 Conclusion

In the 1950s, Dutch government policymakers framed recreational cycling as a public good by linking it with multiple policy goals. The province of Drenthe pioneered this strategy. Its leading politicians looking to combat unemployment and economic crisis as well as attract tourism and industry, saw cycling infrastructure as part of a package of measures. In Zuid-Holland, social democrat politicians held ideas – long-standing and widespread in and beyond the Netherlands – about cities' unhealthy impact on their residents. Ideas about green lungs and garden cities circulated transnationally. In the Interbellum, Dutch social democrat urban planners and public housing experts had emphasized the need for recreation – a notion provincial policymakers in both highly urbanized and tourist regions revisited in the 1950s. This time their motivation for building recreational infrastructure was economic: provinces with considerable nature reserves such as Gelderland or Drenthe were convinced a cycling path network would attract tourists – the kind of national tourism the ANWB promoted in its journal *De Kampioen*. Building on private organizations' earlier work and benefiting from the recreational cycling culture established before 1950, provincial policymakers in turn expanded cycling infrastructure.

Policy beliefs about recreational cycling developed at a time when regional planning ideas were equally strong. The result was a layered structure of national, provincial/regional, and local plans for land use, roads, and recreation. That these beliefs about cycling and governance coincided, helps explain why so many regional policymakers supported and initiated regional cycling plans. During the 1940s and 1950s, the national government was noticeable by its absence from this story, while provincial authorities took the initiative. Despite the lack of national coordination, regional authorities regularly shared best practices. Only in the 1960s did the national state start subsidizing recreational infrastructure on a larger scale.

Unlike in earlier decades, local politicians were the driving force. The ANWB's tourism and recreation lobbies played a background role by connecting people and institutions with relevant knowledge. However, the main initiative came from the ruling elites. Many mayors were personally involved in the organizations, as were leading provincial politicians and officials – a

by far the most important." Archive Eindhoven Regionaal Historisch Centrum, 10091 (Intergemeentelijke organisatie voor rijwielpaden in de Kempen, Valkenswaard), inv. no. 2, "Verslag bespreking studiecommissie heroriëntering taakstelling van de rijwielpaden-organisatie 10-09-1976."

marked difference from the urban cycling activism that would emerge in the 1970s. The large and well-connected group of Dutch politicians who took an interest in recreational cycling also improved the connections for rural commuter cyclists. Mostly appointed and not elected, their longevity in office and continuing support for the policy were crucial. A stable corps of civil servants with whom cycling advocates could forge relationships was significant for urban cycling activism in the 1970s and beyond. Building these cycling infrastructure networks took decades, as competition for funding was a recurring theme. Engineering departments had many tasks and did not always prioritize implementing recreational plans. The piecemeal approach nevertheless produced significant results over time. The key players who made a real difference at the right time were social democrat politicians Stien de Ruyter-de Zeeuw and Joop Borgman in Zuid-Holland, and chief engineer Gerard van den Muyzenberg in Drenthe. Similarly, in a province not discussed here, Gelderland, Public Works chief engineer J.G. Kruimel was also the local representative (consul) for the ANWB. Like De Ruyter-de Zeeuw and Van den Muyzenberg, he had to struggle with a Public Works department that often objected to constructing cycling infrastructure and only reluctantly executed provincial policy.[70] He stimulated subsidizing cycling path organizations throughout his career, so it is not just Gelderland's natural areas but also his support that explains why so many organizations were formed here.[71]

By the late 1960s, policymakers not only expanded recreational but also utilitarian cycling. The next section discusses this development, but the link with recreational cycling policies is important. After 1970, provincial policymakers and engineers remained the key intermediaries between national and local government – for negotiating subsidies, subsidizing cycling, and coordinating (the interconnectedness of) local cycling networks. Designing regional cycling path networks was an invaluable stepping-stone for the future.

70 GA 0039, inv. no. 9488, letter PWS Gelderland to GS, March 28, 1949 and letter GS Gelderland to HID PWS, April 21, 1949.

71 "Fietsen op de paden van de Zuid-Veluwe," 176.

4 Popular or Outdated? National Policymakers' Ambivalence about Bicycles

In 1950, the Netherlands' major roadbuilding trade journal *Wegen* ("Roads") celebrated its twenty-fifth anniversary. For this occasion, the special May issue featured a specific theme. A few years after World War II, in a context of reconstruction, readers might expect a vista on the future of highway building, perhaps inspired by American highway engineering ideas circulating through Europe at this time. Instead, the special issue was dedicated entirely to cycling paths "since bicycle traffic and the cycling path question are at the center of attention at the moment."[1] What does this theme choice tell us about cycling's status in the Netherlands around 1950? Was it the last stand for cycling before automobility took it over entirely? Or was it rather an indication of cycling's continuing presence in practice and governance?

Postwar reconstruction was still in full swing in 1950. Building up industry and roads for trucks and cars was a priority. With the country's low levels of automobility, for many Dutch people, cycling was still the best mobility option. Yet the justification for the bicycle tax to fund cycling infrastructure outside cities had gone: abolished by the occupying German government in 1941 and not reinstated after the war. And with it, the discourse of giving cyclists something in return for their tax money disappeared. Did this also entail a return to the pre-1920s situation in which the state did not consider cycling infrastructure a public good? Or did leading engineers and politicians' 1950s and 1960s technocratic approach still have a place for cycling? The *Wegen* special issue suggests that engineers continued to take cycling seriously. But to what extent could cycling persist alongside advancing automobility in the Netherlands?

In Chapter 2, "Contested Compromise," I argued that cycling's political status had become more firmly embedded in the Netherlands than elsewhere. This suggests that the generated political support and engineering norms were strong enough to have a lasting impact. This chapter investigates that hypothesis. Did the cycling policies created in the 1920s and 1930s set in motion a path-dependent process where older engineering norms still applied, even as ideas about cycling were about to change? Based on

1 *Wegen* 24, no. 5 (1950): 136.

engineering committees' archives and literature produced outside commercial and academic publishing, this chapter explores the lingering role of policies initiated in an earlier period, to explain how cycling survived as part of standard engineering practices.

"Popular or Outdated? National Policymakers' Ambivalence about Bicycles" argues that while national engineers and the ANWB gave lip service to urban cycling as a transport mode, they increasingly left cycling policy to provincial and local policymakers. At the same time, the 1920s and 1930s engineering norms and practices for traffic separation remained relevant nationally. Older policies continued to guide new cycling infrastructure outside the city. And in designing new car infrastructures, national engineers investigated ways to safely combine rising levels of car ownership with (commuter) cycling. Additionally, for the first time, specific groups of cyclists started to catch the government's eye: recreational cyclists and schoolchildren, for instance. In other words, cyclists' status as citizens with equal rights to drivers came under pressure but did not crack with the rise of car mobility. In some ways, that status was even reinforced in unintended and unexpected ways, as we will see.

4.1 Dutch Cycling's Staying Power from an International Perspective

Although beyond the scope of this study, the lack of mobility alternatives, as Adri Albert de la Bruhèze and Frank Veraart already pointed out, partly explains cycling's persistent popularity in the Netherlands. In particular the (late) adoption of the car in the 1960s meant cycling remained popular – although the moped and bus had become rivals in the 1950s.[2] By 1960, Dutch car ownership levels were significantly lower than in other European countries: 45 cars per 1,000 inhabitants, only half the number in similarly small countries like Belgium (82), and Switzerland (89). Meanwhile, the number in France and the UK was over 100, and in the United States almost 350.[3] This huge gap closed by 1970, when Dutch car ownership reached

2 Patrick Bek book with working title "Governing Workers' Mobility: Job Accessibility and the Politics of Movement in Twentieth-Century Netherlands" (PhD diss., TU Eindhoven, forthcoming); Albert de la Bruhèze and Veraart, *Fietsverkeer in praktijk en beleid*.

3 Importantly, even in countries with high car numbers, it was not self-evident that people would own a car around 1960; many people did not, as Simon Gunn points out for the UK in his article, "People and the Car: The Expansion of Automobility in Urban Britain, c. 1955-70," *Social History* 38, no. 2 (2013): 220-37.

similar levels to Belgium and Switzerland (200), although still less than France and West Germany, and especially the United States.[4] The end of Dutch wage control policy in 1963 suddenly increased purchasing power, resulting in a rapid uptake of car ownership. Mom highlights other factors to explain the different development of car culture and ownership in the Netherlands, like the absence of an "aristocracy" of car owners convincing the bourgeoisie to take up the car, so important in other countries. In addition, the country lacked a car industry.[5] Instead, the democratic, favorable bicycle culture – increasingly linked to the train system that the Dutch tourist organization ANWB cultivated, meant cycling was stigmatized far less than elsewhere and remained a crucial transport mode.[6]

Similarly, the role of public transit and urban development in Dutch mobility patterns is beyond the scope of this study.[7] It is, however, important to note that public transit outside the large Dutch cities did not compete significantly with cycling. This is partly why cycling remained so popular in medium-sized cities. Dutch railways ran an inter-city network, linking major cities rather than constructing spiderwebs of lines around cities connecting suburbs and small towns, which was the case in Belgium.[8] The 1960s new urbanization policy in the Netherlands strengthened public transit's role and was called "bundled deconcentration," a term introduced in the 1966 Second Memorandum on Spatial Planning. Planners did not want to spread the population equally over the country because it would stretch public facilities and increase travel distances. Yet concentration was equally undesirable as most people preferred a house with a garden and in a more spacious environment. In typical Dutch planning fashion, bundled deconcentration became the adopted compromise. Planners designated so-called growth centers near existing cities. Public transit played a major role in connecting these new towns to large cities. Inspired by an envisioned population of 20 million, space – always at a premium in the Netherlands – had to be used efficiently. Planners saw public transit along with the bicycle and moped as preferable to cars. However, power struggles between spatial planners

4 Filarski in cooperation with Mom, *Shaping Transport Policy*, 177. Numbers based on Hanna Wolf, "Following America? Dutch Geographical Car Diffusion, 1900 to 1980."

5 Mom, *Atlantic Automobilism*, 67-69.

6 Jan Ploeger shows Dutch Railways (NS) focused on cycling early on as a way to get to and from stations, and facilitated this, strengthening the bicycle-train combination as alternative to commuting by car: Jan Ploeger, "The Influence of the Netherlands Railways on Spatial Planning (1960-1990). How the Bicyclist was Discovered as a Feeder," Working paper, 2021.

7 Patrick Bek, "Governing Workers' Mobility," book manuscript, 2021.

8 De Block, Kool, and Meulder, "Paradise Regained?"

and traffic engineers limited the implementation of this vision, as Mom notes. While these new growth cores did get good public transit facilities, automobility levels also soared.[9] Advocates of driving, cycling, and public transit kept each other in check, avoiding hard choices for any one mode. As a result, Dutch planners and engineers ended up trying to do a little bit of everything – a system which provided opportunities for cycling advocates to carve out their own space.

The popularity of cycling gave rise to much fiercer competition for space in historic Dutch cities when automobility grew in urban centers around the 1960s. Parked and moving cars increasingly obstructed cyclists – yet drivers and some policymakers blamed cyclists for slowing down motorized traffic. The authors of *Cycling Cities* show that urban policymakers and traffic engineers struggled with drivers and cyclists competing for space.[10] Urban authorities were so preoccupied with the traffic problems cars presented, that they lost sight of cyclists' plight in the 1960s: cycle lanes were rarely part of street redesigns in historic city centers.[11] By contrast, planners did include cycling lanes in new residential areas on the outskirts of towns.[12] Dutch planners pioneered the idea of the *Woonerf* in the 1960s – a traffic-calming residential area where walking, playing, and cycling were the norm and cars only could move around as guests.[13] Outside these new residential areas, traffic separation rather than traffic calming remained the Dutch engineering norm.[14] Traffic engineers were rarely willing to sacrifice space for cycling lanes in city centers due to car drivers' ever growing demands

9 Faludi and Van der Valk, *Rule and Order*, 131; Jan Straub, "Binnenstadsbeleid en verkeers- en vervoersproblematiek. Een onderzoek naar het beleid ten aanzien van de verkeers-, vervoers- en parkeerproblemen in binnensteden in Nederland in de periode 1960-1990" (PhD diss., University of Amsterdam, 1991), 32-33; Mom, "The West and the Rest"; Marco Bontje, "A 'Planner's Paradise' Lost? Past, Present and Future of Dutch National Urbanization Policy," *European Urban and Regional Studies* 10, no. 2 (2003): 135-51.

10 Ruth Oldenziel and Adri A. Albert de la Bruhèze, "Amsterdam: World Bicycle Capital, By Chance," in *Cycling Cities: The European Experience*, eds. Ruth Oldenziel, Martin Emanuel, Adri A. Albert de la Bruhèze and Frank Veraart (Eindhoven: Foundation for the History of Technology and Rachel Carson Center for Environment and Society, 2016), 21.

11 Adri A. Albert de la Bruhèze, "Enschede: An Experiment in Cycling," in *Cycling Cities: The European Experience*, ed. Ruth Oldenziel, et al. (Eindhoven: Foundation for the History of Technology and Rachel Carson Center for Environment and Society, 2016), 44.

12 Hans Buiter, "Utrecht: Bicycles Rule – Again," in *Cycling Cities: The European Experience*, ed. Ruth Oldenziel, et al. (Eindhoven: Foundation for the History of Technology and Rachel Carson Center for Environment and Society, 2016), 33; Berkers, "Fietsgebruik en -beleid in Maastricht en Parkstad in historisch perspectief."

13 Van der Cammen and De Klerk, *Ruimtelijke ordening*, 204-6.

14 Veraart, Emanuel, and Oldenziel, "Eindhoven: Engineering a Path for Bikes?," 56-57.

for space. At the same time, many cities invited foreign engineers to design large-scale traffic plans for cars, like the German planner Feuchtinger in Utrecht (1958), Friedrich Lehner in The Hague (1964), or American engineer David Jokinen in Amsterdam (1967), invited by the car lobby *Stichting Weg* ("Roads Foundation").[15] Local politicians ultimately rejected these plans, not least because of citizen protests and concerns over the destruction of historic neighborhoods. Had these plans been implemented, the long-term development of Dutch urban mobility would have looked very different. And because most Dutch (like Danish) cities did not invest in subway construction, cycling had less competition from public transit.[16] The conservation of historic city centers and a fear of cars' destructive effects on American cities already played a role in this reluctance.[17] In other places, redesigning the city for cars was more successful. For the industrial Dutch city of Eindhoven, German engineer Schaechterle drew up a new traffic plan in 1961, which the city did implement. The number of cyclists surprised the German planner: creating a car-friendly city meant finding ways to separate traffic in an urban environment.[18] Involving some separated bicycle infrastructure, the main goal was nevertheless to speed up car traffic by getting cyclists out of the way.[19] And in the 1950s and 1960s, the similarly industrial port city of Rotterdam – as Eric Berkers, Frank Schipper, Patrick Bek and Ruth Oldenziel note – a "modernist urban planning lab ... was not kind to cyclists."[20] Plans for the city center that was destroyed in World War II were American in inspiration, making room for cars while leaving little pockets of space for pedestrians. Yet even here, critical voices were lamenting the marginalization of cyclists already in the 1950s and 1960s.[21]

In many countries and cities, the people who continued to ride their bikes in the 1950s and 1960s were no longer visible to policymakers and engineers. But cyclists were still there, in the streets and on the roads, as Albert de la Bruhèze, Veraart, and later co-authors in *Cycling Cities* have uncovered through painstaking archival work. From a technological point

15 On Jokinen: Oldenziel and Albert de la Bruhèze, "Amsterdam: World Bicycle Capital, By Chance," 21-22. On Feuchtinger: Buiter, "Utrecht: Bicycles Rule – Again," 32-35; Tim Verlaan, "De toekomst van de Nederlandse binnenstad 1960-1978" (PhD diss., University of Amsterdam, 2016), 60.

16 Carstensen and Ebert, "Cycling Cultures in Northern Europe," 45.

17 The demolition for a subway system in Amsterdam sparked major protests in the late 1960s.

18 Albert de la Bruhèze and Veraart, *Fietsverkeer in praktijk en beleid*, 81-83.

19 Veraart, Emanuel, and Oldenziel, "Eindhoven: Engineering a Path for Bikes?" 56-57.

20 Berkers et al., *Cycling Cities: The Rotterdam Experience*, 7.

21 Ibid., 30-34.

of view, this is not so surprising: technologies remain in use long after engineers and policymakers have lost interest in them. Moreover, the car boosters who campaigned for motorizing the masses faced a barrier: there were also urban social democrats promoting public transit for working-class commuters. Engineers' predominant focus on driving as the path towards the future, however, meant that cycling's public status declined precipitously. As Oldenziel argues, by excluding them in traffic counts or otherwise erasing them from official documents, cyclists became invisible to policymakers in many parts of Europe.[22] German traffic planner Feuchtinger, who also designed a controversial urban renewal plan for Utrecht in the 1950s at the invitation of local authorities, is a striking example. A 1956 traffic count in Germany's Saarland region showed two-wheelers had a 16 percent share of road traffic, but nevertheless Feuchtinger argued that cycling levels were so low that they could be omitted in future traffic counts without this "significantly altering the general traffic picture."[23] It was a modernist planning argument he would use again for his Utrecht circulation plan This deliberate decision to render cyclists invisible had major negative repercussions for cycling planning.[24] As geographer and management expert Bent Flyvbjerg summarizes: "Power is more concerned with defining a specific reality than with understanding what reality is."[25] It was only in the 1970s and 1980s that traffic counts would include cyclists and pedestrians again in response to activists call to arms – and in that way their (for planners surprisingly large) share of trips became visible once more.[26] According to Frédéric Héran, a whole generation of French engineers and planners were not interested in cycling and consequently their technical knowhow – and users' tacit knowledge – were lost.[27] Maxime

22 Oldenziel, "Accounting Tricks."

23 Fläschner, "Stahlroß auf dem Aussterbe-Etat," 13-14. Buiter, "Utrecht: Bicycles Rule – Again." For a present-day perspective on counting and calculating procedures: Jens Stissing Jensen, Matthew Cashmore, and Morten Elle, "Reinventing the Bicycle: How Calculative Practices Shape Urban Environmental Governance," *Environmental Politics* 26, no. 3 (2017): 459-79.

24 Similarly, British planners did not include cyclists in future mobility plans, despite still significant numbers in the 1950s: Rorie Parsons and Geoff Vigar, "'Resistance was Futile!' Cycling's Discourse of Resistance to UK Automobile Modernism 1950-1970," *Planning Perspectives* 33, no. 2 (2018): 167-68.

25 Bent Flyvbjerg, *Rationality and Power: Democracy in Practice* (Chicago: University of Chicago Press, 1998), 36.

26 Barbara Schmucki, *Der Traum vom Verkehrsfluss. Städtische Verkehrsplanung seit 1945 im deutsch-deutschen Vergleich*, Deutsches Museum: Beiträge zur Historischen Verkehrsforschung, vol. 4 (Frankfurt: Campus, 2001), 169-72, 93-96.

27 Frédéric Héran, *Le retour de la bicyclette: Une histoire des déplacements urbains en Europe, de 1871 à 2050* (Paris: La Découverte, 2014), 153.

Huré shows how the concerted efforts of non-governmental actors were often required to put cycling back on the political agenda – a struggle that is ongoing.[28]

In the 1950s, the future projections became a self-fulfilling prophecy. Because urban traffic planners believed cycling was on the way out, cycling infrastructure was not worth public funding – even in cycling-friendly cities like Copenhagen. Engineers constructed no new cycle lanes in the city, considering this, like in the Netherlands, an inefficient use of space.[29] Martin Emanuel notes: "National and local policy makers' focus on the car and on public transport had drawn attention away from bicycle traffic. In the 1960s, hardly anyone mentioned the bicycle."[30] Nevertheless, many Copenhagen and Dutch residents kept cycling in the 1950s and 1960s.

Where cycling culture was not as robust as in the Netherlands or Denmark, the practice of cycling – discouraged more actively by policymakers – disappeared much faster.[31] Planners envisioned and eventually built ring roads, parking garages, and other facilities for cars, but very little provision for cyclists.[32] Barbara Schmucki concludes that while Germany focused more on cycling paths right after World War II, the predominant idea – reinforced by the importance of American influences – in the 1950s and 1960s was that the car should be able to reach every (inner-city) destination with as little interruption as possible.[33] To the extent that urban policymakers invested in alternatives to the car, they focused on buses, subways, and light railways

28 Maxime Huré, "Les réseaux transnationaux du vélo. Gouverner les politiques du vélo en ville. De l'utopie associative à la gestion par des grandes firmes urbaines" (PhD diss., Université Lyon 2 Lumière, 2013).

29 Martin Emanuel, "Copenhagen: Branding the Cycling City," in *Cycling Cities: The European Experience. A Hundred Years of Policy and Practice*, ed. Ruth Oldenziel, et al. (Eindhoven: Foundation for the History of Technology, 2016), 81-82.

30 Emanuel, "Making a Bicycle City," 509.

31 On the declining numbers of walkers and cyclists, see Colin Pooley and Jean Turnbull, "Modal Choice and Modal Change: The Journey to Work in Britain Since 1890," *Journal of Transport Geography* 8, no. 1 (2000): 11-24; Colin Pooley, Jean Turnbull, and Mags Adams. *A Mobile Century? Changes in Everyday Mobility in Britain in the Twentieth Century* (Aldershot: Ashgate, 2005).

32 Frank Veraart, Stijn Knuts, and Pascal Delheye, "Antwerp: Cycling Claims a Comeback," in *Cycling Cities: The European Experience*, eds. Ruth Oldenziel, Martin Emanuel, Adri A. Albert de la Bruhèze and Frank Veraart, (Eindhoven: Foundation for the History of Technology, 2016), 93; Martin Emanuel, Frank Veraart, and Peter Cox, "Manchester: Cycling at a Standstill," in *Cycling Cities: The European Experience*, 105; Emanuel, "Malmö: A Center of Cycling Innovation," in *Cycling Cities: The European Experience. A Hundred Years of Policy and Practice*, ed. Ruth Oldenziel, et al. (Eindhoven: Foundation for the History of Technology, 2016), 141-42.

33 Schmucki, *Der Traum vom Verkehrsfluss*, 98-99, 118-53; Albert de la Bruhèze and Oldenziel, *Cycling Cities: The Munich Experience*, 26.

rather than cycling.[34] According to Huré, many French planners saw a mix of "automobility and underground public transit" as the future; planners in the Dutch port Rotterdam had similar ideas.[35] In the 1960s, urban planners in capitals like Stockholm and Amsterdam were also experimenting with strictly separating pedestrians and cyclists from car traffic in their new suburbs.[36] Such safe areas for cyclists and pedestrians were built outside these suburbs, giving them the character of an island reservation. The car ultimately became popular everywhere. Dutch traffic discourse in the 1960s focused on cars too. What needs to be explained is why these high car levels became more compatible with cycling in some countries than others. Policymakers in the Netherlands never lost sight of cycling in the 1950s and 1960s, even though public debate and engineering discourse seemed single-mindedly focused on driving: cyclists remained more visible than elsewhere.[37] What forms of cycling survived in the Netherlands? And which policies supported this?

4.2 Ambiguities and Continuities

In the postwar period, Dutch policymakers' attitude towards commuter cycling became more ambivalent. State engineers in the 1920s and 1930s had been positive about utilitarian cycling's status as a public good worthy of (some) funding and policies, despite an active car lobby. Criticism of cycling did not intensify in the 1950s and 1960s, but cyclists were increasingly ignored by certain policymakers, preoccupied as they were with the problems of the car's massive space demands – whether moving or parked – particularly in historic cities not built for cars. Nevertheless, some engineers and urban

34 Adri A. Albert de la Bruhèze and Martin Emanuel, "Hannover: Sidelining the Bicyclist," in *Cycling Cities: The European Experience. A Hundred Years of Policy and Practice*, ed. Ruth Oldenziel, et al. (Eindhoven: Foundation for the History of Technology, 2016), 118; Emanuel, "Copenhagen: Branding the Cycling City," 82.

35 Maxime Huré, "Lyon: The Bicycle Goes Corporate," in *Cycling Cities: The European Experience*, ed. Ruth Oldenziel, et al. (Eindhoven: Foundation for the History of Technology, 2016), 177; Berkers et al., *Cycling Cities: The Rotterdam Experience*.

36 Martin Emanuel, "Stockholm: Where Public Transit Eclipses Cycling," in *Cycling Cities: The European Experience. A Hundred Years of Policy and Practice*, ed. Ruth Oldenziel, et al. (Eindhoven: Foundation for the History of Technology, 2016), 153. This idea was also considered and partially implemented in the Bijlmer area of Amsterdam and the new Dutch town Lelystad: Petra Brouwer, *Van stad naar stedelijkheid: Planning en planconceptie van Lelystad en Almere 1959-1974* (Rotterdam: NAi Uitgevers, 1997), 35, 56.

37 Buiter, "Utrecht: Bicycles Rule – Again," 33.

planners even recognized cycling was a potential solution for the excessive demands that cars made on urban space. In the interwar period, engineers had translated cycling's status as a middle-class public good into engineering norms that remained in place throughout the 1950s and 1960s, regardless of changing ideas about cycling's place. And cycling's public support continued to be strong – reversing these norms was never a serious consideration. The ambivalence about cycling had no major negative repercussions in policy terms.

We should note that although the evidence of the support for cycling cited in this section comes from the 1950s that continued in the 1960s, the public utterances supporting cycling were rare. Some engineers criticized this, attributing the worsening quality of cycling infrastructure to neglectful engineers, who increasingly swapped the bicycle for the car themselves, one colleague observed.[38] Efforts and investments in cycling reached a low point in the early 1960s. Instead of bicycles, the popular moped along with fear and/or excitement over the future growth of car ownership dominated the public and engineering debates. Behind the scenes, however, some construction of utilitarian and recreational cycling infrastructure continued. Efforts varied from place to place: cycling conditions deteriorated for some cyclists and improved for others. Overall, as with cycling levels, cycling policy attention in the Netherlands dropped in this period but still remained higher than elsewhere.

The continuity of engineering norms mattered. As we have seen, having decided in the late 1930s that cycling should primarily be governed locally, Dutch state engineers increasingly turned their attention to highway planning at the expense of cycling, even as a relatively bicycle-friendly country, it was also at the vanguard of highway planning internationally. Engineers and civil society representatives on the Roads Consultation Committee (COW) had often discussed cycling in the 1920s and 1930s (see 2.3), but during the postwar decades of reconstruction and economic growth, narrowed its focus to determining ideal routes for (new) highways connecting cities, while continuing to implement separate cycle paths along them. Local mobility patterns, where cycling, walking, and public transit outnumbered driving, had been assigned to local government, so state engineers no longer discussed them.

In the 1950s, local policymakers working on cycling did confront state engineers with the increasing conflicts between highway car traffic and local non-motorized traffic. Working in The Hague, the nation's seat of

38 H.B. Bakker, "Opnieuw over rijwielpaden," *Wegen* 38, no. 3 (1964): 70-76.

government, COW officials visited cities or towns impacted by these national new highway plans. As the national COW members discovered, local officials regularly insisted on cycling infrastructure. For instance in Laren, one of Amsterdam's first commuter satellite towns, a planned road would cut straight through its center. Discussing the plan with state engineers in 1961, the municipality demanded the national government built "small tunnels" for pedestrians and cyclists.[39] Laren was home to many wealthy Amsterdam elites, which perhaps explains why such a small place felt it could demand this special infrastructure. Other local authorities specifically requested "bicycle tunnels" to address the danger of crossing the new roads. The traffic futures imagined in national traffic planners' minds collided with the everyday reality of mobility in local communities. In the 1960s, several municipalities demanded cycling facilities (often tunnels under newly planned highways) for tourists and school children, in particular. In 1967, Abcoude, a village near Amsterdam and rapidly becoming a commuter area, requested "a cycling tunnel (for school children)," and in 1970, the mayor of Schaijk, a rural village in the province Noord-Brabant, wanted a bicycle tunnel "for the school-going children of Schaijk and the recreational cyclists."[40] Local policymakers made sure national engineers were aware of cyclists' existence and concerns.

State engineers may have refused a major role in cycling governance, but not necessarily because they thought cycling was outdated. Their neglect probably has more to do with their belief in decentralized governance than any overly negative opinions of cycling. Evidence suggests they started to value urban cycling as a transport mode. While firmly convinced of the (economic) importance of facilitating the car, its disproportional demands on urban space concerned city planners. Tim Verlaan contends that Dutch urban planners recognized these problems earlier than their counterparts in other European cities.[41] Berkers, Schipper, Bek, and Oldenziel make a similar claim even for the car-centered city of Rotterdam: local authorities and planners sought to avoid American models and pushed for public transit. Rotterdam Public Works engineer Theun Brouwer wrote critical pieces about car dominance and its negative effects on cycling in the early 1960s,[42]

39 NA 2.16.11, inv. no. 374. meeting January 26, 1961.

40 NA 2.16.11, inv. no. 375, meetings March 1, 1967 (Abcoude) and May 14, 1970 (Schaijk).

41 Tim Verlaan, "Mobilization of the Masses: Dutch Planners, Local Politics, and the Threat of the Motor Age 1960-1980," *Journal of Urban History* 47, no. 1 (2021): 136-56. Ambivalence about the car also prevailed in Britain: Simon Gunn, "The Buchanan Report, Environment and the Problem of Traffic in 1960s Britain," *Twentieth Century British History* 22, no. 4 (2011): 521-42.

42 Berkers et al., *Cycling Cities: The Rotterdam Experience*, 30-33.

showing how American traffic engineering ideas circulating widely in Europe were appropriated locally in different ways.[43]

The pushback against American traffic engineering and planning came early. The Netherlands' cycling tradition and late automobility led some engineers in the 1950s already to point to the space-efficiency of cycling versus the car in urban contexts. The nation's leading Public Works (*Rijkswaterstaat*) engineer Marius Cosquino de Bussy defended the advantages of cycling for Dutch mobility at the 1956 Dutch Road Congress on city center traffic. Trained at Delft Polytechnical University, he became a Public Works engineer in 1929. As a highway planner, he nevertheless also participated in a 1954 committee to research pedestrians' interests.[44] He also claimed to have a "predilection for cycling [*voorliefde voor het rijwielverkeer*]."[45]

During a working visit, a delegation of French engineers had witnessed rush hour traffic at a major junction in The Hague, where a traffic police officer regulated traffic. As De Bussy reported, the visitors observed how each time the officer signaled for traffic to move, eight cyclists could cross for every car. De Bussy deduced from the statistics that each car had on average two occupants. As "disorderly" as the cycling traffic looked, "one should not forget that it moves around a large amount of people in our old and narrow cities."[46] Possibly the biggest headache for urban car traffic in historic cities – parking space – was also much less of an issue with bicycles. Figure 16 shows how many cyclists could move through a city at rush hour: it was much more efficient than if all those people were to drive a car. At least some arguments for taking cycling seriously as a public good circulated in the 1950s and 1960s.

For some prominent road engineers, urban cycling was even in theory part of the solution rather than the problem. De Bussy claimed road construction outside cities had to be increased, but at the same time wanted to protect (urban) cycling wherever possible.[47] In a 1956 newspaper article, De Bussy admitted cycling was "fast and efficient despite the oftentimes chaotic sight

43 Carla Assmann, "The Emergence of the Car-Oriented City: Entanglements and Transfer Agents in West-Berlin, East-Berlin and Lyon, 1945-75," *The Journal of Transport History* 41, no. 3 (2020): 328-352, here 331.

44 "Belangen voetganger in het verkeer," *Arnhemsche Courant*, 08-01-1954; "Rapport over het probleem van voetgangers in het verkeer: Wettelijke bescherming gevraagd van oversteekplaatsen," *Het Vaderland*, September 15, 1954.

45 "Verkort Verslag van de Congresdag op 14 december 1956 te Utrecht: Maatregelen ter bevordering van de Vlotheid van het verkeer en van de parkeer-mogelijkheden in de binnenstad," *Wegen* 31, no. 7 (1957): 195-96.

46 Ibid.

47 "Nieuwe wegen zullen moeten worden aangelegd," *Het Parool*, February 28, 1956.

Figure 16 In the 1950s, commuter cyclists still dominated urban traffic like near Utrecht's Central Station during rush hour. The image illustrates what engineers also observed: cycling traffic required much less space than car traffic and might therefore be worth safeguarding. [Source: Collectie Het Utrechts Archief, Photographer L.H. Hofland, cat. nr. 43231]

it presents."[48] Traffic counts showed that cyclists still dominated city traffic so "considerable investments in bicycle traffic" were needed "to avoid making cycling too unattractive in cities."[49] It was paramount that "urban planners" would "promote cycling as much as possible, or at least prevent that it is squeezed out by other traffic."[50] Two years earlier, De Bussy had made a study trip to the United States with Rotterdam's chief of police K.J. Müller and ANWB's head of traffic C.A. Kuysten.[51] De Bussy's later defense of cycling proves he did not find the American paradigm an attractive blueprint.[52] However, this fear applied primarily to cities. Increasing roadbuilding on

48 "In Nederland komt een miljoen auto's," *Nieuwsblad van het Zuiden*, February 28, 1956. The author thanks Eric Berkers for this information.

49 M. Le Cosquino de Bussy, "Verkeer – heden en toekomst," *De Kampioen* Special issue '75 jaar ANWB' (July 1, 1958): 38.

50 Ibid.

51 Mom and Filarski, *De mobiliteitsexplosie*, 334. On adopting US highway plans: Bruce E. Seely, Donald E. Klingner, and Gary Klein, "'Push' and 'Pull' Factors in Technology Transfer: Moving American-Style Highway Engineering to Europe, 1945-1965," *Comparative Technology Transfer & Society* 2, no. 3 (2004): 229-46.

52 Dutch officials visited US (suburban) shopping centers. They also traveled to Sweden, where the more controlled, state-led planning made a better impression: Janine Gosseye, "The

the outskirts was not a (spatial) problem, but an economic necessity in De Bussy's view – although new roads had to have cycle paths which were "an important improvement requiring little costs."[53] De Bussy was not alone in this view. In 1953, engineer Willem Valderpoort published a critical book *The Selfish Car,* arguing that the car benefited only a select group while its negative aspects affected everyone.[54]

Dutch urban planners were also concerned about the rising automobility in cities.[55] At the Ministry of Culture, Recreation and Social Work, engineer R. van der Waal cited research in 1961 showing that recreational cycling was still very popular.[56] In sharp contrast to the first half of the 1900s, national policymakers started to take an interest in urban cycling as a public good and potential solution to the automobility issues as well. That these engineers often rode a bicycle may have reinforced their support for measures that enabled cycling alongside car roads. Public Works highest official, director-general August Maris, was not the only official to commute by bicycle.[57] This new belief had little immediate effect: national policymakers had hardly any influence on urban mobility governance. Struggling with automobility in their cities, urban policymakers ultimately also did little to limit its place in the 1950s and 1960s. Consequently, the urban landscape and space distribution changed dramatically.

The logic of traffic separation had not been applied exclusively to cyclists. Developed in the 1910s, it was the outcome of engineers' general distinction between fast and slow traffic. Instead of roads and cycling paths, engineers considered – and in some cases built – parallel service roads for all forms of slow traffic: agricultural or any remaining horse-drawn vehicles, mopeds, and cyclists. Speaking in 1958 at a conference on future traffic development, province of Gelderland engineer W. Mols argued that "in certain cases it can be better to open the road only to motorized vehicles and let slow traffic use a parallel road which also replaces the uni- or bidirectional cycling paths." Wider than a typical cycling path, this would also "largely solve the

Janus-Faced Shopping Center: The Low Countries in Search of a Fitting Shopping Paradigm," *Journal of Urban History* 44, no. 5 (2018): 862-86.

53 "Wij spraken met Ir. M. le Cosquino de Bussy, Hoofdingenieur-directeur Rijkswaterstaat," *Het Vaderland*, 28-02-1958, 3. NA 2.16.11, inv. no. 373, COW meeting January 14, 1954.

54 W. Valderpoort, *De zelfzuchtige personenauto: Beschouwingen over een onderdeel van het verkeersvraagstuk, met een bijzondere toepassing op de stedebouw* (Amsterdam: Van Saane, 1953).

55 Verlaan, "Mobilization of the Masses," 142; Disco, "De bewogen stad, 1945-1980."

56 NA 2.27.5111, inv. no. 60, 28th meeting INCOR-committee, July 5, 1961.

57 "Generaal van een roemrijk leger, in een land vol vredestrofeeën: Ir. A.G. Maris ten afscheid," *Algemeen Handelsblad*, October 19, 1961, 4.

problem of fast mopeds on narrow cycling paths."[58] Provincial engineers like Mols continued to believe in providing cyclists with their own space, even though they no longer contributed to road improvement through the bicycle tax. According to province of Noord-Holland engineer C. Krijn in 1950, the abolition of the bicycle tax "was by no means a reason not to continue the construction of cycling paths."[59] Like De Bussy, these engineers saw the interests of "slow" traffic as important enough to justify investment.

Many authorities built cycling infrastructure. With no coordinating agency collecting data, however, a comprehensive picture of cycling path construction is lacking. Still, the evidence shows that national and provincial Public Works departments kept working towards providing all major roads with separate cycling paths. In 1954, national engineer De Bussy argued that this was a cost-effective way and "important improvement" to the road network.[60] Annual roadworks reports featured many cycling paths. One traffic journal provided an overview of construction works on major roads in 1956. It mentioned the (re)construction of cycling paths on no fewer than thirty-four national or provincial roads.[61] For other years, the pattern is similar. Although the lack of distinction between separate cycling paths and on-road cycling lanes complicates comparing the statistics, the Roads Atlas that main Public Works statistician Van Gils compiled shows a clear increase in the 1950s. The 1949 and 1960 Atlases show how in the intermediate period, national Public Works engineers had provided over 400 km of national roads with cycling paths.[62] In contrast to other countries, where cycling infrastructure stagnated or existing paths became space for cars, Dutch engineers continued to gradually extend the network.

There were significant regional differences. State engineers only administered part of the Dutch road network: provinces played a large and independent role. Decentralized road governance meant that each

58 *Het Nederlandse Wegennet in verband met de toekomstige verkeersbehoeften. Prae-adviezen te behandelen op de Congresdag 1958, Utrecht, 19 december 1958*, 36-37.

59 C. Krijn, "Rijwielpaden langs Provinciale Wegen," *Wegen* 24, no. 5 (1950): 147.

60 NA 2.16.11, inv. no. 373, meeting January 14, 1954.

61 "De Rijkswegen in 1956," *Berichten en beschouwingen van de verkeersafdeling van de ANWB* 8, no. 1 (1957): 9-20. Even the postwar reconstruction years saw 100 km of cycling paths added alongside national roads: *Kamerstukken II 1950/51, 1900 IX B, 9*, December 7, 1950, 48.

62 Not including parallel roads, but counting separate cycling paths and on-road cycling lanes, 1,815 km of nationally administered roads had cycling infrastructure in 1949. By 1961 this was 2,231 km. Source: Rijkswaterstaat: Centrale Dienst voor de Wegen en de Bruggen: Bureau voor de Wegen- en Verkeersstatistiek, *Atlas van de Rijkswegen 1949* ('s-Gravenhage: Rijkswaterstaat, 1950), 4; Rijkswaterstaat Directie Algemene Dienst, *Atlas Rijkswegen 1960* ('s-Gravenhage: Staatsdrukkerij- en Uitgeverijbedrijf, 1962), Table 4.

province could decide whether, where, how, and when to reconstruct roads and provide cycling infrastructure. Unsurprisingly, a 1942 ANWB overview had uncovered the large differences between provinces in types of cycling infrastructure. As no government agency had such an overview, the ANWB took on the task of data collection throughout the twentieth century.

Table 2 Partial overview of the status of cycling infrastructure on the main provincial roads in the Netherlands, 1942

Province	Type of Cycling Path	Extent of Cycling Paths
Overijssel	Adjacent	Extensive
Noord-Holland	Separate	Partial
Zuid-Holland	Separate	Partial
Limburg	Separate	Partial
Friesland	–	Small
Groningen	–	Small
Drenthe	None/Mixed Traffic	–

Source: "Neerlands rijwielpaden en hun bewijzering," *De Kampioen* 59, no. 9 (1942): 65

This partial overview, omitting a few provinces, made clear that engineers chose different solutions. Separate cycling lanes, the ANWB's choice, were more expensive than cycling lanes on the road. Overijssel had more cycling lanes than other provinces. Some had hardly started to construct any cycling infrastructure at all.[63] In 1954, a Friesland newspaper journalist praised the province's initiative with recreational cycling paths but criticized the inadequate efforts to create safe cycling routes alongside major roads. There were barely any "normal 'utility cycling paths'" (*gebruiksfietspaden*) along these roads.[64] Friesland's provincial policymakers only started constructing

63 "Neerlands rijwielpaden en hun bewijzering," 65-66.

64 "Waar blijft het geld voor onze 'gewone' fietspaden?," *Leeuwarder Courant*, June 26 1954. The response to "who had caused this," was: "Clearly it's the national public works administration [*Rijkswaterstaat*]." The reporters emphatically blamed *national* public works as Friesland's provincial branch in Leeuwarden was aware of the problems but received no funding to remedy the situation. "Time and again the item for our cycling paths is dropped from the budget. No money! We'll see again next year." Other departments received what they requested, but Friesland was too far from the centre of power: "Like so many other things, the Frisian cycling paths are located too far from The Hague." There was some justification for this critique as funds were distributed nationally over the provinces. However, the provinces then chose how to spend their allocated roads budget.

these in the 1960s. Others had already been doing this in the 1930s.[65] One consequence of the decentralized Dutch system was that each province operated at its own pace.

Many different stakeholders were responsible for their share of cycling governance. National and provincial engineers provided "their" roads with separate cycle paths in response to the ANWB lobby. Regional networks of provincial and local policymakers too boosted recreational cycling (see Chapter 3). Just after the war, cycling received a lot of policy attention because the war had damaged recreational cycling paths, the popularity of the moped was putting pressure on cycling infrastructures, and the ANWB continued its advocacy. It is hardly surprising that the *Wegen* special issue was dedicated to cycling paths. The five articles on this theme clearly illustrate the Dutch governance network around cycling at the time. National engineer H.B. Bakker authored an article on cycling paths alongside national roads; provincial engineer Krijn of Noord-Holland wrote about cycling paths alongside provincial roads; legal expert and urban planner B. Mees from Rotterdam (critically) described urban cycling, which he found chaotic; ANWB traffic engineer Kuysten discussed rural cycling paths, and another ANWB official, August Boost, contributed on cycling tourism.[66] The first four topics were considered serving utilitarian functions, governed by different authorities. The recreational cycling infrastructure required a separate and equally complicated coalition. Given the patchwork of governance for cycling, it is hard to overstate the ANWB's role in creating a more or less uniform outcome over time.

Despite the lack of top-down coordination, the initiatives described formed an impressive total. By the mid-1940s, the provinces had constructed some 1,350 km of cycling paths alongside roads.[67] By 1965, an ANWB report found this had increased to 1,750 km, despite the missing justification of the bicycle tax after World War II. The rate of cycling path construction slowed down, but did not stop altogether.[68] By 1983, almost all major provincial roads

65 "De opbouw van een goed wegennet, in het bijzonder in Friesland," *Wegen* 36, no. 6 (1962): 126-33.

66 H.B. Bakker, "Rijwielpaden langs Rijkswegen," *Wegen* 24, no. 5 (1950): 137-44; Krijn, "Rijwielpaden langs Provinciale Wegen," 145-48; B. Mees, "Het rijwielverkeer in de grote stad," *Wegen* 24, no. 5 (1950): 149-54; C.A. Kuysten, "Rijwielpaden ten Plattelande," *Wegen* 24, no. 5 (1950): 154-57; A.G.M. Boost, "Rijwielverkeer en toerisme," *Wegen* 24, no. 5 (1950): 158-59.

67 "Neerlands rijwielpaden en hun bewijzering," 65.

68 Koninklijke Nederlandse Toeristenbond ANWB, *Fietspaden en -oversteekplaatsen* ([The Hague]: 1966), 10.

and a substantial share of the secondary roads would have cycling paths (see table 3).[69] Providing major roads with separate cycling paths became part and parcel of Dutch roadbuilding norms and practices outside cities. In short, cycling path construction was clearly an incremental process, and sustained over time, would become a comprehensive network of suburban and rural infrastructure.

Table 3 Extent of cycling infrastructure on the largest provincial (secondary) roads and the network (tertiary roads) in 1983, calculated by the Organization of Provinces (IPO)

Province	Percentage of Secondary Roads with Cycling Infrastructure		Percentage of Tertiary Roads with Cycling Infrastructure	
	In 1983	Planned by 1986-87	In 1983	Planned by 1986-87
Friesland	98	98	54	67
Groningen	86	88	46	50
Drenthe	91	92	62	75
Overijssel	92	92	48	52
Gelderland	–	90	–	58
Noord-Holland	68	68	53	53
Zuid-Holland	92	93	33	34
Utrecht	100	100	70	70
Noord-Brabant	88	98	64	77
Zeeland	94	95	42	49
Limburg	98	100	25	38

Source: HUA 1205, inv. no. 5065, letter IPO Traffic Committee to Minister of Public Works, May 16, 1983

4.3 ANWB Expands its Role as an Expert Organization

The tourist organization ANWB, so important in promoting the practice and political acknowledgement of cycling in the interwar period, expanded its role as a crucial mobility actor. Historians Buiter and Staal characterize the ANWB from 1919 to 1949 as a lobby organization in shaping the new infrastructure – a role the organization continued in the 1950s and 1960s. It also increased its role in disseminating knowledge.[70] As state involvement in

69 Source: HUA 1205, inv. no. 5065, letter IPO Traffic Committee to Minister of Public Works, May 16, 1983.

70 Buiter and Staal, *Het avontuur van de ANWB*, 5.

mobility and tourism grew, the ANWB placed its governance expertise more central by employing its own civil engineers. The expansion entailed not only creating educational platforms, but also producing and disseminating knowledge through brochures and traffic courses. It is hard to determine what the state of affairs on cycling in international engineering research was around the 1950s and 1960s, but the ANWB's extensive expertise on this topic seems to have developed with relatively little international input. This stands in sharp contrast to its road engineering expertise.[71]

The organization's leaders believed the ANWB was possibly the most important actor in Dutch mobility governance in terms of its engineering expertise but also its representation of the user perspective – especially in the absence of strong national governmental coordination. The growth of car ownership created many conflicts between pedestrians, cyclists, and car drivers. In a decentralized administrative system, where provincial and local engineers had considerable autonomy, the uniformity of road design and other car or cycling facilities was by no means guaranteed. ANWB officials found the lack of standardization dangerous: road users were confused by the diverse local conditions, thus increasing the likelihood of accidents. With its good contacts at all government levels and bird's eye perspective of mobility, the ANWB believed it should be pro-actively involved in formulating road engineering standards. Between 1946 and 1977, engineer and member Kuysten headed the ANWB's Roads Section. In an interview in 1977 to mark his retirement, he reflected that "coordination among the more than a thousand road authorities in the design of roads and junctions was an important task, which the ANWB took upon itself without being asked."[72] He presented the ANWB as non-political and neutral technical expert, independent from the policy reversals as a result of governments changing their political color. In a classic trope of technocratic thinking, the ANWB considered itself a more efficient organization than the state.

According to the ANWB, there was an informal division of labor between various mobility actors. The ANWB had to take the lead in the knowledge role. In 1956, Traffic Department director August Boost asserted that the ANWB subscribed to the three Es: Engineering, Enforcement, and Education. The education (of traffic users) "has by gentlemen's agreement for the

71 The World Road Congresses of the engineering association PIARC discussed cycling occasionally in the Interbellum, but rarely if ever after World War II. See https://www.piarc.org/en/activities/World-Road-Congresses-World-Road-Association/Congress-Proceedings

72 "De verkeerskundige moet eenvoudig durven zijn! Gesprek met ir. C.A. Kuysten bij zijn afscheid als Directeur Verkeer en Recreatie van de ANWB," *Verkeerskunde* 28, no. 11 (1977): 506.

most part been awarded to the Traffic Safety Association" (*Veilig Verkeer Nederland*), another non-governmental organization.[73] Enforcement was the government's task, but Engineering was the ANWB Traffic Department's specialty. This was more than an informal collaboration: the ANWB performed "the function of a consultancy bureau officially – i.e. at the request of municipal and provincial authorities."[74] The ANWB enjoyed many personal links with the state. Engineer Frederik Hendrik van der Linde van Sprankhuizen worked for the provincial public works and planning departments, while chairing the ANWB between 1962 and 1980.[75] The ANWB's executive board, as Linders-Rooijendijk shows, were often prominent figures, including mayors, lawyers, and a few (state-employed) engineers. The relationship between the ANWB and the state might not have been formalized, but it was very close.

Indeed, authorities recognized and sought out the ANWB's expertise. When a municipality consulted the Association of Dutch Municipalities (VNG) about a dangerous crossing, it recommended the village turn to the ANWB instead.[76] It is telling that the Association did not refer them to provincial or national Public Works engineers: the ANWB was more visible and easier to contact even though government engineers developed cycling engineering expertise over time. Local authorities considered the ANWB a prominent expert and source of free advice in traffic engineering.[77] Mom and Filarski show that in the 1950s the number of occasions on which the ANWB provided engineering advice to municipalities rose from 100 to more than 250 annually, before it declined in the 1960s.[78] National Public Works not only accepted, but actively supported this ANWB role. Historian Marcus Popkema discovered that after 1950, the ANWB received a secret government subsidy from Public Works to publish a traffic engineering journal for experts. The reason for this secrecy lies in the administrative tradition of decentralized autonomy, where lower authorities considered advice and guidance from national Public Works an encroachment on their autonomy. That is why the Public Works administration, to promote its vision

73 HUA 1007-3, inv. no. 23060, opening speech Boost (ANWB) at 'Eerste Politie-Verkeersleergang' May 9, 1956. Boost succeeded H. de Bruyn, also a central figure in Dutch road and traffic planning. Mom and Filarski, *De mobiliteitsexplosie*, 335.

74 HUA 1007-3, inv. no. 23060, opening speech Boost, May 9, 1956. Linders-Rooijendijk, *Gebaande wegen II*, 576.

75 Ibid., 1032.

76 NA 2.19.185 (Archive VNG), inv. no. 2628, letter VNG to municipality Asten, June 16, 1951.

77 Popkema, "Tussen techniek en planning," 74.

78 Mom and Filarski, *De mobiliteitsexplosie*.

of preferred mobility solutions, chose to work with the seemingly neutral social actor ANWB.[79] The cooperation between the two organizations confirms the close entwinement between state and non-state actors in Dutch mobility governance. Furthermore, it demonstrates that engineers and ANWB officials had very similar core ideas. Had this not been the case, Public Works would not have subsidized the ANWB.

The ANWB used various means to create this monopoly on mobility knowledge. Since the 1920s, the ANWB had published *Wegen*, organized road conferences, and launched a traffic engineering journal.[80] In the 1950s, in collaboration with academics from Delft technical college, it set up traffic courses. As there was no official degree course for traffic engineering in the Netherlands, many Dutch officials responsible for traffic in their region received their training through the ANWB courses. Civil servants and engineers responsible for traffic at the local level often had no specific background in traffic engineering and received further training. Through these courses, local traffic engineers could acquaint themselves with the new traffic expertise. In a time before the establishment of traffic engineering degrees at universities or independent research institutes, the ANWB fulfilled a key role.

The ANWB expanded its educational role for government engineers. To help local traffic engineers design safe and uniform traffic environments, the ANWB wrote manuals. These included sections devoted to cycling paths, pedestrian crossings, intersection design, and so on. The instructions were detailed, technical, and presented objectively to help local officials design a better traffic system. Leading ANWB official Boost described how these had "the intention as well as the effect, fortunately, of achieving standardization of traffic facilities."[81] A memorandum on the proper construction of cycling paths and pedestrian crossings underwent many revisions in the 1960s.[82] In line with cycling policies up to that point, it concentrated on road design in rural areas rather than the city. Dominant in these brochures and other

79 Popkema, "Tussen techniek en planning," 78.

80 Ibid., 75-82. *Wegen* featured traffic planning in the Interbellum but geared towards road-building when increasing specialization led to new traffic engineering journals: *Berichten en Beschouwingen van de Verkeersafdeling van de ANWB* (Reports and Reflections from the ANWB's Traffic Department, 1950-1958), *Tijdschrift voor Verkeerstechniek* (Traffic Engineering Journal, 1958-1959), *Verkeerstechniek* (Traffic Engineering, 1960-1974) and *Verkeerskunde* (Traffic Studies, 1975-now). The ANWB also published *Traffic Law* from 1953, which still exists, and *Recreatievoorzieningen* (Recreational Facilities).

81 HUA 1007-3, inv. no. 23060, opening speech Boost, May 9, 1956.

82 ANWB, *Fietspaden en -oversteekplaatsen*.

knowledge distribution channels, was the belief that traffic flow was the gold standard – a dominant international view increasingly under pressure as traffic volume grew and demanded solutions. A 1961 brochure on urban traffic stressed that cities had to reserve space for the growing car traffic (and parking) because "the (car) traffic function of the street is primary."[83] The brochure did include The Hague's traffic count, demonstrating that cyclists and mopeds outnumbered cars three to one. Consequently, the ANWB proposed alternative main routes for commuter cycling traffic: complete with smooth asphalt and optimized waiting times at traffic lights, these should entice cyclists to avoid major car routes.[84] In effect, the ANWB proposed a more space-efficient way to separate traffic that prioritized cars. As in the 1920s, the car-serving logic is apparent, despite not ignoring cyclists.

As the ANWB's expert role grew stronger and stronger from the late 1940s onwards, how did its stance on cycling and consequently its cycling advocacy develop over the same period? When World War II disrupted travel patterns in the Netherlands, the ANWB had turned its attention more towards cycling than before. At the time, the organization admitted it "was, at least in appearances, primarily oriented toward car tourism," but this made less sense "now that more than ever the entire Netherlands is cycling."[85] In line with its self-imposed task of educating road users, the organization immediately published an article explaining to all former drivers-turned-cyclists how to behave responsibly and orderly in traffic.[86] The ANWB's liberal stance was characteristic – it claimed users were free to choose their preferred transport mode. The ANWB and government should then facilitate that choice. They did not acknowledge that these policies nudged users into certain behavioral patterns.

In the late 1940s, the ANWB had been particularly active in forging new connections in cycling governance between engineering departments, local politicians, and private cycling path organizations. As Chapter 3 showed, local policymakers were keen to boost cycling. The lack of clear cycling governance hindered progress. If a cycling path functioned primarily within a local area, the municipal policymakers were in principle interested in financing it. The trouble started when a path crossed different state level boundaries. It required the ANWB's constant networking to solve the resulting deadlock.

83 Koninklijke Nederlandsche Toeristenbond ANWB, *Verkeersstructuurplan en dwarsprofielen van stadsstraten*, Verkeersmemorandum, vol. 16 ([The Hague], 1961), 22, 24, 30.

84 Ibid., 21-22.

85 "De A.N.W.B. werkt door!" *De Kampioen* 57, no. 12 (1940): 177.

86 "De correcte fietser," *De Kampioen* 57, no. 12 (1940): 179-80.

A 1948 ANWB article encapsulated this: "The statement 'a path is a path and all that matters is that it is a good path' might be true for users, but it is not true for managing authorities. They self-evidently only spend their limited means to the extent they benefit from a path."[87] Getting spatial planners and engineers within the state to work together, or different levels of government, even neighboring municipalities, often required ANWB officials to intervene. That they were highly active in this role in the late 1940s shows how much stock they still put in recreational cycling as a public good.

Throughout the 1950s and 1960s, the ANWB remained firmly in favor of separate cycling paths alongside suburban and rural national roads. Traffic safety research indicated that this significantly reduced accidents.[88] Over the 1960s, the ANWB frequently revised its manual on designing paths, with many diagrams and photographs indicating ideal junction layout. The ANWB recommended the cycling path was built a few meters away from the parallel road, so that drivers turning right would have more time to slow down and check their mirrors for cyclists.[89] Similarly, the ANWB proposed short cycling paths around city bus stops, so that busses would not have to cross cycling traffic when pulling in to stop. These kinds of innovations the ANWB pioneered show its commitment to finding space for different road users. Facilitating motorized traffic was a priority, but also protecting cyclists. Rather than banning cyclists completely, ANWB engineers sought effective compromise solutions.

The ANWB did not approve of separate cycling paths in city centers, however. The main radial routes leading to and from the city should have separate paths, but near the many downtown junctions, it was better "to include bicycle traffic in the traffic flow so that it can get into the lane in front of traffic near junctions [*voorsorteren*]."[90] Separate cycling paths were recommended for the more spacious new neighborhoods on the outskirts of cities. The ANWB gave precise design norms: a 1-meter strip separating the path from the road (clearance to prevent accidents from parked cars opening doors); if there was a cycling path on only one side of the road, it should be 2.85 meters wide: 1 meter for mopeds, two 0.80-meter lanes for bicycles, and 0.25 meters of additional space because cyclists automatically keep a certain distance from the curb, so this space is effectively lost.[91] The

87 "Heden en toekomst van de rijwielpaden," *De Kampioen* 63, no. 10 (1948): 267.
88 ANWB, *Fietspaden en -oversteekplaatsen*, 6-8.
89 Ibid., 31-53.
90 ANWB, *Verkeersstructuurplan en dwarsprofielen van stadsstraten*, 48.
91 Ibid., 48-50.

Figure 17 The Dutch ANWB increasingly promoted driving over the 1950s and 1960s, and sought to maintain cycling, believing it could simultaneously facilitate both. This junction near De Meern, Utrecht in 1975, developed by ANWB experts in the 1950s, has been widely applied since then to prevent collisions between cars turning and cyclists crossing. The ANWB circulated its design ideas among engineers through its trade journals, engineering courses, and good contacts with authorities. [Source: Collectie Het Utrechts Archief, Photograph Provincie Utrecht 847432]

organization's research into cyclists' behavior, experiences, and needs is further proof they were still taken seriously as road users (see figure 17). It also showed its investment in increasing its expertise about cycling.

The ANWB continued its cycling advocacy and lobbying at a local level, where its networks provided extensive knowledge of local road (safety) conditions. Its new traffic journal in 1950 offers numerous examples. The ANWB had asked Public Works to construct a "bicycle groove" (*rijwielgoot*) in the bridge over the river Oude Maas, a major connection on the busy commuter cycling route between Dordrecht and Zwijndrecht. This bridge had steps, and a groove on the side would make it easier for cyclists to transport their bicycle over the bridge. Another example is the cycling path along the national road Haagsche Schouw-Wassenaar. It had a sharp bend which cyclists sometimes missed at night due to bright car headlights. The ANWB asked Public Works to paint white lines on the road to enhance visibility. A final example comes from the southern province of Limburg, where a cycling path route went around a gas station. There was a telephone pole just behind the station, and the ANWB knew that "on multiple occasions, cyclists have crashed into this pole." At the ANWB's request, the pole was moved further away from the cycling path.[92] These examples demonstrate the ANWB's detailed knowledge of local infrastructure and cycling practices. Most state-employed traffic engineers certainly did not have access to this type of information; without the ANWB's role as mediator between user and government, it would have been much harder to resolve problems like these.

The ANWB's cycling expertise was clearly large. What about acting on behalf of cyclists and managing the conflicting interests of drivers and cyclists, or traffic and nature? The ANWB continued to be a large and multifaceted organization. Some defended cyclists (and pedestrians) against car traffic. Leading ANWB official and vice-director of Traffic and Recreation C.A. Kuysten had mixed views. In a 1964 interview, he commented that utilitarian cycling was becoming obsolete. Commuting distances were now too long, and in the future, people would mainly use the bicycle to go to train and metro stations. In the same interview, Kuysten chastised municipal authorities for neglecting cyclists, for example by not building cycling routes to industrial zones, so workers could not go there by bike.[93] In his 1977 retirement interview, he also called the average Dutch driver "aggressive and immature," attributing this partly to the fact that most people had not grown up with the car (hinting at late automobilization in the Netherlands) and regarded driving too much as a race. After retirement, Kuysten became chairman of the Pedestrian Protection Association (*Vereniging Bescherming Voetgangers*). A central ANWB figure like Kuysten held views that did not boil down to exclusive support for one transport mode, but typically avoided hard choices.[94]

Other ANWB officials believed cycling was outdated. According to J. Koolhaas Revers, a veteran car journalist and editor of the ANWB journal *Bromfietskampioen* ("Moped Champion"), the bicycle was about to be superseded by the moped and the car: "our old faithful pedal bicycle," he argued, was fast becoming a "tiny national minority." The last bicycle might even end up as a curiosity in Amsterdam's Rijksmuseum. When he wrote this in 1951, however, cycling was not declining at all.[95] But the succession of transport modes was a survival of the fittest: the moped was a "conqueror,

92 *Berichten en beschouwingen van de verkeersafdeling van de ANWB* 1, no. 3 (1950), 9, 13.

93 "Steden laten de fietsers versukkelen," *Trouw*, February 28, 1964, 11.

94 "De verkeerskundige moet eenvoudig durven zijn!" 507.

95 J. Koolhaas Revers, "Bromfiets de 'Veroveraar': Een beschouwing over het ontstaan van een nieuwe phase in ons gemotoriseerde verkeer," *Bromfietskampioen* 1, no. 1 (1951): 6. The term pedal/push bike (*trapfiets*) distinguished it from the motorized bicycle (*bromfiets*).

who imposes his will, against which there is nothing to be done."[96] For the more car-minded wing of the ANWB that Revers represented, technology's progress was unstoppable. It was best to facilitate it – though not without a certain nostalgia for the loss of the old. Others remained invested in safeguarding Dutch natural and cultural heritage; they saw the encroachment of cars and mopeds on natural areas as a regrettable development. Allied to conservationist and heritage protection movements as part of its promotion of national tourism, the ANWB had some qualms about allowing cars into historic Dutch inner cities.[97] Within and beyond ANWB circles, not everyone believed cycling would disappear. A 1955 ANWB survey among 12,000 bicycle dealers in the Netherlands showed only 22 percent thought the moped would eclipse the bicycle. Other dealers were certain the bicycle would remain the most popular two-wheeled vehicle.[98]

In a sign of the times, the various wings of the ANWB increasingly clashed in the 1960s. While one ANWB branch praised nature and national parks, others encouraged people to enjoy these parks by car and offered cheap deals. Klaas Kroon, a graduate of Wageningen Agricultural University, who became head of the ANWB's new department of Environment and Landscape in 1966 "to promote the interests of nature within the ANWB," had no easy task. He had to negotiate constantly with the ANWB's traffic department, where his "nagging about trees" was not appreciated.[99] This internal tension became manifest for example in the late 1960s, when a new road was proposed through the dunes between the west coast seaside resorts Scheveningen and Zandvoort. Boost, representing the ANWB's roads and traffic interests, supported this road, claiming it was necessary for the future development of traffic. However, ANWB director Blankert and chairman Van der Linde van Sprankhuizen opposed the road because it would damage nature and make the dunes less attractive for recreation.[100] This tension between utility and aesthetics also played out in discussions about American parkways, or the environmental embedding of the Autobahn and autostrada in Germany and Italy.[101] More so than in those countries,

96 Ibid.

97 Buiter and Staal, *Het avontuur van de ANWB*, 132.

98 "Bromfiets zal rijwiel niet overvleugelen," *Bromfietskampioen* 5, no. 4 (1955): 285.

99 Hugo Arlman and Rudie van Meurs, "'De adem van de ANWB stinkt eeuwig naar benzine.' De macht van het vijftiende ministerie [*the ANWB's breath stinks for ever of petrol*]," *Vrij Nederland* (*Bijvoegsel*) 17 (1980), 28.

100 Linders-Rooijendijk, *Gebaande wegen II*, 853-55.

101 On Germany's relationship between roads and driving vs nature: Thomas Zeller, *Driving Germany: The Landscape of the German Autobahn, 1930-1970*, Studies in German history (New

however, the central discussion in the Netherlands was twofold: not just how to embed a road in nature, but whether a road was necessary at all when that nature could also be enjoyed by bicycle, a vehicle still widely owned and used by the Dutch, in contrast to Germany and certainly the United States, even in the 1960s. And now, the onetime lobby organization for cycling tourism in the Netherlands was increasingly considered an organization whose "car face ... wins out too much."[102] This was why around 1970, new activist groups would emerge around cycling: the ANWB was no longer considered a cycling advocate.

4.4 Conclusion

In the two decades when cycling lost much of its public status in many countries, the Dutch did not abandon their bikes. This chapter focused on the continuities and influence of policies initiated in the prewar period. It showed that engineers and policymakers at various levels still believed in the importance of cycling. Engineers continued to extend the network of separate cycling paths alongside major suburban and rural roads. This was neither controversial nor subject to debate. In the major engineering journals and archives consulted here we do not find any serious implications or discussions about ending this policy. It had become unremarkable and normal engineering practice, just as cycling remained an everyday practice.

Within a few decades, traffic separation became so deeply rooted in engineering practices that it developed path dependent characteristics. Traffic separation still did not take place inside cities. Alongside the major rural and suburban roads, it became ever more prevalent. The growing – and controversial – popularity of the moped contributed to this, as I will discuss in the next chapter. The Dutch had already started building cycling paths alongside roads when car ownership levels were so low that there was little need for it.

As a result of path dependent processes, in the suburbs and countryside there far was more continuity between the prewar and postwar situation: the suburban system of traffic separation Dutch policymakers had

York: Berghahn Books, 2007); Christof Mauch and Thomas Zeller, eds., *The World beyond the Windshield: Roads and Landscapes in the United States and Europe* (Athens, OH: Ohio University Press; Stuttgart: Franz Steiner Verlag, 2008).

102 Arlman and Meurs, "De adem van de ANWB," 22.

adopted as a political compromise in the 1920s was still producing cycling paths in the 1950s and 1960s when new roads were built or existing roads reconstructed. The central idea of path dependency theories is that (in this case) mobility and planning policies tend to produce mobility patterns and physical infrastructure such as cycle paths, which are so deeply embedded that changing them is both difficult and expensive.[103] Initial choices, however contingent and small, can have long-lasting and crucial consequences. The contingent decision to levy a bicycle tax in the 1920s put Dutch mobility policy on a unique path. The commitment to traffic separation led to a material infrastructure of suburban commuter cycle paths which supported cycling as a practice, even as car levels started to rise on roads. For one thing, the material costs of going back to mixed traffic were high, requiring the demolition of cycle paths or their transformation into car lanes. Institutionally, engineers would have to change road norms and roadbuilding practices at high costs. And most importantly, drivers who used roads "free" of cyclists would not accept them returning to main roads. This is a positive feedback process which drives path dependent processes. Once traffic separation was firmly embedded, and as long as cycling levels were high, it was prohibitively difficult to change this pattern.[104] The continuity in rural and suburban cycle path construction is easy to overlook, since it had become standard practice for engineers in this period, leaving fewer traces in the historical records than changes in policy direction. These tacit patterns of behavior, deeply rooted in certain institutions, are important in explaining the long-term success of cycling in the Netherlands.

At the same time, it was not inevitable or predetermined that Dutch cycling would remain a public concern throughout the 1950s and 1960s. That this happened also has to do with cycling's cultural and political status, afforded a strong position in the Dutch system. Dutch policy also accommodated the interests of various social groups and a large interest group like cyclists had a certain power. Moreover, with the powerful ANWB lobby on their side, groups' appeals to policymakers who also cycled for recreation, or even for their commute, meant cycling had public support.

If national policymakers discussed cycling less than in previous decades, this lack of attention is not necessarily indicative of hostility towards

103 Hommels, *Unbuilding Cities: Obduracy in Urban Sociotechnical Change*; Hommels, "Studying Obduracy in the City," 323-51.

104 Martin Emanuel, in an article on Copenhagen's cycling culture, aptly pointed to the "obduracy of infrastructure and its capacity to preserve habits and cultures of the past." Emanuel, "Making a Bicycle City," 497, and his "Moments of Unsustainability," 101-21. See: Hommels, *Unbuilding Cities: Obduracy in Urban Sociotechnical Change,* and "Studying Obduracy in the City," 323-51.

cycling. Rather, it shows the ambivalence towards cycling in a period of automobilization. It is also the reflection of deeply rooted Dutch ideas about decentralization and the local management of local affairs. As with cycling culture, the ANWB fulfilled an active role in promoting car culture. At the same time, the organization dedicated a significant amount of time and money explaining to policymakers how they could enable safe cycling alongside rising car numbers. In the process, the ANWB came up with many innovative cycling design solutions and greatly enlarged the knowledge of cyclists' behavior and needs.

Cycling as a public good became more ambiguous as the motorization of commuting increasingly seemed unstoppable. Cyclists refused to give up their bicycles however, and developments in the 1950s and 1960s reinforced cycling's political position, as well as its place on the road. The popularity of mopeds ended up strengthening the commuter cycle paths alongside major roads.

5 An Accident of History: How Mopeds Boosted Dutch Cycling Infrastructure

On holiday weekends in the 1950s, there was an impressive spectacle on Dutch roads and cycling paths leading to recreational areas. A journalist described how "one long procession of cyclists" used the cycling paths next to roads. But among these cyclists something new was visible: "The mopeds have not had it easy ... they certainly did not move forward much faster than the cyclists." These mopeds, in the early 1950s basically bicycles with a small engine, were not popular with the cyclists forced to share their cycling path. Witnessing this, the journalist noted that "cyclists stubbornly kept their ranks closed whenever they heard one of those mopeds coming." An understandable reaction, he noted: "It is almost impossible to get out of the way each time, cycle behind one another, for vehicles that are so much faster than those for which cycling paths are intended after all."[1] What were these mopeds doing on the cycling paths and why were they not using the road?

In 1960, Dutch policymakers were preoccupied with the space demands of the fast-growing automobility park. Yet the true growth spurt of car ownership in the Netherlands only came in the 1960s, later than elsewhere. For much of the 1950s and 1960s, Dutch working- and middle-class people used bicycles and the new moped much more than the car. Many cyclists who eventually (also) became drivers, made this move through an intermediate vehicle: the moped. Research in France, Germany, and Sweden has shown how mopeds were popular across Europe with working-class commuters (1950s) and young people (1960s). This was also the case in the Netherlands. European traffic conventions meant that mopeds were classified as a type of bicycle everywhere. As far as we know, only in the Netherlands did this have a strong impact on cycling, however. The comparatively high levels of cycling combined with an extensive network of suburban and rural cycle lanes, both separate and alongside roads, led to busy and overcrowded cycling paths. Based on (provincial) engineers' archives, parliamentary records, and moped journal articles, I show the intense political and engineering debate about moped's rightful place on the road. It discusses the ambivalent status of the moped as neither a bicycle nor a car, and the consequences of the decision to consign mopeds to cycling paths. Why were moped riders forced onto the same narrow cycling paths? How did national and provincial

1 "Tienduizenden trokken naar bos en water," *Utrechtsch Nieuwsblad,* May 26, 1953, 2.

engineers and policymakers make and defend this decision? And why did it also, paradoxically, benefit cyclists?

"An Accident of History: How Mopeds Boosted Dutch Cycling" argues that the Dutch decision regarding where mopeds could ride added an unintended impetus to the construction of cycling paths. This process happened in a car-oriented context. By prioritizing the status of car drivers, moped riders were forced onto cycling paths. This decision overburdened the already congested paths at a time when roads were not yet heavily used. It also convinced engineers that cycling path construction should remain a priority. This argument is new in the international scholarship, which discusses the moped as a technical artefact or looks at its user groups. An underexamined topic is how road engineers reacted to the moped. This merits attention as part of cycling policy, not only for its impact on cycling path construction. It also provides a unique perspective of ongoing debates about cycling's utilitarian or recreational value as a public good.

5.1 Mopeds Widen Citizens' Action Radius

International experiments with motorized bicycles were common in the 1930s. They generally failed for technological reasons: light bicycle frames could not sustain the weight and power generated by auxiliary engines.[2] After World War II, European firms started producing bicycles with more reliable auxiliary engines. As cars were still too expensive for many people, the moped was an alternative with a greater level of comfort and action radius than the bicycle.

Mopeds became highly popular in a very short time span around the 1950s and 1960s in several countries, including the Netherlands, before ownership stagnated. After the moped's introduction in 1947, purchases rocketed to two million by 1969. Experts predicted this: in 1956, the Dutch Traffic Institute (NVI) estimated that by 1960 there would be one million mopeds in the Netherlands, a total that had already been reached by 1949.[3] Combined moped ownership exceeded car ownership significantly,

2 Héran, *Le retour de la bicyclette*, 57-58. In France a 1925 decree stated that bicycles with an auxiliary engine (*bicyclettes à moteur auxiliaire, BMA*) should (1) not weigh more than 30 kg; (2) not go faster than 30 km/h; (3) and be propelled by pedals. But no registration or driver's license was needed, nor was there a maximum cylinder capacity or minimum age.

3 "Miljoen bromfietsen in 1960?" *Algemeen Handelsblad*, March 27, 1956, 7; "Bromfiets: op naar het miljoen...!," *Trouw*, February 25, 1959.

causing Jan Meijer, journalist with the social democratic newspaper *Het Vrije Volk* to characterize the moped in 1950 as the "car-of-the-little-man" (*auto-van-de-kleine-man*).[4] When Dutch wages rose rapidly in the second half of the 1960s, cars came within reach of the working-class men who had represented most moped use in the 1950s.[5] As table 4 shows, Dutch moped ownership was among the highest in Europe in the 1950s. Although numbers dropped after 1970, mopeds remained part of everyday Dutch mobility. A 2013 report by the Dutch Statistics Agency shows that 58 of every 1,000 Dutch residents still had a moped, the highest number in Europe.[6]

The moped became part of a Dutch mobility system alongside continuing bicycle and increasing car use. Cycling did not disappear: Dutch people still owned millions of bicycles, outnumbering the (growing) moped and car park in the 1950s and 1960s.[7] Not everyone who owned a bicycle used it, however: traffic counts on provincial roads that actually distinguish bicycles and mopeds (many do not) verify the moped's popularity.[8] Even in urban traffic it was popular, as numbers in Rotterdam's city center show. At the height of the moped's popularity in 1968, moped riders outnumbered cyclists by between 40 to 60 percent crossing the Nieuwe Maas river on their daily commute in Rotterdam.[9] Traffic counts in Gelderland province in 1973 showed almost 4,000 mopeds compared to 4,750 cyclists.[10] Given the larger distances mopeds could travel, they probably showed up more often in traffic counts than cyclists. These numbers mean that mopeds were not as popular as the bicycle, but they did form a very large minority. Photographs from this period confirm that the moped's share in traffic was significant. Policymakers and experts were aware of this and felt compelled to take action.

4 [Jan Meijer], "Spreeuwenpraat," *Het Vrije Volk*, July 31, 1950.

5 Mom and Filarski, *De mobiliteitsexplosie*, 268.

6 Doreen Ewalds, Ger Moritz, and Michel Sijstermans, *Bromfietsen in Nederland* (Den Haag/Heerlen: Centraal Bureau voor Statistiek, 2013), 7.

7 ANWB archive, inv. no. 1241, FNRV meeting May 9, 1956.

8 HUA 1210, inv. no. 1408 and 1409: data is available for 1955, 1960 then annually in the 1960s. It shows the moped goes from a minority share to almost equal use in 1960, and majority use in 1968. Traffic counts on Zuid-Holland's provincial tracks in 1969 show a similar 50:50 division: NA 3.02.27, inv. no. 1560, letter Provincial Public Works and Planological Office to Provincial Executive, Oct 28, 1969.

9 The author is grateful to Eric Berkers for sharing this data.

10 Memorandum "Fietsverkeer," *Kamerstukken II* 1982/83, no. 17745, sub-no. 2, p. 47. https://repository.overheid.nl/frbr/sgd/19821983/0000146649/1/pdf/SGD_19821983_0005626.pdf, accessed 27-11-2019.

Table 4 Moped ownership in Europe in 1958

Country	Number of mopeds	Population	Mopeds per 1000 people
France	2,600,000	44,243,000	59
Netherlands	503,421	10,759,000	47
West Germany	1,130,000	53,518,000	21
Italy	752,000	48,634,000	15
Belgium	135,169	8,869,000	15
Switzerland	59,201	4,981,000	12
United Kingdom	227,712	50,947,000	4

Source: Article in ANWB journal Wegen (Roads): "Het probleem van de rijwielen met hulpmotor," *Wegen* 32, no. 8 (1961): 191

The lack of (comparative) research on mopeds has led historians to claim they were only popular in certain countries. According to Frank Steinbeck, the use of mopeds or light motorcycles was an important intermediate step to full motorization in Germany.[11] Frédéric Héran argues that France is the only country "to have known a considerable amount of moped trips" in the immediate postwar years.[12] Blomkvist and Emanuel have recently shown the moped's high use levels in Sweden.[13] The moped fulfilled a similar role in the Netherlands. It seems mopeds have played a major but underexplored role in many European countries.

The Dutch moped was primarily a working-class commuter vehicle in the 1950s before young people adopted it in the 1960s. Mom and Filarski show that about 65 percent of riders used the moped to go to school or work.[14] In the 1950s, mainly adult men rode a moped, before they switched, in the 1960s, to the car. Then the young generation took over. The percentage of young moped-owners rose rapidly as the moped became part of a burgeoning youth culture.[15] The moped's expansion between 1960 and 1970 came almost

11 Frank Steinbeck, "Rückenwind für motorisierte Fahrräder," *Kultur & Technik*, no. 2 (2013): 34-39, here 34 and his *Das Motorrad: ein deutscher Sonderweg in die automobile Gesellschaft* (Stuttgart: Franz Steiner Verlag, 2012), 132-70 and 239-55.

12 Héran, *Le retour de la bicyclette*, 60.

13 Pär Blomkvist and Martin Emanuel, "Regulating Innovation: Utility vs. Leisure in Swedish Moped History, 1952-1961," *Technology and Culture* 61, no. 3 (2020): 815-842.

14 Mom and Filarski, *De mobiliteitsexplosie*, 270.

15 On the moped's role in Dutch youth culture: ibid., 250. For a more international perspective: Becky E. Conekin, "Fashioning Mod Twiggy and the Moped in 'Swinging' London," *History and Technology* 28, no. 2 (2012): 209-15.

Figure 18 In the 1960s, the moped, initially a working man's vehicle, became more popular with young people and women. As we see at Utrecht's Janskerkhof square, bicycles were still around. [Source: Collectie Het Utrechts Archief, Photo Gemeentearchief Utrecht, no. 810884]

exclusively from sixteen- to twenty-one-year-olds, and to a lesser extent from twenty-one- to twenty-four-year-olds.[16] Popular moped types like the Austrian and German Puch, Kreidler, and Zündapp were part of rival youth cultures.[17] In the Netherlands, sales reached nearly a million for the French Solex moped, which was slower and initially (marketed as) the workman's moped before becoming more popular with women.[18] The "modest and functional" Berini was associated with white collar workers. More "sturdy" mopeds like Union or Magneet carried connotations of blue-collar workers.[19] Mopeds' dominant user groups changed over time, but still constituted a sizeable chunk of the Dutch population.

16 Stichting Wetenschappelijk Onderzoek Verkeersveiligheid SWOV, *Snorfiets, veilig of niet? Te verwachten consequenties van de eventuele invoering van de snorfiets voor de verkeersveiligheid* (Voorburg: SWOV, 1976), 14. Between 1960 and 1970, ownership in the 25-49 age group dropped from roughly 750,000 to 600,000, while in the 16-21 age group it went from almost zero to 700,000, and in the 21-24 bracket doubled from just below 100,000 to 200,000.

17 Wim de Jong and Bas van Kleef, *De Puch-story en andere brommerverhalen* (Bussum: Uitgeverij Thoth, 1994).

18 Ibid., 29.

19 Ibid., 40, 122. The Berini, despite its Italian-sounding name, was a Dutch product: Servaas van der Horst, *Berini: Geschiedenis van een Hollandse brommer* (Amsterdam: Uitgeverij Thoth, 1991).

In the 1960s, more women adopted the moped and the gender distribution of moped users also changed completely: initially there was a strong gender gap as most moped users were men. This gap closed in the 1960s: the male/female ratio changed from 6:1 in 1963 to 2.5:1 in 1968, and 2:1 in 1970 (see figure 18). And while in 1970 there were still two men riding mopeds for every woman rider, the ratio of new moped users had become 1:1, meaning that just as many men as women now bought mopeds.[20] According to industry organization RAI, in 1968, about 30 percent of moped riders were women. Their numbers more than doubled from 250,000 to 570,000 between 1964 and 1968.[21] The ANWB promoted the sale of mopeds to women right from the very first edition of its moped journal in 1951, in a special column "How women ride mopeds" (*Hoe de vrouw bromt*). Summing up, large numbers of the Dutch (working-class) population adopted the moped in the 1950s and 1960s. This was no surprise to traffic experts after the moped's quick uptake in the late 1940s. However, they had to decide how to classify the moped and consequently its place on the road.

5.2 Sharing the Cycling Path

How did cyclists and moped riders experience sharing cycling paths in the 1950s and 1960s? Letters to newspapers and governance actors provide some insight. I focus on three related issues: the difference in speed, congestion, and whether these paths were considered social spaces.

First, the issue of speed differences was – and remains to this day – a major bone of contention between cyclists and moped riders. While the average cyclist travels at 15-20 km/h, mopeds were allowed 40 km/h. And many went faster: moped builders as well as bicycle repairers and mechanics tinkered with mopeds within the legal maximum 50cc cylinder capacity. Tuned-up mopeds could reach speeds of 60 km, while still legally obligated to use cycling paths.[22] However, when the government announced a lower speed limit, in anticipation, many producers limited the maximum speed of their mopeds to 40 km.[23] This did not stop moped users from tinkering and tuning up their engines.[24] This became a key argument by cyclists against framing the moped as a type of bicycle and in favor of its removal from cycling paths.

20 SWOV, *De bromfietser en de verkeersveiligheid: Een beschrijving van de groep bromfietsbezitters en van de onveiligheid van bromfietsers* (Voorburg: SWOV, 1973), 11.

21 GAA 1302 (RAI Archive), inv. no. 401, "Historische ontwikkeling," 16.

22 De Jong and Van Kleef, *De Puch-story*, 16.

23 Ibid., 25.

24 Ibid., 96-101.

National legislators discussed lowering the moped speed limit further. In the moped subcommittee of the national advisory committee for Road Traffic Safety, the ANWB advised an informal 20 km/h speed limit. The ANWB framed this as a "decency norm" (*fatsoensnorm*) that would not become law – moped riders could go faster on a cycling path when it was quiet, but had to slow down during rush hours.[25] This fitted in with the ANWB's self-image as traffic educator to the public, but also its liberal ideology of individual freedom: the state should not restrict or regulate motorized traffic more than necessary for safety reasons.[26] Committee members C. Krijn, chief engineer and director of Public Works in Noord-Holland, and Albert van Walraven, director of Public Works in Amsterdam, supported the ANWB's idea.[27] However, the committee decided to stick to the 40 km/h speed limit at all times and re-evaluate at a later point. The only change came in the late 1950s, when Minister of Public Works Jacob Algera agreed to lower the speed limit within built-up areas to 30 km/h.[28] The various committees discussed safety measures for mopeds at length, ranging from speed limits to mandatory number plates, and from mandatory helmets to riding tests for all moped users.[29] For the short term, the committees only succeeded in introducing the 40 km/h speed limit.[30]

The second bone of contention was the width. Given the problem of speed, the narrow, separated cycling paths could get very busy, leading to irritation between cyclists and moped riders. I started this chapter with a 1953 reporter observing that mopeds got stuck in busy cycling traffic, especially on public holidays. The only advantage of this bicycle and moped congestion was that the slow speeds reduced the likelihood of accidents. After another busy holiday weekend in 1955, the same newspaper noted that

25 "Tussentijdse balans," *Bromfietskampioen* 2, no. 1 (1952): 8; J.W.A. [Ankersmit], "Divergerende lijnen," *Bromfietskampioen* 6, no. 8 (1956): 524.

26 Mom, Schot, and Staal, "Civilizing Motorized Adventure," 139-58.

27 NA 2.16.11, inv. no. 1363, meeting Subcommittee Mopeds, December 27, 1955.

28 NA 2.16.11, inv. no. 1363, meeting Committee for the Safety of Road Traffic, March 19, 1956, when 10 of the 17 members supported the idea. The Minister did not think the lower speed limit necessary initially (letter minister Algera to Committee, March 20, 1957) but changed his mind a year later (letter minister Algera to Committee, June 6, 1958). He cited the new speed limit for motorcycles in built-up areas (to 50 km/h), making it hard to distinguish motorcycles from mopeds.

29 Accident numbers with mopeds were high. In the 1960s, every year c. 20 percent of all traffic deaths were moped riders (SWOV, *De bromfietser en de verkeersveiligheid*, 42.). Number plates for mopeds were introduced in 2005, though an insurance plate was mandatory since 1966.

30 Third-party insurance became mandatory in 1965, helmets only in 1975, which promptly led to a slower type of moped (*snorfiets*), limited to 25 km/h and a helmet was not compulsory.

special moped tracks were urgently needed, indicating that both groups were sizeable enough to deserve their own lane.[31]

In many letters to the editor, cyclists and moped riders expressed both anger and resignation or presented potential solutions. One "passionate cyclist" found the shared tracks a "misery" but also noted with resignation that mopeds would not disappear and that "car drivers might drive them off the road in a less sportsmanlike fashion."[32] His proposal, given that the shared paths were now a fact of life, was that mopeds should switch off their engines when overtaking. Other readers rejected this idea and "floated their own trial balloons."[33] Moped riders acknowledged that their presence on cycling paths could be a nuisance. One moped rider, claiming he drove 35 km/h on average, said: "Concerning the use of the cycling track, I am aware of being a danger to myself and others."[34] Another moped rider wrote to the ANWB that as an experienced motorcyclist, it was not dangerous for him to use the road on his moped. Restricting mopeds by consigning them to the cycling track hindered progress and was dangerous as well: moped riders "stuffed away on the cycling track between honorable fathers and mothers with kids in the front and backseats, it is going to be a mess (*janboel*), I am telling you."[35] Many users, both cyclists and moped riders, were more inclined to frame the moped as a light motorcycle than as a type of bicycle.

The third issue concerned the social practice of cycling abreast, allowing cyclists to converse with each other. This practice became controversial as it obstructed mopeds who were overtaking cyclists. In contrast to car drivers, who have little or no opportunity to interact, enclosed as they are in their cars, cyclists can and do ride abreast because it is possible to communicate while cycling. Mobility scholars have analyzed this sociability as an advantage for slow modes like cycling and walking.[36] Riding home from work (or school), cyclists could chat with others by riding next to them. With the congestion caused by consigning moped riders to already jam-packed cycling paths, engineers became concerned about efficiency. Some tried to apply the car logic to cycling paths and argued for prohibiting riding abreast.

31 "Compacte massa auto's en fietsers," *Utrechtsch Nieuwsblad*, May 31, 1955, 2.

32 Th.C.M.L., "Bromfietsenwel," *Utrechtsch Nieuwsblad*, June 6, 1953, 3.

33 "Lezers-tribune: Ook bromfietser moet aan z'n trek komen," *Utrechtsch Nieuwsblad*, June 10, 1953, 4.

34 D.H.G., "ANWB Hoe zit dat?," *Bromfietskampioen* 1, no. 4 (1951): 71.

35 "ANWB Hoe zit dat?," *Bromfietskampioen* 1, no. 7 (1951): 123.

36 Marco te Brömmelstroet et al., "Travelling together alone and alone together: mobility and potential exposure to diversity," *Applied Mobilities* 2, no. 1 (2017): 1-15.

To allow for uninterrupted speed, moped riders wanted cyclists to ride in single file. George Oversteegen, a well-known local bookseller and anarchist, wrote to an Utrecht newspaper, suggesting a ban on cycling abreast. Oversteegen characterized himself as both a cyclist and a moped rider but noted there was little civility on cycling paths. Cyclists did not give way to overtaking mopeds, while moped riders act "as if the cycling track belongs to them alone." This lack of civility was an issue: "Everyone will admit that even the widest cycling tracks are still too narrow for the continually increasing number of cyclists and moped riders."[37] If wider cycling paths were not available, cyclists should ride behind each other and a maximum speed for moped riders introduced. As the newspaper editor replied, cycling abreast was more sociable (*gezellig*) and would not disappear. In a letter to the *Moped Champion* journal, a moped rider complained that cyclists refused to acknowledge times had changed: they no longer had a "monopoly" on the cycling path and should stop seeing mopeds as "pariahs." Cyclists stubbornly kept cycling two abreast, talking and having fun without making way for overtaking moped riders. The moped rider wanted a traffic rule which prohibited cycling abreast, believing it would eliminate 80 percent of cycle path accidents.[38] The assumption behind his plea seems to be the same as many car lobbyists: like the road, the cycling path primarily had to provide "flow," in other words the fastest way from A to B. Socializing and using the cycling track as a space not primarily dedicated to speed did not match this vision. However, such a ban was never seriously considered. Cyclists and moped riders had to negotiate their narrow strip of road space every day.

Cyclists viewed the issue differently. The large numbers of cyclists had to share their already limited space with faster vehicles emitting noise and exhaust fumes. Critical commentators noted how this decision privileged a still small minority of car drivers. A letter published in *Bromfietskampioen* asked: "How would you like it if the 200,000 moped riders – possibly half a million in two years – would decide where cars had to drive?"[39] Another letter from a cyclist and moped rider argued that it was drivers who caused accidents, and their responsibility to drive more carefully rather than remove all other road users from the car lanes. "If we continue to let drivers have their

37 George Oversteegen, "Fietspaden zijn veel te smal," *Utrechtsch Nieuwsblad*, May 16, 1956, 7.

38 G. Pouderoyen, "Van de lezer voor de lezers: De bromfietser heeft 't gedaan," *Bromfietskampioen* 5, no. 9 (1955): 661. This strongly worded letter went too far for the ANWB, who pointed out that cyclists' annoyance at the new presence on cycling tracks was understandable and that statistics showed that mopeds riders were often to blame for accidents.

39 "ANWB Hoe zit dat?" *Bromfietskampioen* 1, no. 11 (1951): 195.

way, little will remain of the public character of the roads."[40] In response, the ANWB argued that its genuine and primary concern was traffic safety. For some, however, the ANWB had shown its true colors by promoting the moped as a faster type of bicycle. In a 1975 letter to the new activist group Cyclists' Union, a former ANWB member wrote that "slow traffic" was a seemingly neutral technical term which "had clearly been thought up from behind a car window." Drivers and their interest groups thought: "everything that is slower than us, all that inconvenient stuff [*lastige spul*] that we have to overtake all the time, it's all the same, it has to leave our fast lane, we'll throw it out altogether."[41] Increasingly, cycling advocates criticized the compromises as prioritizing drivers and not giving cyclists enough space. In the 1970s this would lead to widespread activism, as Chapter 6 will discuss.

5.3 Framing Mopeds and Cycling Paths

The moped's arrival presented a challenge to the national politicians and engineers who had agreed to the cycling path compromise. That system was based on two transport modes: the car and the bicycle, both competing for space. As Chapter 2 showed, for various reasons utilitarian commuter cycling infrastructure came to be considered a public good and national and provincial government had to provide cycling paths. When a sizeable share of these commuters started to travel by moped in the 1950s, they needed a place in the dual system. Adding a third lane on or next to roads for mopeds was an option engineers never really considered. Sometimes suggested, this solution was immediately rejected as too expensive and space consuming. The alternative was to consign mopeds to the road or the cycling path. In other words, this required answering the question: was the moped more car-like or bicycle-like? The early moped looked like a bicycle with a small engine mounted on it and was an ambivalent vehicle. Everyone writing about the moped in the 1950s noted the ambiguity. It led to intractable and still ongoing debates about its exact status.

International governance played a role in framing the moped. Organized by the United Nations, the 1949 Geneva Convention on Road Traffic decided mopeds should be classified as bicycles. The Netherlands had signed this convention that stipulated: "Cycles fitted with an auxiliary combustion

40 "ANWB Hoe zit dat?" *Bromfietskampioen* 1, no. 12 (1951): 224.

41 IISG ARCH01969 (Cyclists' Union Archive), inv. no. 1, letter H.D. Coster to Werkgroep 2000, May 31, 1975.

engine having a maximum cylinder capacity of 50 cm^3 (3.05 cu. in.) shall not be considered as motor vehicles, provided that they retain all the normal characteristics of cycles with respect to their structure."[42] This formulation was ambivalent (not specifying "normal characteristics") but mopeds with an engine were added as category to the 1953 Dutch Road Traffic Regulations.[43] Problems arose because this formulation was based on cylinder capacity rather than speed: soon technological improvements (and tinkering at home) would enable mopeds to achieve much higher speeds than policymakers anticipated in 1949. A new version of the convention in 1968 stated that vehicles only counted as mopeds if they had a 50 km/h speed limit.[44] This was, of course, still much faster than cyclists went.

In Dutch governance, the leading stakeholders at Public Works and the non-governmental organization ANWB collaborated to give mopeds a place in the Dutch roads system. In 1953, the Minister of Public Works consigned the moped to the cycling path.[45] The ANWB got involved in the policymaking process as representative of moped riders. This was not self-evident. Smaller lobby groups for car drivers (KNAC) and motor cyclists (KNMV) could also claim to be representing moped riders. Car journalist and editor of ANWB journal *Moped Champion*, Koolhaas Revers, asserted in his first editorial that new moped riders would "instinctively seek the support of the largest traffic federation in the Netherlands."[46] Koolhaas Revers was walking a tightrope.[47] While he stressed the moped's resemblance to the (old-fashioned) bicycle by evoking moped riders' earlier cycling experience, he also emphasized the moped's modernity. He asserted that the ANWB

42 United Nations, "Convention on Road Traffic," (Geneva United Nations, 1949), 35. The full text is at: https://treaties.un.org/doc/Treaties/1952/03/19520326%2003-36%20PM/Ch_XI_B_1_2_3.pdf accessed June 19, 2019.

43 *Staatsblad van het Koningrijk der Nederlanden*, 1953, no. 301-399; January 1, 1953, no. 337, 599-618, accessed through Delpher on December 24, 2018, https://resolver.kb.nl/resolve?urn=KBDC001:003028005:00001

44 United Nations, "Convention on Road Traffic," (Vienna: United Nations, 1968), 3. https://treaties.un.org/doc/Treaties/1977/05/19770524%2000-13%20AM/Ch_XI_B_19.pdf, accessed 19-06-2019. The 1968 text left it to the discretion of member states to classify mopeds as bicycles or motorcycles.

45 Sweden made a similar decision in 1955. Blomkvist and Emanuel, "Regulating a Freedom Machine."

46 J. Koolhaas Revers, "Bromfietsers van Nederland, de A.N.W.B. heet u welkom," *Bromfietskampioen* 1, no. 1 (1951): 3.

47 Koolhaas Revers was editor of multiple car journals, including the ANWB's *Autokampioen* since 1937. His true love was the motorcycle and in the moped's success, he saw an indirect victory for the motorcycle over the car as mass transport mode: Vic Sniekers, "Bekende auto-journalist J. Koolhaas Revers (75 jaar): Heimwee naar motorfiets," *De Tijd*, 19 February, 1966, 7.

"adjusts immediately to present-day circumstances" and framed the moped as a modern transport mode.[48] Whatever its motivations, the ANWB was a very powerful actor and had the government's ear. Unsurprisingly, the national government agreed with the ANWB.

Public Works Minister Spitzen explained his position in a 1949 report to Parliament. According to Spitzen it was too dangerous for relatively slow mopeds to share the road with cars. It would be safer to treat them as bicycles.[49] He acknowledged that there were "certain objections in admitting large numbers of these ambiguous [*tweeslachtige*] vehicles on the cycling tracks which are often already overcrowded."[50] The ANWB and state engineers were in favor of mopeds riding on cycling paths and not using main roads.[51] Defending this decision, the ANWB argued that the difference in speed between cars and mopeds was higher than between bicycles and mopeds. Therefore, the safest, although less than ideal solution was to refer mopeds to cycling paths. Given the policymakers' commitment to provide car drivers on main roads with a "flow" uninterrupted by cyclists, this might have been an unspoken reason for the decision. Although "aware this is not the most pleasant solution," the ANWB upheld its seemingly neutral, technical alternative: "the mentality of the really fast traffic is simply less receptive to the presence of road users, who are hard to see and slower in speed."[52] Engineers and the ANWB mostly used the safety argument. The more implicit argument that drivers did not want to share the road with mopeds, having managed to clear the roads of cyclists, certainly influenced the decision.

Car interest often trumped the bicycle's in (utilitarian) traffic debates. While there was broad consensus on framing the moped as a type of bicycle when it came to traffic roads (to justify relegation to the cycling path), the issue was more complicated regarding recreational infrastructure. The ANWB's ambivalent position manifests itself here once more. The organization wanted to guarantee an obstacle-free road for drivers, but also reserve certain nature areas for non-motorized vehicles only. The first aim required framing mopeds as bicycles, the second framing mopeds as motorized vehicles. To resolve this dilemma, the ANWB exploited the ambivalence of the moped as a technology: either highlighting the moped's bike-like

48 Koolhaas Revers, "Bromfietsers van Nederland," 3.

49 Spitzen assumed mopeds could not go faster than 20-25 km/h: *Kamerstukken II* 1948/49, 1000 IX B, 9, 20.

50 *Kamerstukken II* 1949/50, 1400 IX B, 11, 20.

51 Koolhaas Revers, "Bromfiets de 'Veroveraar,'" 6.

52 ANWB archive, inv. no. 1241, Memorandum 'De Bromfiets', attachment to letter ANWB (Boost) to FNRV, October 30, 1951.

qualities when trying to convince policymakers to remove it from the road; or emphasizing the moped's motorcycle-like character when discussing its place on (recreational) separate cycling paths. These tracks in the countryside were not connected to the road network or beside unpaved roads. The earlier discussed cycling path organizations were often responsible for their construction. For strategic reasons, they framed the paths as mixed-use: serving both tourists and local commuters. For the ANWB, some of these paths were exclusive recreational routes that should ban fast and noisy mopeds because they would spoil cyclists' experience of nature. In this case they brought to the fore the motor-like characteristics of the moped which they minimalized in the discourse on utilitarian mobility.

Politicians also often invoked the distinction between recreational and utilitarian (or "traffic") cycling. In 1949, liberal MP Roelof Zegering Hadders pleaded against allowing mopeds on rural cycling paths (*landelijke rijwielpaden*).[53] He distinguished "cycling tracks, constructed to improve traffic safety" and "cycling tracks which have been constructed for quiet enjoyment of the most scenic spots of our country."[54] In line with the ANWB's stance, he agreed with the minister's plan for mopeds on traffic cycling paths, but objected to the presence of mopeds on recreational routes. In response, the Public Works Ministry legally distinguished traffic and recreational paths.[55] The Ministry basically framed separated cycling paths alongside roads as utilitarian, while cycling paths with an independent trajectory (not following the road network) were recreational. It forced moped riders to use the utilitarian paths and to switch off their auxiliary engine when using designated recreational paths.[56] Two different signs were created for these cycling paths. This final stipulation implied that the moped was after all considered more like a motorized vehicle than a bicycle – it was too fast and noisy and had to become more "bike-like" by turning off the engine.

Provincial policymakers, now often responsible for classifying cycling paths as either one or the other, were in a quandary: in 1962, politicians on

53 *Handelingen II* 1949/50, 34 (December 13, 1949), 970. See parliament records online: https://zoek.officielebekendmakingen.nl/uitgebreidzoeken/historisch.

54 *Handelingen II* 1950/51, 34 (December 19, 1950), 956.

55 Recreational tracks had a different sign: not the round blue one for cycling paths alongside roads to prohibit cycling on the road, but a rectangular black sign, which meant that all non-cyclists were officially not allowed on the path. This was to protect the (basic) paving from damage by horses, cattle, or motorized vehicles.

56 *Handelingen II* 1950/51, 35 (December 20, 1950), 965. These early mopeds still had functional pedals and could be used without the engine. Later mopeds, while keeping the pedals, could not be used as bicycles and were effectively banned from these paths.

Figure 19 This rectangular sign, unlike the round sign, denotes a cycling path banning mopeds because their noise and exhaust fumes were not compatible with the recreational function of these paths. The sign reads "Cycling path" above "so no mopeds [brommers]." The verb "brommen" for riding a moped originally means humming or droning, a reference to the moped's sound. [Source: Donald Trung Quoc Don (Chữ Hán: 徵國單) – Wikimedia Commons]

Friesland's Provincial Board represented various standpoints in a typical fashion. One member squarely opposed allowing mopeds on cycling paths as "the mopeds of today are basically motorcycles" and even on wider cycling paths "it just does not work." Another politician pointed out that cycling paths were also used "for commuter traffic; by men who go to work in neighboring places by moped." The compromise, a third member suggested, was an arrangement distinguishing "touristic cycling paths from cycling paths for commercial traffic [*economisch verkeer*]."[57] Policymakers regarded cycling and moped riding from very different perspectives. Some had more of an eye for tourism, others were more concerned with everyday mobility. Coupled with the various moped appraisals, debates continued long after the formal solutions in 1953.

In practice, mopeds and rural cycling paths were used both for commuting as well as recreation. Cycling track organizations often argued that their rural tracks served local needs as the shortest distance between villages. The ANWB agreed: if the Minister's plan to prohibit moped riders from using these paths became law, many moped riders would face unnecessary detours. The existence of unpaved sand roads meant that all cyclists and

57 "Harlinger haven was nog niet 'haalbaar'daarom uitdieping van grote kanelen. Fietspaden: gemeenten voeren plan uit en GS coördineren," *Leeuwarder Courant*, July 4, 1962, 7.

moped riders preferred riding on cycling tracks that had some rudimentary kind of paving.[58] Nor was it only tourists who used the tracks. So did "farmers going to the market ... postmen, district nurses, the minister and the priest, as well as countless workers."[59] In its letter, the ANWB claimed that only 800 of the 5,000 km of rural cycling track (*plattelandsrijwielpad*) were "of purely touristic interest ... The remainder, including the whole of Twente and the Achterhoek, as well as Drenthe and Noord Brabant, is primarily used by the regions' residents."[60] Elsewhere, it called the independent paths in Twente that farmers used "working tracks" (*werkpaden*).[61] The provincial government of Noord-Holland agreed with the ANWB, noting that cycling tracks unconnected to roads were all used for commuting as well as recreation.[62] The decision to keep the non-mandatory cycling paths free of moped users was a problem for the rural residents who commuted via these paths.[63]

Moped riders were not amused. They complained about no longer being allowed to use the cycling paths that cycling path organization UMO constructed southeast of Utrecht. These paths were often the shortest connections available but could now only be used by bicycles. In the early 1950s, many moped riders applied for an exemption to this rule. One resident of the village Amerongen bought a moped in 1953 and wanted a permit for his 10 km commute to Veenendaal.[64] A local journalist argued that he missed all the local scoops because every time he encountered a UMO path, he had to dismount and walk alongside his moped.[65] Another moped user wrote to the UMO club that the cycling path alongside the route to his work as a mill-hand was next to an unpaved road: if he was not allowed to use this cycling path, he had to take a 10 km detour.[66] The Utrecht cycling path organization routinely refused these requests because it did not have the legal authority (as a private organization) to

58 "Nieuwe verkeerswet sluit duizenden kilometers rijwielpad voor bromfietsen," *Berichten en beschouwingen van de verkeersafdeling van de ANWB* 1, no. 5 (1950): 1-2.

59 Ibid.

60 HUA 1201, inv. no. 25539, letter ANWB to Ministry of Public Works, November 2, 1950. As proof, the ANWB showed that the Twente Chamber of Commerce gave the local cycling path organization a yearly 1,000-guilder subsidy.

61 "A.N.W.B. Hoe zit dat?," *Bromfietskampioen* 1, no. 1 (1951): 14.

62 HUA 1201, inv. no. 25539, letter Provincial Government Noord-Holland to Ministry of Public Works, June 27, 1951.

63 "Open deze paden voor ons!," *Bromfietskampioen* 2, no. 5 (1952): 170-72.

64 GAZ 8020, inv. no. 52, letter L. Blankestijn to UMO, June 20, 1953.

65 GAZ 8020, inv. no. 52, letter C.B. Dekker to UMO, undated.

66 GAZ 8020, inv. no. 52, letter E.J. Reemst to UMO, June 5, 1959.

allow these exemptions; only the Provincial Executive did.[67] As a result, some moped riders decided to use the paths anyway and risk a fine since police controls were not very strict. The moped riders' disobedience and the authorities' failure to enforce the rules then led to conflicts with regular cyclists: one long-time UMO member cancelled his membership because he was often almost pushed off the rather narrow paths by mopeds overtaking. He suggested either making the paths twice as wide, or actually enforcing the moped ban.[68] However, the UMO did not have the money to do the former, and the police did not have the capacity for the latter. In short, the popularity of mopeds in a system based on the well-functioning binary system of traffic separation presented unsurmountable problems in practice.

5.4 How Mopeds Boosted Cycling Path Construction

While the outcome of the moped debate was in many respects a decision that prioritized drivers over commuter cyclists, it did have a positive, albeit unintended, impact on cycling path construction in the long run. After national legislators classified the moped as a bicycle, it was up to the engineers governing roads and cycling infrastructure to deal with the consequences. Many were convinced that the network of separate cycling paths was not adequately constructed to deal with this influx of moped riders. Aware of large numbers of suburban cycling commuters, and the fact that moped riders were faster and would constantly overtake cyclists, engineers felt that wider cycling paths were necessary.

Politicians, responding to their working-class constituency, also urged Public Works engineers to continue investing in cycling-and-moped paths. For instance, social democrat MP Siep Posthumus (PvdA) argued that the moped issue required a "radical solution," like "the construction or improvement of cycling tracks or secondary roads in good condition, which are suitable to both the slower moped riders and ordinary cyclists."[69] Like other stakeholders, Posthumus noted the "tensions" and "problems of a psychological character" cyclists experienced as they had to share paths with mopeds. It was to relieve this tension that engineers should "rapidly widen existing cycling tracks and construct new ones with a greater width

67 GAZ 8020, inv. no. 52, letter UMO to H.J. Olie, August 13, 1960.

68 GAZ 8020, inv. no. 52, letter M. Zieleman to UMO, March 3, 1956.

69 *Handelingen II* 1951/52, 21 (November 27, 1951), 634.

than has been the case up to this point."[70] In 1962, Christian democrat Cor van der Peijl repeated Posthumus's call for continued construction of cycling infrastructure: he advocated the "permanent" interpretation of mopeds as "slow traffic" and demanded more and wider cycling paths.[71] Communist MP Jan Haken echoed claims for separate moped-only paths, but like other proponents did not believe this was a feasible option.[72] Politicians were aware of the commuter needs of moped riders and pressured state engineers to take steps to address it. At the highest political level, calls for more cycling infrastructure could be heard throughout the 1950s and 1960s.

Engineers in provincial Public Works departments agreed cycling paths needed widening, even if they disagreed on cycling's status and future. At a 1956 meeting of the private cycling path organization platform, the Federation of Dutch Cycling Path Organizations, two engineers, C. de Groot from provincial Public Works in Utrecht and S.J. Faber from Gelderland, differed on cycling's future. According to De Groot, the moped was "the transport mode of the future" and cycling would only be an urban activity.[73] The suburban and interurban cycling paths in this scenario would not be used by cyclists but only by mopeds, not requiring modifications. Conversely, Faber did not believe the moped's rise was directly correlated with a decline in cycling: overall mobility levels were increasing: "one should not state that the normal bicycle is about to disappear. It is not like that: traffic has become more intensive."[74] Both agreed, however, that engineers should construct more and wider cycling paths.

These provincial engineers formulated new design norms for cycling infrastructure. Faber explained in 1956 that the province of Gelderland had "decided to extensively construct cycling tracks along the roads in the Provincial Roads Plan. Those tracks will be 2.40 m wide on each side of the road. So, in the long run, all the roads in Gelderland's Provincial Roads Plan will have two cycling tracks, dedicated to traffic on two wheels."[75] In 1959, engineers in the province Drenthe also adjusted their design standards for cycling paths. Calling this "the logical consequence" of the "stormy development" of moped use, cycling paths "that had been initially designed with

70 *Handelingen II* 1953/54, 42 (December 22, 1953), 3629.

71 *Handelingen II* 1961/62, 40 (January 17, 1962), 3425.

72 *Handelingen II* 1951/52, 21 (November 27, 1951), 637. Van der Peijl also called for these special moped paths in 1968: *Handelingen II* 1968/69, 2 (September 18, 1968), 24.

73 ANWB archive, inv. no. 1241, FNRV meeting May 9, 1956.

74 Ibid.

75 Ibid. A 1957 report shows that cycling paths in Gelderland were often constructed 2.40 m wide: "De Rijkswegen in 1956," 17.

a hardening width of 1.50 meters, will now be given a width of 2 meters."[76] Public Works engineer Hendrik Baltus Bakker in the province Zuid-Holland described road reconstruction: "Taking into account the possibility that cycling paths will become more important due to the motorized bicycle, their width has been designed spaciously ... at 2.30 meters."[77] For its recreational cycling paths which the province had controversially opened up to mopeds in certain cases, its engineers even decided the cycling paths should be 3 meters because of this.[78] Gelderland widened some of the Zuid-Veluwe organization's major paths around 1960 because of moped traffic.[79] The lack of national guidance is clear from the varying norms that provinces adopted. Engineers and politicians widely agreed, however, that the rise of the moped necessitated investments in cycling infrastructure.

The ANWB attempted to achieve some unity in these norms. As discussed in Chapter 3, by publishing guidelines, the ANWB provided additional training in traffic engineering for local engineers.[80] The booklet on cycling paths said that cyclists needed at least 0.75-0.80 meters in width and mopeds 1 meter. To allow cyclists to overtake, paths should be 1.50 meters wide; for a moped to overtake a cyclist, 2 meters was advisable. Ideally, the ANWB advised 2.50 meters as a minimum so that two cyclists could ride abreast and a moped could still overtake them.[81] The ANWB and cyclists valued this social aspect of being able to ride side-by-side (see section 5.2). Besides the brochures, the ANWB also used its traffic engineering courses to promote the idea of wider cycling paths. In 1954, provincial engineer D. Mathlener's lecture stated that 2.50 meters was a minimum width for unidirectional cycling paths, and those with intensive moped traffic should even be 3

76 DA (Drents Archief) 0923, inv. no. 6258, letter Public Works Drenthe to Ministers of Education, Arts and Sciences, Economic Affairs and Social Affairs, December 4, 1959.

77 H.B. Bakker, "Over het saneren in het algemeen en over het saneren van een oud Rijkswegvak in het bijzonder," *Wegen* 25, no. 1 (1951): 4.

78 "Nieuwe rijwielpaden in Zuid-Holland," 2.

79 "Fietsen op de paden van de Zuid-Veluwe," 176.

80 ANWB archive, inv. no. 1241, letter Boost (ANWB) to FNRV, May 7, 1956 with attached Memorandum no. 4 *Rijwielpaden langs verkeerswegen* ("Cycling tracks alongside traffic roads"), published in 1956. It was reissued in updated form in 1966 and 1970 as *Fietspaden en -oversteekplaatsen* ("Cycling tracks and crossings").

81 The ANWB even argued that the cycling track surface had to be smoother because of mopeds' higher speed (and different road-holding): "Vlakke rijwielpaden," *Berichten en beschouwingen van de verkeersafdeling van de ANWB* 5, no. 2 (1954): 6. An engineer involved in Zuid-Holland's cycling track plan made the same point in 1964: tile tracks were too lightly constructed for mopeds, and basic asphalt paving had to be provided: J.C. Fekkes, "Provinciaal rijwielpadenplan voor Zuid-Holland," *Tijdschrift voor Verkeerstechniek* 15, no. 8 (1964): 362.

meters wide.[82] The ANWB's many means of knowledge dissemination and prestige as a knowledge actor likely helped convince engineers to adopt these new norms.

The construction of suburban cycling paths did indeed continue in the 1950s and 1960s, partly because of the developments sketched in this chapter. The ANWB traffic engineering journal's 1956 overview of all construction work showed that cycling paths had been (re)constructed on thirty-four national or provincial roads in that year.[83] One specific example: In Zeeland in the far south-west of the Netherlands, "a separate cycling path has been constructed between the built-up areas of Oostburg and Aardenburg ... The path is 2.40 meters wide and in total 7.1 km long."[84] In general, where this article mentions width, cycling paths are 2.50 or 3 meters, significantly wider than before. For other years, this overview of road works shows the same pattern (see figure 20).

In conclusion, in its motivation as well as its unintentional outcome, this compromise is reminiscent of the 1920s bicycle tax. In both cases, a major decision concerning a vehicle's place on the road disadvantaged cyclists. In both cases, politicians and engineers realized this and decided that cyclists had a right to compensation. There are also differences: the position of cyclists was weaker in the 1950s than it had been three decades earlier: in the 1920s, cycling was on an upward trajectory and had broad support as vehicle of the masses. In the 1950s, although internationally more so than in the Netherlands, an increasingly large group of experts doubted the future of cycling as a transport mode. Ultimately, though, the long-term development of cycling in the Netherlands benefited in unintended ways from these key compromises. This is another manifestation of the Dutch "*poldermodel*" whereby compromise and the (somewhat) equal distribution of benefits and burdens are key beliefs. As Public Works Minister Spitzen noted in 1949: the moped "is dangerous on the highway and less dangerous, but not entirely undangerous, on the cycling path," while people "are anxious about the arrival of this minor noisemaker on the quiet heath lands and forests." He said: "We hope to be able to find a solution which reconciles all these interests, that are not all in line."[85] That cyclists had to give up a little more than drivers went unmentioned, however.

82 "De bromfiets: het snelverkeer van het rijwielpad: Nationale Verkeerstechnische Leergang 1954," *Bromfietskampioen* 4, no. 3 (1954): 142.

83 "De Rijkswegen in 1956," 9-20.

84 Ibid., 12.

85 *Handelingen II* 1949/50, 35 (December 14, 1949), 1002.

Figure 20 Provincial engineers constructed cycling paths in part to accommodate moped riders. They believed widening and expanding the cycle path network was crucial given the considerable cohort of Dutch commuter cyclists. This path was built between 1955 and 1960 near Rhenen, Utrecht province, at a time when only a few countries were still constructing (utilitarian) cycling infrastructure [Source: Photographer Provincie Utrecht, Collectie Het Utrechts Archief, no. 837959]

5.5 Conclusion

The moped's role has been unjustly neglected in the (cycling) historiography, considering its major role in Dutch mobility as an intermediate step between cycling and driving. In that role, it impacted both the Dutch cycling system of separation – and helped keep cycling high on the political agenda – as well as the comparatively late adoption of the car. Many working-class commuters and young adults took up the moped enthusiastically between the late 1940s and late 1960s. This posed a challenge to the engineers and the ANWB's traffic-separation paradigm. The framing of the moped was crucial to the solutions: both its status as an intermediate vehicle between bicycles and cars, but also its status as either a utilitarian or recreational vehicle. These motorized bicycles were neither bicycles nor cars, not fast enough to use the road, but too fast to use the cycling path. The chosen option meant that cyclists had to make way for those who had left the bicycle behind and adopted the moped. Leading state and ANWB engineers justified their decision by arguing that car drivers and moped riders sharing roads posed a greater danger than moped riders and cyclists sharing cycling paths. Over time, however, this became less and less true, for engineering improvements enabled mopeds to go ever faster while still legally classifying as a moped. Cyclists were the ones who had to adapt to this new mode, and engineers chose to ensure car users an unobstructed drive.

A key finding of this chapter is that the decision to consign moped riders to the cycling path had unintended, positive consequences for cyclists because it positively reinforced the status of cycling paths as a public good. It was not uncommon for roadbuilders outside the Netherlands to pave over existing cycling paths and make them part of car infrastructure. They did this inspired by a vision of the future, assuming that everybody would be able to afford and own a car. In this visualization of future mobility, there was no place for cyclists and consequently no reason to build new cycling lanes or even maintain existing facilities. Cycling levels in the Netherlands were higher than elsewhere, and cycling was much more embedded both in Dutch culture and in mobility institutions at a national, provincial, and local level. Without speculating about the development of Dutch cycling without mopeds, it is fair to say that thanks to the moped's popularity, engineers regarded constructing cycling infrastructure as more justified. It also reinforced the existing separation that engineers saw between utilitarian and recreational cycling, by formalizing the distinction in traffic legislation. As a result, the category of recreational cycling became quite distinct in governance, with separate funding streams and governance coalitions, as the next chapter will demonstrate.

Path dependency shaped the moped story in several ways. Without the prior commitment to traffic separation (of drivers and cyclists) on major roads there would not have been so many cycle paths. Without these, the question of the moped's place on the road would not have escalated to the same extent as there was only one option: the road. The decision to assign mopeds to cycle tracks was shaped by the particular path Dutch politicians and engineers had chosen in the 1920s. Additionally, the decision reinforced engineers' commitment to their chosen path. Given the popularity of cycling, it is not very likely they could go back from a system of traffic separation to mixed traffic, but the developments around Western Europe at this time do suggest cycling was in peril. Engineers' reinforced vigor to construct more and wider cycle paths meant the traffic separation system remained strong throughout the car-centered 1960s. This chapter has not studied moped drivers' experience of the cycle paths. While speculative, it is therefore possible that moped riders were also 'path dependent' in the sense that separate paths gave them safer and more comfortable options than moped riders who had to share the roads with cars. While not at its 1960s levels anymore, the continuing popularity of mopeds in the Netherlands, as with bicycles, might have something to do with its place in cycle lanes.

From a governance standpoint, the Dutch history of mopeds illustrates how not just multiple stakeholders, but also multiple levels of government shaped everyday mobility. From the international UN Convention on Road Traffic to the national adoption of legislation, then provincial and municipal implementation, each level played a role in shaping the Dutch system of cycling and moped riding. In the Netherlands, politicians, influential non-governmental actors like the ANWB, and engineers called for widening existing cycling tracks and accelerating the pace of constructing new ones. Those same actors saw the few suggestions of dedicated moped tracks as unviable and settled upon a sort of combined moped-and-cycling path. Such paths were wider, smoother, and more extensive than the cycling infrastructure for cyclists alone. This does not detract from the fact that cyclists now had to share their separated cycling infrastructure with faster, noisier, and smellier vehicles, making cycling less pleasant. For those who could share the cycling path – or had no other mobility options – the many moped users made it possible to ride alongside major roads in the Netherlands, without having to mix with the 1960s fast-rising tide of car traffic.

Conclusion Part II

Motorization everywhere meant that cycling's share of everyday trips declined precipitously in the 1950s and 1960s. This decline occurred in the Netherlands as well, but to a much lesser extent. A sizeable cycling culture remained, and by the late 1960s, far from having virtually disappeared, cycling was still a part of Dutch everyday life – whether for commuting, getting around town, or recreation. Multiple factors explain this. Many commuting distances remained cyclable, sometimes in combination with public transit. Cycling was also more compatible with driving in the Netherlands: partly because traffic separation enabled safe cycling outside cities, partly because late automobilization made cycling a necessary and still viable urban mode, although increasingly beleaguered. Finally, the cultural framing of cycling developed before 1950 showed its staying power, so that cycling did not develop the same stigma as elsewhere. Politically, engineers across the world increasingly ignored cycling while they were grappling with or actively promoting automobilization.

Cycling infrastructure's status as a public good remained strong or even increased in the period which has been described as one of overall decline. To whatever extent cycling did decline, we can conclude it was partial and uneven. The cycling experience varied significantly among different cyclists. For some, it was very eventful – particularly urban cyclists. Others saw the safety and comfort of their cycling routes improve. Tourists had ever more options. My assessment is more positive than earlier cycling histories of the Netherlands because they focused on urban cyclists, the group who suffered the worst consequences of automobilization. My analysis of provincial and regional policymakers and the coalitions they formed with the ANWB to facilitate suburban commuter as well as recreational cycling shows that pro-cycling policies initiated in the 1950s were still being implemented in the 1960s. Traffic policy thus had a positive effect on cycling's development in the Netherlands at this time.

In short, in discussing cycling as a public good in the 1950s and 1960s, we have to distinguish the focus on utilitarian cycling from the focus on recreational cycling. The moped and recreational cases in previous chapters show that politicians and ANWB officials applied this discourse opportunistically: sometimes it was more expedient to stress commuter cyclists' needs, while at other times the tourism potential was the key to funding. Politicians and engineers did this consistently, resulting in diverse, yet overlapping governance and material networks. In practice, this distinction did not hold up.

Utilitarian cycling's status in this period is ambiguous because it was rediscovered in the late 1960s. Provincial policymakers began to realize how many people still used the bicycle for commuting. We have seen how policymakers in the 1950s still valued cycling as an efficient (urban) mode of transport, perhaps allaying some policymakers' qualms when thinking about automobility's explosive growth. At most we can say that for a period of five to ten years from the late 1950s to the mid-1960s, utilitarian cycling was rarely discussed. Even at this low point in cycling's public status, however, the construction of separate cycling paths alongside roads never stopped. This was a legacy of an older conception of cyclists as (tax-paying) citizens deserving cycling paths, enshrined in traffic engineering norms and reinforced by the need to construct more and wider cycling paths for mopeds.

In the same decade, recreational cycling became more important as a public good. The large and successful private governance coalition around recreational cycling was incorporated in provincial and regional government. This was not so much because these policymakers saw cycling as inherently good. They viewed it as a means to an end, such as public health or the economic benefits of tourism. New regional state bodies for recreational cycling proliferated, as did cycling path plans and subsidies. Ultimately this also had a bearing on utilitarian cycling, as the revived focus on this type of cycling around 1970 did not require new administrative procedures or institutions but could build on the provincial officials' existing expertise.

Cycling policies in the 1950s and 1960s were governed by various national, provincial, and local coalitions in which the ANWB often participated. However, cycling policy was not centrally coordinated but decentralized. It is revealing that, when politicians asked engineers in the province Zuid-Holland to provide a comprehensive overview of cycling paths in the late 1950s, no one could. It illustrates how much effort it took for politicians to convince engineers to implement their cycling vision. The national government's role was limited to that of enabler. It funded provincial and municipal traffic and recreation departments. As the moped debate showed, for example, it also formulated laws. The government emphatically shied away from coordinating cycling policies or infrastructure networks, however. That task was left to the provinces, and other regional forms of cooperation between municipalities. It is no accident that the overall cycling infrastructure data used here mostly comes from ANWB sources: this non-governmental organization had representatives in the various government bodies, was in a better position to compile statistics than any state agency, and functioned as a de facto policy coordinator between the different layers of government.

There was also a vertical separation between the agencies involved in road construction and in recreation. The agencies all saw cycling in a different light. Public Works officials were responsible for traffic and tried to give cyclists a place, while also protecting the flow of car traffic. Spatial planners and civil servants, responsible for recreation, failed to contact each other on a regular basis – hindering knowledge sharing and coordination. At the same time, it is also a sign of how widely cycling planning was spread within Dutch bureaucracy. Officials at different levels of government and in various branches or sectors were all making cycling policies. This dispersed and decentralized rather than concentrated approach grew organically over time.

Most importantly, the role of local citizen initiatives, so crucial before, was assimilated by the provinces. Now the formal control in building rural cycling infrastructure was entirely in the hands of provincial actors. Some of the citizen organizations continued privately, but financial limitations gradually lessened their influence.[1] The national organization ANWB, however, successfully kept – and even expanded – its control of engineering knowledge and education.

To sum up, in the 1950s and 1960s, Dutch cycling governance became more embedded within various government circles. Recreational cycling was policymakers' most visible concern. Policies aimed at utilitarian cycling were still implemented behind the scenes. Both in policy and in practice, various types of cycling persevered throughout the 1950s and 1960s.

1 Around the 1940s, there was more focus on the meso level of government in Dutch politics: Toonen, "Dutch Provinces and the Struggle for the Meso," 123. The provinces were deemed key actors for the major domains of land use, spatial planning, and infrastructure.

Part III

Dutch Model: How Urban Cycling Became a National Political Demand after 1970

Figure 21 This junction in Utrecht, 1968 illustrates cyclists' increasingly daunting battle in urban traffic. According to urban engineers, cycling infrastructure was impossible in city streets. In practice, road experts were not willing to limit the space for vehicles when faced with fast rising car levels. As a result, cyclists and drivers had to negotiate their place on the road every day and cycling was not for the faint-hearted. The situation led to growing protests from cyclists no longer willing to face these street battles. [Source: Collection Het Utrechts Archief, no. 810895]

Until the 1970s, Dutch urban cyclists found their own way through cities, with little protection from other road users and even less policy focus, as figure 21 illustrates. For urban policymakers, the priority was dealing with the car's place in dense historic cities. As one traffic planner admitted to a journalist in 1964, while he "knew everything about cars" in his city, "maybe we have forgotten about cyclists a little due to all the worries about the car."[1] Like elsewhere in Europe, car ownership in the Netherlands grew in the 1960s. Welcoming (suburban) car drivers into historic city centers made life hard for downtown cyclists. National and provincial authorities might have addressed the needs of (utilitarian) suburban, small-town, and recreational cyclists in earlier decades, but ignored urban cyclists. By the 1970s, urban cyclists decided they would no longer accept their status as second-rate citizens. Across the Western world, social movements emerged, protesting the car's negative impact on their cities. Other countries' attempts to transform the city had less long-time success than their Dutch

1 "Steden laten de fietsers versukkelen," 11.

counterparts, who managed to form a coalition with urban policymakers and engineers that has made Dutch cities safer for cycling over the past fifty years. This may be a familiar argument. Surprisingly, we know relatively little about these Dutch activists' strategies, the role of national and provincial policymakers, and how they worked together to put cycling higher on the agenda. While the activist decade has been described as radically different from the previous period, we have no assessment of the policy continuities and discontinuities. This section aims to address these omissions.

According to political scientists, in countries with more consensual and corporatist traditions, social movements are more likely to gain access to policymaking circles.[2] In these countries, policymakers tend to acknowledge that certain interest groups represent significant portions of society and form valuable intermediaries for consensus-driven policymaking. In the context of cycling activism, Emanuel, Balkmar, and Kriesi have made this claim for Denmark, Sweden, and Switzerland respectively, though US historians like Furness, Friss, and Longhurst have shown how much more antagonistic this situation was in America.[3] With a consensual political tradition like in Denmark and Switzerland, Dutch cycling action groups could operate in a similar way. Indeed, cycling activists had reached policymakers' ears, but we know surprisingly little about their cycling advocacy and fight for a redistribution of space. Albert de la Bruhèze, Veraart, and Oldenziel found examples in major Dutch cities. But a comprehensive analysis of Dutch cycling activism's roots, social background, and action strategies is lacking. How did activists establish a working relationship with state officials? And what was the activists' input in policymaking? In Chapter 6, I explore these questions.

Dutch cities provided little dedicated cycling infrastructure before the 1970s. Over the last fifty years, they have now undergone a major transformation in catching up with the countryside. This required not just citizen activism but also a concerted effort by urban policymakers. Research by Albert de la Bruhèze, Veraart, Oldenziel, and Emanuel demonstrates how

2 Andeweg and Irwin, *Governance and Politics of the Netherlands*; Van den Berg, "Dynamics in the Dutch Policy Advisory System," 63-84; Kickert, *The History of Governance in the Netherlands*; Van Veen, "From Neo-Corporatism to Regulatory Governance," 85-102.

3 Dag Balkmar, "Cycling Politics: Imagining Sustainable Cycling Futures in Sweden," *Applied Mobilities* 5, no. 3 (2020): 324-40; Emanuel, "Making a Bicycle City," 493-517; Hanspeter Kriesi, "Federalism and Pillarization: The Netherlands and Switzerland Compared," *Acta Politica* XXXI, no. 4 (1996): 539-56; Pete Mijnssen-Hemmi and Florian Boller, *30 Jahre Veloförderung in der Stadt Zürich, 1975 bis 2005* (Zürich: Tiefbauamt der Stadt Zürich, Verkehrsplanung, 2006); Friss, *On Bicycles*; Furness, *One Less Car*; Longhurst, *Bike Battles*.

urban politicians and officials, pressured by activists, started to take cycling more seriously in the 1970s and beyond. These cities operated in a larger regional, provincial, and national context, on which they partly depended for funding. In addition, higher government levels formulated the policy goals that cities had to take into consideration – a dimension largely ignored in the history of cycling. Scholars and historians of policymaking have shown that this multi-level structure is complicated. We do not know how this dynamic impacted urban efforts to redesign the city. My aim in Chapter 7 is to see what role, if any, provincial and national government played in cycling governance. The time frame is the 1970s and 1980s, a period when stakeholders sought a governance arrangement that worked for the new policy goal of protecting and promoting urban cycling. These chapters paint a fuller picture of the multi-stakeholder coalition that shaped cycling policy in the Netherlands than earlier studies.

Curtailing the place of the car in the city and returning some of this space to cyclists and pedestrians has been a long and arduous process. It was by no means finished in the late 1980s. In the early 1990s, national cycling policy received new impetus with the implementation of the national policy document Bicycle Master Plan (*Masterplan Fiets*). My final chapter shows how this (internationally) relatively well-known policy document was an outcome of two decades of debates.[4] The overarching theme is the government's attempt to decentralize cycling governance.

To understand the mobility context of the 1970s and beyond, we have to take a closer look at mobility developments in the Netherlands. The Dutch population's overall mobility rose sharply in the 1960s in line with households' increased purchasing power. It is well known that car sales more than doubled within just a decade (182,000 in 1963 to 430,000 by 1973). Over this same period, bicycle sales also soared, from 555,000 to 1,068,000 – and outnumbered cars in absolute terms.[5] Social trends, like growing awareness of environmental problems (reinforced by the Oil Crisis and Car-Free Sundays) but also the potential health benefits of cycling, increased cycling's

4 An English summary of its results and historical context was one of the earliest publications on Dutch cycling in English: Public Works and Water Management: Directorate-General for Passenger Transport Ministry of Transport, *The Dutch Bicycle Master Plan: Description and Evaluation in an Historical Context* (The Hague: Ministry of Transport, Public Works and Water Management: Directorate-General for Passenger Transport, 1999).

5 Centraal Bureau voor de Statistiek, *Statistisch zakboek* 1974 ('s-Gravenhage: Trio, 1974), 181; Henk Maurits, "Energiecrisis leidt tot recordverkoop van fietsen," *De Tijd*, January 3, 1974; "Gouden tijden voor rijwielindustrie," *NRC Handelsblad* November 8, 1973; "Fietsenmaker H. Kind Jr.: 'Ze zeggen weer u tegen me'," *Het Vrije Volk*, November 8, 1973.

popularity and status across the world and in the Netherlands.[6] Having lagged behind in car ownership in comparison to other European countries, in the late 1960s the Netherlands also became a car-oriented country – yet with an atypical number of cyclists.

The rising critique of automobility coupled with roadbuilding around 1970 led to a cycling revival – or to put it more accurately, stopped the declining trend. Public transit also flourished. Between 1960 and 1990, Dutch Railways (NS) added 110 stations to its network, leading to a total of 389 by 1990. New suburbs or satellite towns were connected to this railway network.[7] Since the 1980s, the expansion led to a sharp increase in train use, although overall kilometers traveled by car also rose in an ever more mobile society.[8] In this revival, the bicycle – privately owned or more recently the Dutch Railways' shared bicycle system – has played an important part as before-and-after transport in people's daily public transit commute.[9] Moreover, there was no competition from subway systems. Like in Denmark, few Dutch cities operate subways (only Rotterdam since 1968 and Amsterdam since 1977) or streetcars, making the bicycle a necessary alternative.[10] Urban planning policy, for instance in the 1976 Urbanization Memorandum (*Verstedelijkingsnota*), did stipulate that new neighborhoods must be within 4 km distance of train stations to curb automobility growth.[11] At the same time, the 1960s spatial planning policy to concentrate population growth in new satellite towns only partially succeeded.[12] Government policy could not entirely prevent urban sprawl and automobility.

In the late 1960s, coupled with rising car levels was "an unprecedented highway building fever."[13] National and provincial governments constructed hundreds of kilometers of highways, including many expensive tunnels and crossovers.[14] According to Mom and Filarski, municipalities also extended

6 The health benefits of cycling were explained in a Dutch bicycle (industry) lobby brochure: Stichting: fiets!, *Fit door fietsen: hoe fiets je voor je gezondheid?: wat zegt de dokter?: conditietraining, trimmen, trainingsschema's*, Fietsotheek, vol. 8 (Amsterdam: Stichting: fiets!, 1974).

7 Mom and Filarski, *De mobiliteitsexplosie*, 366-67.

8 Ibid., 379-81.

9 Jan Ploeger and Ruth Oldenziel, "The Sociotechnical Roots of Smart Mobility: Bike Sharing since 1965," *Journal of Transport History* 41, no. 2 (2020): 148-53.

10 Carstensen and Ebert, "Cycling Cultures in Northern Europe," 45.

11 Van der Cammen and De Klerk, *Ruimtelijke ordening*, 278. Jan Ploeger's forthcoming book will shed more light on this development.

12 Bontje, "A 'Planner's Paradise' Lost? Past, Present and Future of Dutch National Urbanization Policy," 145.

13 Mom and Filarski, *De mobiliteitsexplosie*, 316.

14 Gijs Mom, "The West and the Rest," 172; Hans Buiter and Kees Volkers, *Oudenrijn: geschiedenis van een verkeersknooppunt* (Utrecht: Matrijs, 1996).

and densified road networks. By the early 1970s, the Netherlands had "a road network with a density almost unique in Europe."[15] The radical shift in a context where cycling and moped riding were still popular modes of transport created a dramatic conflict. It partly explains why citizen resistance against automobility was so intense, widespread, and sustained in the 1970s.

The government's facilitation and rise of automobility had many negative effects. The increase in traffic deaths was perhaps the most dramatic. Pedestrians, cyclists, and playing children had ever less space. The lack of restrictions on (urban) automobility coupled with little government effort to protect urban cyclists from the resulting dangers, led to ever-increasing traffic deaths. At cycling's lowest point in the late 1960s, hundreds of cyclists and children were killed every year. As figures 22 and 23 show, the already high levels of road fatalities in the 1950s rose even further in the 1960s. Measures taken since then to improve traffic safety have been effective, although the number of cyclist deaths has not decreased further over the past twenty years.

Dutch cities had always been cycling cities but despite, not because of state policies. Together with the powerful tourist organization ANWB, policymakers had promoted and facilitated recreational and suburban commuter cycling in the Netherlands throughout the twentieth century. Urban cyclists had been relatively ignored until the 1970s. The rising anger over traffic safety, coupled with growing concerns over environmental degradation and the car's space demands in such a densely built nation triggered a wave of broad-based activism. This acted as a catalyst for growing concerns about precisely these issues among certain policymakers and engineers. The social pressure led to a wide coalition of actors aiming to curtail the car's place in the city and redistribute urban space. In the 1970s, wide social unrest, a peak of activism, and a receptive government all aligned to create a strong pro-cycling coalition and a shift in cycling policy.

15 Mom and Filarski, *De mobiliteitsexplosie*, 316-17.

Figure 22 This graph shows how hundreds of children died in traffic every year after mid-century. Many of these cases concerned drivers hitting playing children, leading to widely shared anger and activism in the 1970s. Data from Dutch Road Safety Institute SWOV (Stichting Wetenschappelijk Onderzoek Verkeersveiligheid). See: https://www.swov.nl/feitenencijfers/verkeersveiligheidscijfers-verkeersongevallen and https://theseus.swov.nl/single/?appid=5dbac35a-5fbd-401f-b711-682176941688&sheet=jvpBF&opt=currsel%2Cctxmenu for the data from 1996 and later.

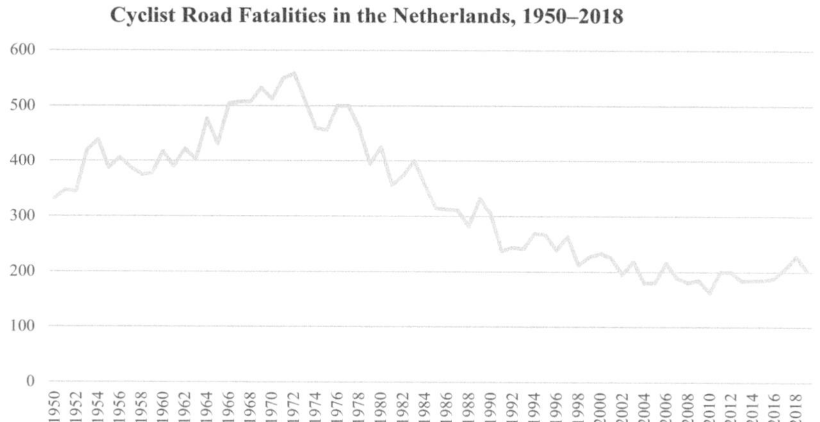

Figure 23 Increasing automobility in the 1960s coupled with insufficient measures to protect cyclists, led to dramatic accident rates peaking around 1970. Separate cycling infrastructure existed alongside major roads, but very little within cities, making urban cycling ever more dangerous. Data from SWOV and CBS.

6 Citizen Expertise: Urban Activism Shapes Local Cycling Policy in the 1970s

"When it comes to cycling, there is only one expert: *the cyclists themselves*. The cyclist knows best which situations are dangerous, which short cuts are needed, which barriers should be removed ... Where the government is doomed to fail given its technocratic approach, the Cyclists' Union can be strong."[1] With this resounding statement, the Dutch Cyclists' Union (*Fietsersbond*) claimed that successful cycling governance was impossible without the input of everyday cyclists. Rather than making policy "from behind the car window" as cycling activists often suspected, urban engineers should work with cycling groups representing everyday cyclists. With their detailed knowledge of urban roads and the dangers of everyday mobility, they had the expertise engineers lacked. To curtail the car's place in the city, fight the record high numbers of traffic casualties, and make cycling a safer and more attractive option in Dutch cities, engineers and activists had to work together. The new organization of young activists also questioned the traditional, mediating, and monopolistic role of the ANWB in cycling governance. It was part of a larger critique.

By 1975, the technocratic policymaking of the 1950s and 1960s had come under fire. A rising chorus of activists took aim at both the undemocratic way engineers and policymakers operated, and the negative effects of their decisions: pollution and rising road accidents, for example. Citizens took to the streets to protest and demand change. Urban policymakers' lack of interest in cycling meant that urban engineers were unaware of the problems that cyclists faced daily in cities where engineers gave cars free rein. Instead, they argued, these cyclists were the true experts when it came to turning back the clock to when bicycles ruled the streets. A new generation of young, educated, and left-wing activists put cycling back on the agenda and demanded that urban cyclists, neglected up to that point, would finally be taken seriously.

In the early 1970s, an international wave of environmental activism was sweeping over the world. Central to this activism was a fear that society

1 IISG, Archive Cyclists' Union (ARCH01969), inv. no. 1, Christiaan la Poutré, "Inventarisatie knelpunten," 2. Underlined in original.

was losing (democratic) control over large technical systems. Many opposed what they considered technocracy gone awry, critiquing the waste of natural resources. The Oil Crisis only reinforced this conviction. The activist world was heterogeneous. In the area of innovation and governance, for instance, it split over the question of how ordinary people could develop counter-expertise and participate in government. Some activist groups opposed building their own knowledge base and preferred extra-parliamentary activism with an eye to long-term mentality changes. Other groups wanted to achieve results pragmatically in the shorter term.[2] These themes also permeate the Dutch environmental movement and cycling activism. Due to high cycling levels and rapid motorization, cycling activism in the Netherlands was a large subset of environmental activism, more so than in other countries. As such, it played an important role in putting urban cycling on the political agenda. This role has been acknowledged but not yet comprehensively researched.

In *Cycling Cities*, the authors present social movements' achievement in putting cycling higher on the political agenda as a key reason for successful Dutch urban cycling since the 1970s.[3] They show how in various cities, activists were the ones who pushed policymakers into action and turned thinking about public space, cars, and bicycles on its head. Due to its approach – "biographies" of urban cycling in various cities – the studies do not analyze the entire Dutch Cyclists' Union. In this chapter, I offer a more comprehensive analysis of the precursors and origins, social background, goals, and action strategies of cycling activism in the Netherlands. In addition, this and the following chapters will delve into the new governance arrangements for (urban) cycling of the 1970s and beyond. Through (local) policy coalitions, politicians, planners, engineers, and activists redistributed urban space somewhat towards cyclists and pedestrians. In this process, cycling activists played an important role.

My analysis builds on the environmental activism scholarship, which observes a split between activists' more oppositional and cooperative stances vis-à-vis the state. A key argument in this chapter is that from the start, besides engaging in large street demonstrations and other public protests,

2 Jacqueline Cramer, *De groene golf: geschiedenis en toekomst van de Nederlandse milieubeweging* (Utrecht: Van Arkel, 1989), 9-13; Andrew Jamison et al., *The Making of the New Environmental Consciousness: A Comparative Study of the Environmental Movements in Sweden, Denmark, and the Netherlands* (Edinburgh: Edinburgh University Press, 1990).

3 Ruth Oldenziel et al., eds., *Cycling Cities: The European Experience* (Eindhoven: Foundation for the History of Technology and Rachel Carson Center for Environment and Society, 2016), 11-12.

cycling activists also pragmatically sought cooperation with state officials. The main Dutch cycling activist organization (Cyclists' Union) cooperated with local policymakers to make cities safer for cycling by applying bottom-up user knowledge, for instance about bottlenecks and blackspots. These activists were knowledgeable, media savvy, and capable of establishing a fine-meshed network of local chapters.[4] They engaged with a progressive government that came into power during the period and a bureaucracy willing to experiment in order to make cities more livable. In his study on the role of French NGOs in cycling governance since the 1970s, political scientist Maxime Huré shows how (transnational) networks of interest groups and market parties sought public-private partnerships to stimulate cycling at the local level in France.[5] As I show, activism in the Netherlands was bottom-up and broadly supported. In France, the entrepreneurial rather than non-governmental actors like (advertising) companies and a smaller cadre of lobbyists played a role. Dutch activists had different concerns and tactics, aiming to improve the situation (by redistributing space or traffic calming) for an existing group of cyclists. In France, activism sought to increase the very low cycling levels. Different cycling levels led to different types of cycling activism.

International political science scholarship on activism indicates why this type of cycling activism was more likely to work in the Netherlands and similar small democracies like Denmark. It argues that the Dutch political situation is more favorable to social movement participation than surrounding countries for several reasons.[6] The Netherlands has a low electoral threshold, allowing new parties to gain political influence quickly and forcing established parties to adopt a social movement platform in order to win votes or preempt the foundation of rival parties.[7] This political system was beneficial to cycling activism. A second factor is Dutch professional bureaucracy: civil servants do not change significantly after each election. It guarantees a measure of stability and enables activists to build long-term relationships with officials or participate in the civil

4 Young engineering and architecture students from the major Dutch technical universities or architecture academies in Delft, Eindhoven, Enschede, and Rotterdam were an important part of this movement.

5 Maxime Huré, "Les réseaux transnationaux du vélo."

6 J.W. Duyvendak and R. Koopmans, "Protest in een pacificatiedemocratie. Nieuwe sociale bewegingen en het Nederlandse politieke systeem," in *Tussen verbeelding en macht: 25 jaar nieuwe sociale bewegingen in Nederland*, ed. J.W. Duyvendak, et al. (Amsterdam: SUA, 1992), 39-57.

7 James C. Kennedy, *Nieuw Babylon in aanbouw: Nederland in de jaren zestig* (Amsterdam: Boom, 1995).

service. Lastly, the informal political culture in the Netherlands embodies facilitation, assimilation, and cooperation. There is little tendency towards repression, confrontation, or polarization. In countries like the US, France, and Germany, by contrast, the absence of this consensus-driven political culture makes it much harder for social movements to succeed.[8] All these factors have largely to do with access to politics. While this access proved comparatively easy to achieve, the implementation of actual pro-cycling or car-limiting policies was a slower process which often frustrated activists.

In so-called corporatist democracies, policymakers acknowledge certain groups as representatives of social interests or groups (*corpora*). The preferred mode of interaction between the state and society is through the mediation of these groups to reach a consensus. Such consensus democracies differ from majoritarian democracies where minority interests are given less consideration.[9] Such a consensual tradition can benefit cycling activists who argue that the state has prioritized cars too much at the expense of cyclists and pedestrians. In countries where the balancing of interests is valued and cyclists form sizeable and well represented groups, the conditions of success seem better. This was the case for Dutch activism and seems to apply to Danish or Swiss cycling activism as well.[10] From a governance perspective, in other words, it is probably no accident that cycling action groups were most successful in the Netherlands, Denmark, and Switzerland. The political culture of consensus building in these countries seems to accommodate citizen participation more easily and has produced some of the most cycling-friendly cities in the world.

Corporatist democracies like the Netherlands and Denmark also differ from other countries such as Sweden, in that they still had significant cycling levels when activism emerged.[11] This shaped activism, increased the grounds for recruiting both active volunteers and passive supporters, and gave them access to user knowledge as well as a basis of legitimacy. In many other countries, pro-cycling policies were much more politicized. In

8 Duyvendak and Koopmans, "Protest in een pacificatiedemocratie."

9 Arend Lijphart, *Patterns of Democracy: Government Forms and Performance in Thirty-Six Countries* (New Haven: Yale University Press, 1999); Rudy B. Andeweg, "Consociational Democracy," *Annual Review of Political Science* 3 (2000): 509-36.

10 Emanuel, "Copenhagen: Branding the Cycling City," 89-99; Emanuel, "Making a Bicycle City," 493-517; Mijnssen-Hemmi and Boller, *30 Jahre Veloförderung in der Stadt Zürich, 1975 bis 2005*.

11 Balkmar, "Cycling Politics: Imagining Sustainable Cycling Futures in Sweden," 324-40. Dag Balkmar and Jane Summerton, "Contested Mobilities: Politics, Strategies and Visions in Swedish Bicycle Activism," *Applied Mobilities* 2, no. 2 (2017): 151-65.

the Netherlands, a consensus emerged that cycling is a public good: hence the notion that it is the government's responsibility to provide a safe traffic environment with fast, direct, and comfortable cycling routes, as well as parking facilities. This happened in a context where space remained a scarce commodity: conflicts over distribution of public space are common in the Netherlands like elsewhere. Still, cycling is not framed as a fringe leftist project. And therein lies the remarkable difference between the Netherlands and other, primarily Anglo-Saxon countries.

To sum up, this chapter aims to refine and corroborate the *Cycling Cities* claim about the importance of social movements. It also offers the political history scholarship a particularly compelling "*poldermodel*" case study as explanation. It underlines the uniqueness of Dutch cycling activism in an international context. I do so by using the as yet unstudied Cyclists' Union archives to examine how this key activist organization was structured, how it functioned, and its types of activism, to analyze its cycling governance role in the 1970s and 1980s.

6.1 Early Cycling Activism: Goals and Methods, 1965-1975

Widespread activism marks the period from 1965 to 1975, ranging from countercultural initiatives to environmental protests and peace demonstrations. This included various groups protesting the car's dominant place in society and advocating more space for cycling. It led in 1975 to various groups coming together to establish a new action group for cycling: the Cyclists' Union (*Fietsersbond*). The organization was a federation of local cycling groups, uniting existing traffic, cycling, and environmental protest organizations. It employed a discourse that argued urban cycling was a neglected public good, and experimented with different action repertoires. Much of this had been pioneered by earlier groups who helped found the Cyclists' Union. To better understand its history and later success requires first taking a deeper look into the origins of these groups. They include countercultural movements like Provo and Gnome (*Kabouter*), unorganized (parent) protests around children's safety, Stop the Child Murder (*Stop de Kindermoord*), and alternative planning group the Dooievaar. Each contributed in its own distinctive way to subsequent protest movements: Provo's discourse returned in Stop the Child Murder. This movement's lobbying of national policymakers was efficacious because well-organized local activists and traffic safety protesters succeeded in politically conveying the idea that millions of Dutch citizens supported the movement. And once those groups had put the issue firmly on the agenda, the young Dooievaar architects contributed by developing an innovative form of participative policymaking – or what we today would call co-design. Many Cyclists' Union chapters then were able to use this widely published user-based design process to develop a robust cycling path plan. Armed with such a plan, the Cyclists' Union activists lobbied local – and often progressive – authorities. Despite not always producing results as quickly as the activists wanted, it did set in motion a process of urban transformation that has played out over the last fifty years.

The early countercultural movements already supported cycling in their trenchant critique of car culture.[12] The Provo (1965-1967) and Kabouter movements (1969-1974) criticized capitalism and the materialist consumer culture it had created. The car served as a powerful symbol of all they opposed and car culture had to be provoked. The city had become a place of (large-scale) consumerism rather than a living environment. The car, with its large demands on space and polluting exhaust fumes, irritated these urban activists. Unlike later activists, their anarchist actions were typically provocative and playful. Their early interventions contributed to a mentality change rather than actual transformations of urban space.

Provo activists were among the first to draw attention to traffic safety problems in the Netherlands.[13] Their White Bicycle Plan critiqued the car's polluting and space-consuming role in the city.[14] This was a happening (rather than a plan) that entailed Provo activists painting a number of bicycles white and leaving them throughout Amsterdam free to use. A photograph of John Lennon and Yoko Ono with a white bicycle became iconic. Besides politicizing the bicycle, this also marked the invention of the

12 On Provo and other action groups of the mid-1960s: Roel van Duyn, *Provo. De geschiedenis van de provotarische beweging 1965-1967* (Amsterdam: Meulenhoff, 1985); Virginie Mamadouh, *De stad in eigen hand: provo's, kabouters en krakers als stedelijke sociale beweging* (Amsterdam: Sua, 1992); Kennedy, *Nieuw Babylon in aanbouw*; Niek Pas, *Provo! Mediafenomeen 1965-1967* (Amsterdam: Wereldbibliotheek, 2015); Marko Otten, *Provo: de ludieke opstand tegen bom en regentendom* (Doetinchem: Het Boekenschap, 2017). On the Kabouter movement, see Coen Tasman, *Louter Kabouter: Kroniek van een beweging* (Amsterdam: Babylon-De Geus, 1996).

13 Bicycle historian Zack Furness traces Provo's inspiration to French Situationist thinking of the 1950s. Zack Furness, "Biketivism and Technology: Historical Reflections and Appropriations," *Social Epistemology* 19, no. 4 (2005): 401-17, here 402.

14 Van Duyn, *Provo*, 75-76; Mamadouh, *De stad in eigen hand*, 67-68, 73; Furness, *One Less Car*, 55-58; Fred Feddes and Marjolein De Lange, *Fietsstad Amsterdam: Hoe Amsterdam de fietshoofdstad van de wereld werd* (Amsterdam: Uitgeverij Bas Lubberhuizen, 2019), 58-62.

shared bicycle.[15] Another Provo member, Thom Jaspers – described in the press as the "safe traffic magician" (*veiligverkeermagiër*) – organized traffic safety events in 1965. He was one of the first people to publicly draw attention to worsening safety conditions. Shortly before, Jaspers had witnessed a fatal accident, motivating him to denounce traffic as "a pagan god" to which "we Dutch people sacrifice seven people every day."[16] Jaspers staged a happening at an art gallery in Schiedam near Rotterdam; it involved his Japanese friend and (performance) artist Yoshio Nakajima pouring water and red paint over himself while standing on a wrecked car, to symbolize the blood and tears of traffic victims. Jaspers and others then demolished the car with metal poles.[17] For the artists, the traffic-themed performance was an isolated incident, yet their framing inspired Luud Schimmelpennink's focus on mobility and foreshadowed the traffic safety group Stop the Child Murder's stark language of death. The art performance was rooted in social upheaval in the street. Street accidents skyrocketed in the early 1970s. Anger over traffic accidents reached a critical mass, prompting a broad-based action group.

After the demise of Provo, in 1969 to 1970 a similar movement crystallized: the Kabouter (Gnome) movement – a name indicating small people fighting giant technocratic forces. Far-reaching in its ambition to transform society, the movement's vision of traffic revolved around (free) public transit. With multiple activities in 1970, its most active year, it also supported cycling, like its cycling club (*Kabouterfietsclub*) promoting the cargo bike as alternative to the car.[18] In early 1970, the Kabouters also organized multiple protest rides with regular and cargo bikes.[19] Their key complaint, beyond the excessive use of space, was air pollution; at one demonstration, Kabouters pretended to faint and need oxygen because of car exhaust fumes.[20] The Car

15 Ploeger and Oldenziel, "The Sociotechnical Roots of Smart Mobility: Bike Sharing since 1965," 134-59.

16 Joop Daalmeijer, "We moeten bewuster leven ook met het verkeer," *Het Nieuwe Stadsblad*, May 19, 1965. Jaspers placed a piece of cardboard that he was carrying around under the victim's head. This bloodied cardboard he then wanted to exhibit as a "national monument for traffic victims".

17 "Auto kapot rameien voor 'veilig verkeer'," *Rotterdamsch Parool/De Schiedammer*, May 22, 1965; "Happening voor veilig verkeer Schiedam," *Het Vrije Volk*, May 22, 1965; "Hysterische 'kunstrakkers' slopen Nieuwlandse galerie," *Het Nieuwe Stadsblad*, May 25,1965. Jaspers' interest in traffic safety was short-lived; he later moved to Groningen to join the squatters and peace movements.

18 Tasman, *Louter Kabouter*, 61, 154-58; Mamadouh, *De stad in eigen hand*, 105.

19 Tasman, *Louter Kabouter*, 155.

20 "Zieke kabouter weggebracht: In drie steden acties," *De Volkskrant*, April 13, 1970, 3.

Elimination Service (*Auto-Eliminatiedienst*) organized protests against the car's dominant place in the city in 1970. With sit-ins, they blockaded traffic in busy Amsterdam streets like Leidsestraat, where exhaust fumes and parked cars were a major headache. Despite confrontations with angry drivers and police arrests, the protesters ultimately convinced the city to experiment with closing this shopping street to car traffic.[21] By 1970, the action groups had already managed to extract some concessions from urban government.

In the long run, the countercultural movements' primary contribution to cycling activism was a reframing of urban streets as public places rather than monofunctional car spaces. Across Europe, such movements aimed to make the bicycle what historian Zack Furness calls "a technological embodiment of environmentalism."[22] As elsewhere, the works of urban theorist and activist Jane Jacobs (*Death and Life of Great American Cities*, 1961), economist Ernst Schumacher (*Small is Beautiful*, 1973), and social critic Ivan Illich (*Energy and Equity*, 1974) inspired these activists.[23] The car symbolized everything that was wrong with capitalism; the logical alternative, the bicycle, its opposite of a different economic and ecological system. Protests like the Gnome movement's in Amsterdam introduced the idea of making certain streets car-free and giving cycling and public transit a larger role in urban mobility. Their successor groups incorporated these key values: self-governance, livability, the small (neighborhood) scale, walking and cycling, opposition to bureaucracy, unrestricted economic growth, and the redevelopment of inner cities to accommodate big businesses and cars.[24] In short, these urban social movements reframed the bicycle as a green symbol. They promoted cycling as an antidote to many of the problems plaguing the city.

These groups' relationship with the government was more confrontational than the ANWB had ever been. It was also more contentious than later action groups. Even so, some relationships existed. Policymakers in many left-leaning cities shared the activists' concerns, if not their style and discourse.[25] While cycling activists since the mid-1970s had found it beneficial to work with the government, such dialogue was highly controversial within the

21 Tasman, *Louter Kabouter*, 157-58.

22 Furness, *One Less Car*, 59.

23 Ibid., 66. A translation from a section of *Energy and Equity* appeared in a 1977 edition of the Cyclists' Union journal: Ivan Illich, "Gradaties in het voortbewegen op eigen kracht," *De Vogelvrije Fietser* 2, no. 6 (1977): 18-19.

24 Geert Mak, "Anatomie van een romantische stadsrebellie," *NRC Handelsblad*, October 3, 1992.

25 Kennedy, *Nieuw Babylon in aanbouw*, 121.

Provo and Kabouter movements, which had been based on oppositional and anarchistic traditions. Most preferred provocative, extra-parliamentary activism. Nevertheless, the Kabouter movement scored successes in Dutch cities, winning four seats in Amsterdam's 1970 municipal elections. In the city council, Kabouter Roel van Duyn, a former history and political science student, submitted numerous policy proposals. Sociology student Henk Bakker, a key member of the group The Obstinate Amsterdammer (*De Lastige Amsterdammer*) proposed making the entire city center car-free. He became disillusioned with the Kabouters' playfulness and accused them of lacking "ready knowledge" and being unwilling to acquire expertise.[26] In turn, activists accused The Obstinate Amsterdammer of being "alternative technocrats," pointing to a split over the extent of government involvement.[27] The city council refused to discuss many of the Kabouters' unorthodox policy ideas.[28] Still, they became part of the municipal council and interacted with establishment politicians on a regular basis. If radical groups could work with urban government to this extent, more pragmatic groups had good prospects of establishing a working relationship with local engineers and policymakers.

The anger over the negative effects of automobilization was not limited to Amsterdam or specific action groups. Children everywhere in the Netherlands tended to cycle rather than walk to school, riding on or crossing highly dangerous roads. By the late 1960s, these roads had the same high car levels as other Western countries. Where separate cycling paths were lacking, or crossings were unprotected, this meant danger for cycling children. Their parents in cities, towns, and villages all over the country shared a widely expressed anger over worsening traffic safety and took to the streets to protest the lack of cycling infrastructure and safe conditions. The dangerous journeys to school sometimes led parents to protest by keeping children at home. In 1963, after a twelve-year-old boy died, hit by a driver at a dangerous major crossroads near the harbor city Rotterdam, parents refused to let their children go to

26 Lisette Lewin, "Lastige Amsterdammer Henk Bakker: 'Alles is te regelen als eerst die auto maar weg is'," *Algemeen Handelsblad*, March 18, 1970. While broadly accurate, there were exceptions. Notably, the Rotterdam Kabouter Pax Kroon wrote a serious report on alternative transport in Rotterdam in 1970. Berkers et al., *Cycling Cities: The Rotterdam Experience*, 41. "Rotterdam moet kiezen: auto of mens," *NRC Handelsblad*, October 31, 1970, 20.

27 Jitske Bosma et al., *"... de beste aktiegroep ter wereld ...": 40 dorpsverhalen uit de Nieuwmarkt* (Amsterdam: Stichting Uitgeverij De Oude Stadt, 1984), 59. While the squatter movement was more confrontational in later years, it nevertheless maintained regular contact with city officials. Herman De Liagre Böhl, *Amsterdam op de helling: De strijd om stadsvernieuwing* (Amsterdam: Boom, 2010).

28 Tasman, *Louter Kabouter*, 313-14.

school.[29] In 1968, parents in Holten also refused to allow their children to attend school after a sixteen-year-old girl died on a dangerous road. Until cycling paths came along, people demanded a lower speed limit and a ban on overtaking.[30] A group of 100 six-year old school children protested in Amsterdam in 1970 against cars parked on the pavement.[31] In the same year, 400 children in Hilversum blocked the ring road, demanding a cycling path to their school, while in the village of Stompwijk, parents threatened to block roads with tractors if their children did not get a safe cycling path to schools in suburban Zoetermeer.[32] In 1972, parents in border town Doetinchem kept their children at home when the subsidized school bus was abolished. The parents considered the local roads too dangerous to allow children under ten to cycle.[33] In 1974, parents in a Noord-Holland village kept their children at home in protest against the dangerous school route; after an accident, two children had ended up in hospital with serious injuries.[34] In 1975, concerned parents wrote to Utrecht's city government about the children's school route along the busy Amsterdamsestraatweg.[35] In 1976, seventy schoolchildren spontaneously blocked a dangerous provincial road in Friesland the day after a child was killed in a traffic accident.[36] The construction of separate cycling paths was still not complete in provinces like Friesland or Drenthe, leading to dangerous situations for schoolchildren.[37] These protests took place across the country: anger over the car spilled over into the streets.

29 "Gevaarlijk punt in Bergschenhoek eist wéér mensenleven," *Het Vrije Volk*, July 9, 1963, 1.

30 "Kinderen blijven thuis: Telegram aan minister, motie aan de raad," *Tubantia*, February 22, 1968, 5. The dangers of cycling in the 1960s are well illustrated by the fact that this road was part of the international route to Germany until a highway was constructed south of Holten from 1971 onward. "Rijk legt weg aan naast E 8 bij Holten," *Volkskrant*, February 22, 1968, 11.

31 "'Weg met auto's': 100 kleuters demonstreren," *De Telegraaf*, April 21, 1970.

32 "Amsterdams Dagboek: Notities," *Het Parool*, September 7, 1970, 2.

33 "Ouders houden hun kinderen van school uit protest," *Tubantia*, September 27, 1972, 7. The conflict dragged on for ages; the municipality considered cycling safe enough, the parents did not, since their children had to cross two busy streets which were not adequately protected: "Bus-stop, school-stop," *Algemeen Dagblad*, November 1, 1972, 14.

34 "De Noord heeft schoon genoeg van de onveilige weg naar school," *Het Parool*, April 23, 1974, 9.

35 HUA 1338 (Gemeentebestuur van Utrecht, 1970-1989), inv. no. 12559, letter E. Graafstal-Paasche (parents' committee) to mayor and aldermen Utrecht, December 15, 1975. This busy shopping street had a lot of bus and truck traffic which had to stop or park at the side of the road, crossing the cyclists' path.

36 "Politie steekt stokje voor blokkade Marnezijl door schooljeugd," *Leeuwarder Courant*, September 3, 1976, 15.

37 Gert van Engelen, "'Je ziet hier geen dikke schoolkinderen'," *De Vogelvrije Fietser* 6, no. 1 (1980): 16-17 and 37; Gert van Engelen, "'Actie is hier een vies woord'," *De Vogelvrije Fietser* 8, no. 6 (1983): 10.

Figure 24 Primary school children cycled to their local town hall to protest cycling conditions in the Haarlemmermeer region in 1978, carrying a slogan "We'd better go by boat, because cycling you end up in a moat" ("We kunnen beter met de boot want fietsend kom je in de sloot"). Narrow country roads like these along waterways became increasingly dangerous when cars dominated traffic. [Source: Nationaal Archief, Bert Verhoeff/Anefo, no. 929-7231]

This local activism revolved around the claim that safe cycling for children was a public good which the local and national government failed to provide. The gradual and implicit reframing of the street as an exclusive space for cars had to be undone.[38] As one schoolteacher in Noord-Holland province commented: "every child has the right to protection, and so do little five-year-olds who ride their little bicycle to school every day on the busy Middenweg where there are no cycling paths."[39] The lack of cycling infrastructure everywhere had endangered cyclists since the introduction of the car. The fast rise in car ownership combined with the critical mass of cyclists made this danger even more acute: as commuters switched to driving, those who had to keep on cycling objected to the worsening conditions. It was the state's duty to protect its cycling citizens against this danger.

The step from citizen protests to reconstructing the streets required a commitment from both activists and government officials, the protesters

38 Norton, "Street Rivals: Jaywalking and the Invention of the Motor Age Street," 331-59; Norton, *Fighting Traffic*; Peter Norton, "Four Paradigms: Traffic Safety in the Twentieth-Century United States," *Technology and Culture* 56, no. 2 (2015): 319-34.

39 "De Noord heeft schoon genoeg van de onveilige weg naar school," 9.

believed. In 1978, for example, a group of primary school children held a cycling demonstration in Haarlemmermeer to protest drivers' behavior and demand more cycling paths on the narrow country roads (see figure 24). The chief motivation was to ensure (rural) children could cycle safely to high schools in towns. The group cycled to the town hall in Hoofddorp to hand over a petition and a bicycle they had dredged from a canal, symbolizing the apparently regular occurrence of school children being forced off the road into the water by passing cars.[40] The pressure from citizens led to the promise of a municipal cycling path plan, but funds were lacking to actually build it. The activists knew how the underlying link between finance and governance worked. As a letter from Nieuw-Vennep's community council (*Dorpsraad*) to parliament shows: "for the funding of this plan, the municipality is of course almost entirely dependent on higher authorities."[41] The activists also knew that their province Noord-Holland received 2.5 million guilders in cycling subsidies in 1979 and would receive 4 million in 1980 and 1981, to be distributed among the municipalities. The pittance that would remain for the Haarlemmermeer area made "realization of these plans a hopeless matter," they wrote.[42] So the activists wrote to parliament requesting extra funds. Multi-level government relations complicated citizens' simple demands. Local residents sometimes understood, but were frustrated by the bureaucratic complexities standing in the way of quick solutions.

Both the Provo rhetoric and the broad social unrest over unsafe school routes led to a high level of organization. Stop the Child Murder (*Stop de Kindermoord*), officially established on March 22, 1973, had been active a year earlier. In 1972, L. Rutgers-Naaijkens, a mother from Eindhoven campaigning for safe school routes, together with other concerned parents, approached Vic Langenhoff, a journalist for *De Tijd*, who covered environmental issues and traffic safety.[43] Personal experience motivated Langenhoff: his youngest daughter, six years old, had been killed in an accident in 1971; not much later, another of his children and her ten-year-old friends were hit by a car, this time "only" resulting in concussion.[44] Langenhoff's article received many

40 "Fietsdemonstratie naar Hoofddorp," *Trouw*, May 16, 1978, 12; "Fietsers willen paden," *De Telegraaf*, May 18, 1978, 7.

41 NA 2.02.28 (Archief Tweede Kamer), inv. no. 6270, letter Dorpsraad Nieuw-Vennep to Vaste Kamercommissie voor Verkeer en Waterstaat, February 5, 1979.

42 Ibid.

43 NA 2.19.198 (Archief Stop de Kindermoord), inv. no. 33; Huib Goudriaan, "Een kind blijft ook in het verkeer een kind," *Trouw*, October 21, 1972, 21.

44 Vic Langenhoff, "Pressiegroep Stop de kindermoord" *De Tijd*, September 20, 1972, 9; Joop van Schie, "Stop kindermoord op de wegen," *De Volkskrant*, October 11, 1972, 11.

reactions: parents were now so afraid when their children came home late from school, that he compared it to a "war psychosis."[45] Mothers were particularly active in this protest group.

The group's strong language was deliberate: they spoke consistently of child murder.[46] It was not implying that car drivers killed children intentionally. Rather, it believed the government was not doing enough to address the high annual fatalities. In this sense, traffic casualties were "premeditated" (*met voorbedachte rade*).[47] The government, in its apparent acceptance of this sacrifice, was guilty of murder. Founding mother Rutgers-Naaijkens told Langenhoff in 1972 that for children, the Netherlands was one of the least safe countries in Europe: "In 1970, 8,614 children up to fifteen years old were injured, 3,772 of them pedestrians, 2,709 cyclists. Society accepts this as a kind of natural disaster – what is worse: we don't put up with natural disasters but come up with a Delta plan [the nation's mega-project against flooding] for them."[48] Rutgers-Naaijkens addressed this apathic government response locally by organizing teams of parents to safeguard crossings. She also tried to lobby local government – not always easy because they did not take her seriously: if she requested official data, local officials "wanted to subject her to a kind of exam to test her knowledge of traffic issues," Rutgers-Naaijkens told Langenhoff.[49] Activists accepted this challenge and became experts in their own right.

Representing the broad critique against technocratic governance, Stop the Child Murder distanced itself from the well-embedded Traffic Safety interest group (*Veilig Verkeer Nederland,* VVN), similar to how the Cyclists' Union would later square up against the well-established tourist organization ANWB. Like the ANWB, the highly institutionalized VVN collaborated closely with – and was even accused of being "coopted by" – the government. With an annual government subsidy of 4 million guilders, it could employ a staff of eighty-five.[50] Stop de Kindermoord accused the safety group of only treating the symptoms, not the root cause: the car had been given so much space. Consequently, the number and speed of cars were in direct contact with

45 Van Schie, "Stop kindermoord op de wegen."

46 German citizens also wrote about the car in terms of violence and death: Jens Ivo Engels, *Naturpolitik in der Bundesrepublik: Ideenwelt und politische Verhaltensstile in Naturschutz und Umweltbewegung, 1950-1980* (Paderborn: Ferdinand Schöningh, 2006), 385-86.

47 Ibid.

48 Vic Langenhoff, "Waarom laten wij kinderen bij honderden van de weg rijden?," *De Tijd*, September 20, 1972, 9. The Delta Plan refers to the large engineering works constructed in the aftermath of the devastating 1953 floods.

49 Ibid.

50 Henk Bakker and William Rothuizen, "'Elke dode die er niet valt is er voor ons één'," *Haagse Post* 59, no. 51 (1972): 6.

cyclists and pedestrians.[51] The group blamed the VVN for failing to address the danger drivers posed to children and instead focusing on introducing periodic vehicle inspections or seat belts. Stop the Child Murder labelled VVN "hostile to children."[52] Its activists believed the system needed to change radically.

The Stop the Child Murder organization played a crucial role in lobbying for traffic calming and cycling infrastructure as the key means to improve traffic safety for children. They expanded on the concept of traffic-calming measures for residential areas that had been emerging in urban planning circles, the *Woonerf*, pioneered around 1960 by urban planner Niek de Boer in the town of Emmen. Cyclists and pedestrians needed completely car-free routes when leaving their neighborhood to go to school. In 1974, when the village of Nuland in Noord-Brabant planned a road construction without cycling paths, Stop protested that this was an "almost undreamed-of" plan when so many stakeholders were insisting on separate cycling paths. The reconstruction's primary goal should be "the protection of the non-motorized, especially children; which of course can only be done by guaranteeing car-free cycling routes."[53] In October 1972, Stop the Child Murder wrote to Parliament's Permanent Committee for Public Works, setting out its three-phase demands: first, immediate police protection at intersections for school children and free school buses for children who had no safe walking or cycling route to school; second, more separate cycling and pedestrian facilities, traffic lights, and speed bumps at every intersection, plus a ban on mopeds faster than 25 km/h on cycling paths; third, each municipality should complete a network of pedestrian and cycling facilities, connected to those in nearby communities. The group also demanded that the government financed these from the highways budget.[54] At a national level, Stop the Child Murder lobbied for coordination between various ministries to achieve a more efficient and uniform cycling policy.[55] This petition, as discussed in the next chapter, failed. By the late 1970s, the parent group collaborated with the newly established Cyclists' Union (1975) and the older Pedestrian

51 NA 2.19.198 (Archief Stop de Kindermoord), inv. no. 232, document Vic Langenhoff, "Stuk nr 1 voor rubriek lezers in VARA-gids dd 27 jan."; inv. no. 236, document "'Stop de Kindermoord' mars naar Den Haag."

52 NA 2.19.198, inv. no. 233, letter SdK to Council of Ministers, May 14, 1973. See also Paul Kouwenberg, "Veilig Verkeer Nederland houdt niet van kinderen," *De Volkskrant*, May 12, 1973, 27.

53 NA 2.19.198, inv. no. 233, letter SdK to city council Nuland, February 18, 1974.

54 NA 2.19.198, inv. no. 233, letter SdK to Permanent Committee for Public Works, October 11, 1972.

55 NA 2.19.198, inv. no. 233, letter SdK to Minister of Public Works, January 13, 1974; letter SdK to Minister of Public Housing, January 13, 1974

Figure 25 The dangerous situation for children playing in the street or cycling to school was a powerful motivation for civic activism. Protest group Stop the Child Murder delivers a painting to Public Works minister Tjerk Westerterp on February 6, 1974, illustrating that politicians were receptive to their demands. [Source: Nationaal Archief, Bert Verhoeff/Anefo, no. 926-9884]

Association (1953) to promote lowering the maximum speed in built-up areas from 50 km/h to 25 or 30 km/h. Somewhat to their pleasant surprise, in 1983 the ministry of Public Works introduced legislation to that effect.[56]

Because of the widely shared perception that traffic had become unacceptably dangerous by the early 1970s, Stop found easy access to policymakers. Here social anger aligned with activist work and government concern. Former chair Maartje van Putten, at the time a young Amsterdam mother and a founder of the organization, recalled that on their multiple visits to Parliament, the protesters were received by MPs (see figure 25).[57] On another occasion, the group organized a bicycle tour to Prime Minister Joop den Uyl's home, presenting to him and his wife – they had seven children – a brochure protesting the dangers for children playing outside. Stop the Child Murder believed in and was adroit in generating media attention for its public actions.

56 Interview Steven Schepel, November 13, 2019. The cooperation operated under the heading "Comité 50 is teveel" (Committee 50 is too much): "Voetganger in actie," *Algemeen Dagblad*, November 3, 1979; Han Van den Berg, "50 is teveel, vindt comité," *Het Vrije Volk*, August 29, 1979.
57 NTR/NPO Radio 1, *De Spaak*, podcast audio, Stop de Kindermoord, August 1, 2019. https://www.nporadio1.nl/podcasts/de-spaak [accessed September 2, 2019]; Interview Maartje van Putten with author, June 6, 2020; "Kind moet veilig kunnen leven," *Het Parool*, January 20, 1973.

At the same time, the group recommended to other activists in its *Manual for Participation and Action* (1975) that they establish close contacts with local policymakers. With "strong, well-formulated arguments" one could influence politics, and "diverse contacts" with politicians could give early access to policy proposals.[58] It believed personal relationships with politicians were beneficial and effective. Like young architects' Dooievaar's citizens guide (1974), Stop the Child Murder explained in this manual what happened when an action group addressed specific institutions – warning, for instance, that a letter to the mayor and aldermen would probably never reach the council members' desks.[59] While protesting on the street and building public support for traffic safety, the groups also explored how to influence local politics early on – and were surprised how seriously policymakers took them. Some demands were quickly adopted. As Chapter 7 shows, in response to protests, several MPs championed the traffic safety issue in the 1970s. The pressure that citizens and politicians put on engineers was considerable and successful. Because so many people, including children, rode a bicycle in the Netherlands, traffic safety was a particularly pressing political problem.[60] By the mid-1970s, politicians and civil servants were receptive to their requests after a decade of Provo and Stop the Child Murder-led activism.

Another crucial but insufficiently recognized influence on the Cyclists' Union is the Dooievaar group. Concerned about established urban planning practices, four young architecture students, Hans van Beek, Leo Hamer, Jan Ledderhof, and Arij van der Stelt, founded Dooievaar in the city of The Hague in 1972.[61] The group's name was a play on words: contracting the Dutch words for "dead" and "stork" – the white stork is the symbol of The Hague.[62] The

58 Stop de Kindermoord, *Handleiding voor Inspraak en Aktie* (Amsterdam: Pressiegroep Stop de Kindermoord, 1975), 19.

59 Ibid., 21-22.

60 In contrast, protests against the car in UK cities like Birmingham targeted air pollution caused by congestion rather than safety. Indeed, a serious issue, but as the negative effects were less immediately visible and pressing, it was a less potent reason for activism. According to Simon Gunn, citizen activism against the car mattered, but equally factors like economic recession and rising oil prices led to a revaluation of the car's place in British cities: Simon Gunn, "Ring Road: Birmingham and the Collapse of the Motor City Ideal in 1970s Britain," *The Historical Journal* 61, no. 1 (2018): 227-48; Norton, "Street Rivals"; Norton, *Fighting Traffic*; Norton, "Four Paradigms."

61 Around this time, young architecture students around the world began promoting a new mode of urban planning, influenced by Jane Jacobs and Jan Gehl among others. For instance in Munich: Karl Klühspies, *München nicht wie geplant: Stadtpolitik, Bürgerwille und die Macht der Medien* (München: Münchner Forum, 2015).

62 On a symbolic march, the group carried a stuffed stork as if they were about to bury it (Interview Hans van Beek and Leo Hamer, The Hague, February 5, 2019). On the cycling history of The Hague, see Berkers, Botma, and Oldenziel, *Cycling Cities: The Hague Experience*.

students liaised with Eisse Kalk, an advocate for democratization in politics from the action group Werkgroep 2000, who had studied social geography and political science.[63] A group of like-minded thinkers from progressive Christian circles founded this group in 1965 in order to increase citizen participation in (local) politics.

The emphasis on grass-roots politics fits in a larger international context of technocracy critique and participatory democracy, as we see with the Students for a Democratic Society in the US and the wider '68 movement worldwide. Werkgroep 2000 and Dooievaar, echoing this critique of technocratic rule, believed that behind-the-scenes expertise and the executive's discretionary power made it impossible for legislators to exercise true democratic control. To counter this tendency, Kalk asserted in 1971, "action groups have taken over parliament's monitoring task."[64] Local officials prepared plans in a non-transparent way, only allowing participation once most decisions were done and dusted – only details could be tweaked.[65] In 1973, the municipality of Rijnmond paid Kalk's organization to supervise citizen participation in devising its regional plan (*streekplan*). Earlier, Werkgroep 2000 had worked in Helmond. Kalk appreciated these local policymakers' efforts to involve citizens but disagreed about the procedure – citizens should not just be able to comment on already developed plans, but should be included earlier on in the process for an open discussion about a plan's underlying assumptions and objectives. These implicit aspects, often couched in technical terms, hid fundamental views about society, planning, and mobility, he believed. In short, Kalk wanted to politicize local politics, ensure the trade-offs and priorities were visible for everyone, and utilize citizens' knowledge about their everyday living environment in plans to improve that environment.[66]

Dooievaar applied these grass-root political ideas to design and planning. As staunch defenders of citizen participation, Dooievaar demanded an end to what they considered "authoritarian plan making" in The Hague.[67] According to the four students, architects and planners should be more

63 The 4 students thesis project – the social responsibility of architects and planners – was inspired by Kalk.

64 H. Smits, "'Inspraak': nog minder dan meedenken," *NRC Handelsblad*, June 12, 1971, 7. See also E. Kalk, "Verkeerd verkeer en ruimtelijke wanorde," in *Milieu-aktie in Nederland*, eds. E. Tellegen and J. Willems (Amsterdam: De Trommel/Vereniging Milieudefensie, 1978).

65 "Burger wordt dupe van groeibeleid," *Trouw*, November 27, 1972, T12; N.H.J. de Bruijn, "Wie mee wil praten moet het nu vertellen," *NRC Handelsblad*, February 5, 1973, 8.

66 Paul Kouwenberg, "Inspraak-begeleider Eisse Kalk: Democratie verbeteren kost tijd," *Volkskrant*, March 8, 1975, 19.

67 Werkgroep Dooievaar, "Weg fiets? Nee fietswegen!," *Tijdschrift voor Verkeerstechniek* 24, no. 11 (1973): 552-53.

politically and socially engaged, instead of blindly executing the planning and mobility vision of a small group of policymakers. The engineering ethos should be "a process of planning preparation with as much user/resident influence as possible."[68] Current Dutch transport planning was, in short, "one-sidedly technical."[69] Citizens had to be involved.

Echoing the critique of American urban planner Jane Jacobs, the activists claimed that car-centered planning made cities unlivable and cycling an unattractive option and feared that The Hague would become an American-style car city.[70] City officials devised policies without consulting citizens, resulting in car-dominant planning at the expense of space for walking, cycling, and parks, and presented such options as only technical and politically neutral. They were not. Dooievaar criticized The Hague's city planners for presenting improved cycling routes as technically impossible and obscuring the fact that the car was an ideological choice.[71] Everywhere one looked in the city, cars had priority and the shortest route. Cyclists had to cope with many detours, barriers, and long waiting times, not to mention the danger of cycling in a car city.[72]

Dooievaar urged the city government to ensure citizen participation earlier on in the decision-making process. Because many citizens were cyclists, whose voices went unheard, the activists expected this process would generate a more cycling-friendly city, and by example encourage local Cyclists' Union activists. Dooievaar activists generated attention for the cyclists' ideals, and having established a public platform, set about devising alternative traffic plans.

As a first step they grabbed the attention of the politicians and the media. In June 1973, 1,500 protesters attended a Dooievaar demonstration.[73] The attendees included MP and former Minister of Public Works Willem Drees Jr., as well as The Hague city councilor Nuy and many council members.[74] Excellent contacts with journalists helped the activists to publicize

68 GADH, Archive Dooievaar (0772-01), inv. no. 3, letter E. Kalk to F.J. Smits, July 20, 1972. Kalk (of Werkgroep 2000) explained to Smits, director of the Rotterdam Architecture School, why the students' unorthodox project was relevant for the profession. Kalk and Werkgroep 2000 were increasingly involved in local policymaking experiments and formed an early partnership with Dooievaar.

69 Werkgroep 2000 in association with Werkgroep Dooievaar, "Fietser… kop op!," *Katernen 2000* 1974, no. 9-10 (1974): 1-29, here 6.

70 Interview Author with Hans van Beek and Leo Hamer, February 5, 2019; Furness, *One Less Car*, 53. Jacobs preferred the term vitality. See H. Kaal, "A Conceptual History of Livability: Dutch Scientists, Politicians, Policy Makers and Citizens and the Quest for a Livable City," *City* 15, no. 5 (2011): 532-47, here 537.

71 Werkgroep 2000 in association with Werkgroep Dooievaar, "Weg fiets? Nee fietswegen!"

72 "Dooievaar wil Hagenaar weer laten fietsen," *De Volkskrant*, February 24, 1973, 39.

73 "Protest tegen slechte klimaat voor fietsers," *NRC Handelsblad*, June 12, 1973, 10.

74 Ibid.

their ideas.[75] The group's exhibits and events informed the general public about the harmful effects of government plans. They worked together with a progressive group of Leiden University environmental biologists to (successfully) block the road construction between Leiden and The Hague. Here they practiced their democratization ideal by informing contracted road workers that some of their own homes would be demolished for the building project.[76] Dooievaar knew how to build public support for its ideas.

Once the Dooievaar activists had built support for their goals and established contact with the city government, they preferred pragmatic and expertise-based methods. A shining example is how they designed a cycling plan for The Hague. In 1972, the students had criticized the city planners' "Exploratory Memorandum for Slow Traffic" for containing too few robust measures for cyclists, and being narrowly inspired by traffic safety concerns that "in fact emphasize cycling's subordinated position."[77] Even if they made cycling safer – something they seriously questioned – the proposed measures would also make it less attractive, despite the fact that promoting cycling was in the best interest of cities' living environment.[78] When The Hague authorities deemed a cycling route network technically impossible, the activists said the assessment was inaccurate; all that was required was to upend their car-centric priorities.[79] Ambitious cycling plans could be drawn up and implemented if policymakers chose to do so.

To prove an alternative was feasible, the activists devised a step-by-step plan, published in the form of a manual (see figure 26).[80] This plan demonstrated how to draw up an ideal cycling route network with the input of ordinary cyclists. In 1974, Dooievaar applied this to The Hague, before publishing the

75 GADH Archive Dooievaar (0772-01), inv. no. 4 contains many articles on Dooievaar by journalist Jek Kleijn. Similarly, journalist Lisette Lewin gave the Amsterdam action group De Lastige Amsterdammer multiple long interviews in *Algemeen Handelsblad* and later *NRC Handelsblad*: Lewin, "Lastige Amsterdammer Henk Bakker,"; Lisette Lewin, "Fielt," *NRC Handelsblad*, November 11, 1972; Lisette Lewin, "Henk Bakker: metro is Lammers' struikelblok," *NRC Handelsblad*, November 20, 1972; Lisette Lewin, "Amsterdammer is weer lastig: metro geen comfort," *NRC Handelsblad*, January 26, 1973.

76 Ibid. On the Leiden protest: H. Udo de Haes and W. ter Keurs, "De Leidse Baan," in *Milieu-aktie in Nederland*, eds. E. Tellegen and J. Willems (Amsterdam: De Trommel/Vereniging Milieudefensie, 1978).

77 GADH Archive Dooievaar (0772-01), inv. no. 5, 'Samenvatting studie 1970-1972', 20.

78 GADH Archive Dooievaar (0772-01), inv. no. 5, letter Dooievaar to The Hague city council, November 1, 1972.

79 GADH Archive Dooievaar (0772-01), inv. no. 5, attachment to letter Dooievaar to The Hague city council, November 1, 1972.

80 In 1975, Stop de Kindermoord also published a brochure with tips and best practices for activists: Stop de Kindermoord, *Handleiding voor Inspraak en Aktie*.

Figure 26 Making alternative plans, based on input and expertise from everyday cyclists, was an important action strategy for activist group Dooievaar and the Cyclists' Union. Dooievaar founding members (left to right): young architects and urban planners Leo Hamer, Hans van Beek, Ary van der Stelt and Johan Ledderhof, working on a cycling path plan for the city of The Hague [Source: Photographer: Stokvis, Gemeentearchief Den Haag, no. 1.53120]

guide in Werkgroep 2000's magazine *Katernen 2000*.[81] They recommended that activists first indicate on the map the main work, recreation, and living areas, and secondly determine which main cycling routes were required between these functions. The third step was to map those main routes onto actual streets, creating a network on paper. Next, an article in the local newspapers would elicit feedback from cyclists. In The Hague, their tactic was very successful: the activists' plan garnered many responses from "ordinary [*doodgewone*] cyclists who turn out to be knowledgeable about this because they cycle themselves."[82] With this user input, the fifth step was to create a final plan,

81 Werkgroep 2000 in association with Werkgroep Dooievaar, "Fietser… kop op!" A digitized version can be accessed at https://fietsstadamsterdam.nl/wp-content/uploads/2019/05/1974FI1.pdf. Dooievaar also made a video demonstrating their process. See https://haagsgemeentearchief.nl/mediabank/beeldcollectie/detail/dd534bb6-9178-4e5e-b193-4882ce95ce54/media/55bcf303-e594-5da2-8a9f-3bb898ad7f00?mode=detail&view=horizontal&q=Dooievaar&rows=1&page=1&fq%5B%5D=search_s_entity_name:%22Film%20%5C%26%20Video%22&sort=order_i_sort_date%20desc [accessed September 4, 2019]

82 "Het probleem van de fiets: het opbergen," *De Tijd*, February 24, 1973.

including technical drawings of the ideal cycling path. This report then served to enter into dialogue with local policymakers. Dooievaar included a detailed breakdown of city bureaucracy and politics' inner workings. Activists were often young, not well-connected, and inexperienced in politics.

Though the young architects were not experienced politically, they soon caught up. In contrast with the experienced and well-connected figures in the ANWB or the cycling path organizations, who knew how to make a case for state funding, the Dooievaar activists had to learn this "game" from scratch. Inviting local policymakers for an "informal conversation" was encouraged, although Dooievaar also warned activists should not be naive in dealing with politicians, who knew how to apply "delaying tactics."[83] This gave other activists valuable pointers on how to engage the local citizenry and politicians in a conversation about better cycling facilities. The final result – cooperation between educated traffic planners like the Dooievaar activists and cyclists – had to be a cycling route plan based on the knowledge of both users and experts.[84] City officials transformed Dooievaar's plan into technical drawings that a council committee reviewed positively.[85] If the plan was not implemented immediately, Dooievaar accused the city of delaying tactics.[86] Nevertheless, in 1975 the council selected two city-center streets for experiments with traffic calming and cycling lanes, showing Dooievaar's good relations with the city.[87] Its ideas were taken seriously, for instance after the council's majority supported Dooievaar's alternative plans for a new tramline bridge. One dissenting councilor nevertheless praised the activists for producing "playful insights" as alternatives.[88] In short, enough overlap existed between Dooievaar's goals and ideas and a section of urban politicians for them to remain in conversation.

The level of activists' collaboration with authorities – seen as selling out by the cause – continues to be an issue in the historiography. In their book on Amsterdam's cycling history, Feddes and De Lange recently characterized Dooievaar's activism as "depoliticization," an expertise-based approach suggesting concrete improvements but eschewing open confrontation or

83 Werkgroep 2000 in association with Werkgroep Dooievaar, "Fietser... kop op!" 27.

84 The emphasis was on cycling paths. As architects, The Hague's activists were also very interested in cycling parking, both at home and at public venues (Interview Hans van Beek and Leo Hamer, The Hague, February 5, 2019); "Het probleem van de fiets: het opbergen," 3.

85 GADH Archive Dooievaar (0772-01), inv. no. 10, letter Dooievaar to The Hague city council, November 23, 1975.

86 "Groene Eland en Dooievaar vormen reddend tandem," *De Posthoorn*, January 10, 1974.

87 "Den Haag neemt proef met fietsroute," *Trouw*, June 17, 1975, 6.

88 "Dooievaar krijgt steun voor alternatief: PvdA vraagt raad uitstel 'Koekamp'," *Trouw*, April 17, 1975, 6.

a fundamental system critique.[89] The discussion above suggests this is not quite correct. The emphasis on democratization and increased participation in planning procedures confronted an older way of doing politics. What is more, Dooievaar's bone of contention was the depoliticization of local planning documents, where planners followed "facts" without considering their desirability or other potential policy goals. Dooievaar believed in a co-design process for city planning aimed at consensus through deliberation, with the city's entire population explicitly discussing policy goals and trade-offs. Its tactic formed an influential blueprint for Cyclists' Union activism.

6.2 User Expertise and Cycling Infrastructure: Cyclists' Union Activism, 1975-1985

The activism of the early 1970s culminated in 1975 in the foundation of the Cyclists' Union amid a politically favorable context. A number of smaller political parties had already tried to put democratization and the environment on the agenda in the late 1960s and early 1970s: the Pacificist Socialist Party (PSP, 1957-1991), Democrats 66 (1966-), and Radical Political Party (PPR, 1968-1991).[90] These young political parties also influenced the ideas of the best-established and largest social democrat party, the Labor Party (PvdA). From 1973 to 1977, the Den Uyl government was the most progressive coalition government to date. Its central promise was a fairer "distribution of income, knowledge and power."[91] Faced with the typical coalition politics of the Netherlands, and the worsening economic situation following the 1973 Oil Crisis, this government could not implement its program remotely to the extent it had envisioned.[92] Nevertheless, this was a ruling coalition receptive to the ideas of cycling activists. Promoting cycling fitted the Den Uyl promise to give more space to neglected and underserved groups like cyclists.

These cyclists were represented by one action group when, after a decade of grassroots politics, activists took the initiative to create the national cycling platform, the Cyclists' Union (*Fietsersbond*) in 1975. Importantly, its original

89 Feddes and De Lange, *Fietsstad Amsterdam*, 73.

90 The PSP and PPR merged with the Dutch Communist Party and Christian left EVP party into Groenlinks.

91 Remieg Aerts, Herman de Liagre Böhl, Piet de Rooy and Henk te Velde, *Land van kleine gebaren: Een politieke geschiedenis van Nederland 1780-1990* (Nijmegen/Amsterdam: Uitgeverij SUN, 1999), 310.

92 Ibid., 309-12.

name was intentionally oppositional. Initially founded as ENWB (*Eerste Enige Echte Nederlandse Wielrijders Bond*, "First Only Real Dutch Cyclists' Union"), the activists deliberately adopted this acronym for its similarity with the existing tourist organization ANWB. The name not only pointed the finger at rival ANWB for neglecting its original role as cycling club, it was also designed to goad the ANWB into taking legal action, creating media exposure, and "because possible legal difficulties with the ANWB would enable the ENWB to clarify what it was all about."[93] In 1979, the name became ENFB in response to pressure from the ANWB before eventually settling on Fietsersbond, or Cyclists' Union, the name I will use throughout. The 1970s Dutch cycling activism came from a new generation and a new cycling organization.[94]

Table 5 The initiators of the Cyclists' Union [Fietsersbond]

Name	Founding Year (approx.)	Scope	Main action goal
Werkgroep 2000	1965	National	Democratization
Vereniging Milieudefensie	1971	National	Environment
Nivon	1960	National	Environment
Strohalm	1970	National	Environment
Stop de Kindermoord	1973	National	Traffic safety
Rover	1971	National	Public transport
Amsterdam Autovrij	1974	Local	Traffic/urban planning
Verkeerd Verkeer Amsterdam	1973	Local	Traffic/urban planning
Verkeerd Verkeer Utrecht	unknown	Local	Traffic/urban planning
Dooievaar	1972	Local	Traffic/urban planning
Werkgroep Fiets Den Haag	1974	Local	Cycling
Werkgroep Fiets Hilversum	unknown	Local	Cycling
Werkgroep Fiets Amersfoort	unknown	Local	Cycling
Werkgroep Fiets Leiden	1973	Local	Cycling
Werkgroep Fiets Kampen	unknown	Local	Cycling
ENWB Utrecht	unknown	Local	Cycling
ENWB Helmond	unknown	Local	Cycling
Fietsgroep Rotterdam	unknown	Local	Cycling

Source: IISG, Archive Cyclists' Union (ARCH01969), inv. no. 1. The founding date denotes the earliest year when there is evidence of activity

93 IISG, Archive Cyclists' Union (ARCH01969), inv. no. 1, meeting minutes May 13, 1975, 3.

94 In Denmark, in contrast, the longstanding preeminent cycling organization DCF (*Dansk Cyklist Forbund*) reinvented itself internally: a new generation of (also) young board members were influential in shaping Danish cycling policy. See Emanuel, "Making a Bicycle City," 512. This change took place in the mid-1970s. Denmark's cycling activism took off in the second half of the 1970s, later than in the Netherlands.

Many, predominantly local and some national action groups formed the Cyclists' Union, as table 5 shows. Coordinated by urban planning student Jan Wittenberg and Werkgroep 2000, the platform aimed to increase citizen participation in policymaking. At the founding on October 18, 1975, in RASA, a music venue in Utrecht, former Public Works Minister Willem Drees Jr. (of the social democrat DS'70 party) stressed the importance of selective car use. Other speakers included Leiden biologist-activist Helias Udo de Haes, The Obstinate Amsterdammer (*De Lastige Amsterdammer*) activist Marten Bierman, and others from environmental and traffic protest movements. The program included cycling working groups sharing local activism experiences and ended with a cycling demonstration through Utrecht.[95]

The Cyclists' Union members were young (in their twenties or thirties), urban, university-educated, and left-leaning politically. Despite the dominance of women in the foundational Stop the Child Murder, a 1978 membership survey showed 72 percent were men, and 70 percent were aged between sixteen and thirty.[96] These members' backgrounds confirm what political scientist Hanspeter Kriesi concluded about 1980s New Social Movements: their support was broad, but the active core was more radical and progressive middle class.[97] Support for action groups like the Cyclists' Union ENFB and Stop de Kindermoord was broad-based, while their active core members were young, educated, male and left-wing, even mockingly describing themselves as "a rather elitist club."[98]

Sympathy for their action goals and support for traffic safety and cycling was much more widespread. Especially outside the major cities, local Cyclists' Union chapters were often staffed by a small volunteer group of older, less clearly left-wing men. To illustrate, fifty-five-year-old businessman F.H. Markerink volunteered in 1975 to establish a Cyclists' Union branch around Enschede. His only daughter had been killed in a car accident and, although he did not want the Cyclists' Union to become "too political," he also believed the traffic safety group VVN's lobby was too cautious and weak.[99] Similarly, the driving figure behind the Helmond branch of the

95 Patricia Penrhyn Lowe, "Weg fiets? ... Fietsweg. Oprichting van de Eerste, Enige, Echte Nederlandse Wielrijders Bond," *Jonas: tijdschrift voor sociale en kulturele bewustwording* 6, no. 5 (1975): 10-11.

96 Ien Deltrap, "Waar blijven de dames?," *De Vogelvrije Fietser* 4, no. 2 (1979): 4.

97 Hanspeter Kriesi, "New Social Movements and the New Class in the Netherlands," *American Journal of Sociology* 94, no. 5 (1989): 1078-1116, here 1102.

98 Dirk Dragtstra and Paulien Osse, "Het ENWB-bestuur doet een boekje open: 'De fiets hoort juist in de stad!'," *De Vogelvrije Fietser* 1, no. 5 (1976): 18.

99 NA 2.19.198 (Archief Stop de Kindermoord), inv. no. 241, letter F.H. Markerink to ENWB, October 8, 1975.

Cyclists' Union was sixty-two-year-old retired bus driver Janus Meulendijks, and a key figure in Ermelo was fifty-nine-year-old Hessel van Hoorn, a former PR employee at Shell.[100] Cycling had its most prominent advocates among smaller left-wing political parties, but the more tacit support of a much broader political spectrum. These men, with time on their hands and active in local politics, defended the many ordinary cyclists' interests as a non-partisan issue. As a result, the Cyclists' Union could credibly claim to represent a large group of people. The Cyclists' Union could justify that claim, thanks to its diverse groups, establishing numerous local branches, and a decade of cycling-related activism.

The new group had support, also organizationally, from other action groups. Organized environmental activism in the Netherlands predated the Cyclists' Union foundation by a few years, for instance in the Environmental Defense Foundation (*Vereniging Milieudefensie*, 1971) and Nature and Environment Foundation (*Stichting Natuur en Milieu*, 1972).[101] Two members of the Environmental Defense Foundation, Jos van Gerven and Peter Kramer, joined the Cyclists' Union finance committee, and Kramer even coordinated that side of the organization.[102] Paul Tilanus and Maartje van Putten of Stop the Child Murder regularly attended Cyclists' Union general meetings.[103] Van Putten was also on an advisory committee along with members from environmental action group Straw (*Strohalm*, Bert Kerkhof) and anti-car activists from Amsterdam Car-Free (*Amsterdam Autovrij,* Ello Paul).[104] These contacts were valuable but could also be problematic. In 1976, active member Dick Arendshorst left the Cyclists' Union, unhappy with the direction Stop the Child Murder, Amsterdam Car-Free, and the Utrecht branch were pushing the organization. He implied that this wing was too principled and unwilling to compromise. According to Arendshorst, the Cyclists' Union should be an organization where all or most cyclists could feel at home.[105]

Echoing Stop the Child Murder's belief that the road was a public good and the government was liable, The Cyclists' Union alleged the Dutch

100 Dick Arendshorst, "Janus Meulendijks regelt het wel met de hoge heren," *De Vogelvrije Fietser* 1, no. 2 (1976): 8-9; Gert van Engelen, "Beschaafd, constructief en tóch lastig," *De Vogelvrije Fietser* 10, no. 2 (1984/85): 10-11. Both Meulendijks and Van Hoorn were well-known in local politics.

101 Cramer, *De groene golf*; Mom and Filarski, *De mobiliteitsexplosie*.

102 IISG, Archive Cyclists' Union (ARCH01969), inv. no. 1, "Werkgroepen van de ENWB."

103 IISG, Archive Cyclists' Union, inv. no. 1, "Presentielijst ledenraad ENWB, 13 december, Utrecht, 1975"; 'Presentielijst ledenraad E.N.W.B. dd. 1 mei 1976'.

104 IISG, Archive Cyclists' Union, inv. no. 1, "Werkgroepen van de ENWB."

105 IISG, Archive Cyclists' Union, inv. no. 1, letter Dick Arendshorst to Jan Wittenberg, April 28, 1976.

government was negligent in protecting "weaker" road users against the hostile environment of car-centered cities. In 1978, the Cyclists' Union defined its role as "the catalyst for people who think traffic and the living environment should be more geared to the weakest groups in society."[106] Its members stood up for what they called the "traffic paupers" [*vervoersarmen*], stating "it is no accident that the weakest in society are also "captive cyclists" [*gedwongen fietsers*] and also the weakest in traffic."[107] School children, housewives, those who could not afford a car, or did not drive for other reasons, fell into the category of having to use a bicycle and needing protection. This was in line with Den Uyl's motto of distributing social costs and benefits more equally. Some Dutch politicians agreed with this analysis. Labor socialist (PvdA) MP Jaap van der Doef saw that the same "immobility" and "modern pauperization" caused by a society that provided one-sided access for car drivers, excluded all social groups who could not (afford to) drive.[108] The Dutch urban traffic environment failed to protect this group.

Like Dooievaar and other activists, the Cyclists' Union used protest to get its message across. In the early 1970s, activists in the US, the UK, and France also organized bicycle rides in protest against the car.[109] In the build-up to the Cyclists' Union's official founding in 1975, Dutch cities held a series of mass demonstrations. Under the heading "Amsterdam Autovrij" (Amsterdam Car-Free), three protest rides in 1974 and more in the following years drew increasingly large crowds.[110] Famously, in June 1977, during "Wereldfietsdag" (World Bicycle Day), protesters in Amsterdam lay on the ground at Museumplein, portraying traffic victims. Over time, this type of activism diminished but did not disappear entirely. In 1982, the Cyclists' Union Amsterdam branch protested the proposed closure of a cycling path on a bridge. Since then, the public protest culture around the bicycle has declined in the Netherlands, in sharp contrast with countries like the US, where the protest movement Critical Mass emerged in 1992, using the same tactics.[111]

Once these protests convinced policymakers that the problem needed action, the dialogue between activists and government could begin. Next on

106 IISG, Archive Cyclists' Union, inv. no. 1, "Notitie de E.N.W.B. en de bierkaai," April 5, 1978.

107 Ibid.

108 Smaal, *Politieke strijd om de prijs van automobiliteit: de geschiedenis van een langlopend discours, 1895-2010*, 245.

109 Furness, *One Less Car*, 60-63.

110 Feddes and De Lange, *Fietsstad Amsterdam*, 78-80.

111 On Critical Mass: Furness, *One Less Car*, 78-107.

its agenda, the Cyclists' Union sought to address the engineers' knowledge gap regarding cycling because it found that data on cyclists and their travel behavior or preferences was much scarcer than the corresponding data about drivers. If they existed at all, government cycling reports gave no insight into (subjective) dimensions like safety, comfort, and route choice. Engineer Van Gurp, director of the National Traffic Academy in Tilburg, drew attention to this gap at a 1977 congress on cycling and moped facilities. He noted the interest in cycling from the increasing stream of cycling memoranda (local) authorities were publishing, but also the lack of specific information. "A good picture of the most important cycling flows [*fietsstromen*] and bottlenecks is usually missing."[112] In addition, "there is insufficient understanding of the traffic safety problems which cyclists experience and undergo."[113] This was precisely the knowledge gap the Cyclists' Union had set out to address since its foundation two years earlier.

Adopting a philosophy reminiscent of Dooievaar's, the Cyclists' Union believed that the true cycling experts were the actual cyclists. By riding in the (urban) streets every day, cyclists developed qualitative, experiential knowledge about the entire cycling experience, ranging from the (subjective) safety of the route, to the quality of the road surface, to parking facilities at public buildings. A true pro-cycling policy should be based on this type of knowledge, not on quantitative traffic counts and traffic-flow studies. By virtue of its membership base of ordinary cyclists, the organization believed it could provide local policymakers with this type of knowledge and secure cycling's rightful place in the governance network. In a guideline on bottlenecks, discussed later, the Cyclists' Union emphasized the danger of a quantitative approach. It was impossible to state "objectively" how much motorized traffic was acceptable on a road that cyclists also used. The dangerous apparent objectivity that numbers provided could easily misconstrue everyday cyclists' lived experience, the Cyclists' Union warned: "Numbers cannot lie, but liars can play with numbers! Do not let yourself be tempted into a quantitative approach in your confrontation with the government. It is of much greater importance to make clear that when it comes to cycling, there is only one expert: *the cyclists themselves*."[114] Rather, as quoted at the start of this chapter, "the cyclist knows best which situations are dangerous,

112 P.B. van Gurp, "De (brom)fietser: zijn plaats in de plannen en planvorming," in *Congres fiets- en bromfietsvoorzieningen Tilburg 1977* (Den Haag: ANWB, 1978), 18.

113 Ibid.

114 IISG, Archive Cyclists' Union, inv. no. 1, Christiaan la Poutré, "Inventarisatie knelpunten," 2. Emphasis in original.

which short cuts are needed, which barriers should be removed ... Where the government is doomed to fail given its technocratic approach, the Cyclists' Union can be strong."[115] The Cyclists' Union preferred participation and bottom-up user expertise to technocratic top-down planning.

The Cyclists' Union was convinced that making cycling policy without talking to cyclists was doomed to fail. This was also why some Cyclists' Union branches queried local councilors about their preferred mode of transport. It was highly problematic if urban policymakers did not ride a bike because "only actual cyclists know the problems of cycling through the city."[116] Apart from street demonstrations, the Cyclists' Union also strongly believed in developing counter-expertise as a tool to deal with official planners and engineers.[117] Armed with their own research and data, these activists were convinced it would be much harder for a reasonable government to reject them. At the time, the development of counter-expertise was a popular strategy among Dutch environmental action groups, believing their ability would help raise environmental awareness among citizens and create a stronger negotiating position with the state.[118] While some action groups deplored and refused to do this, in the 1980s, environmental action groups increasingly cooperated with governmental stakeholders, for instance by "developing alternative scenarios for environmental policy."[119]

Lay experts developing legitimacy as a governance partner by gaining credibility is, internationally speaking, not unique to the Dutch case.[120] On the transnational circulation of French cycling activism, French political scientist Maxime Huré discusses action groups' self-legitimization. He concludes that the role of user associations in local French politics was important, particularly in providing cycling expertise which municipal policymakers lacked.[121] Huré focusses on how expertise from countries

115 Ibid.

116 Archief Eemland, Archive ENWB Amersfoort (0242), inv. no. 3, letter Cyclists' Union to Mayor and Alderman, January 18, 1976.

117 A journalist at the Cyclists' Union 1977 annual meeting noted that a cabaret item was a welcome interruption to a program filled with "the thieves' slang [*boeventaaltje*] of officials and planners" making him "dizzy with all the policy switches, extensive literature reviews, coordination, and stimulation." J.W.E. Metselaar, "De vuist van de gemangelde fietsers," *NRC Handelsblad*, December 3, 1977, 5.

118 Cramer, *De groene golf*, 54.

119 Ibid., 99.

120 Steven Epstein, "The Construction of Lay Expertise: The Construction of Lay Expertise: AIDS Activism and the Forging of Credibility in the Reform of Clinical Trials," *Science, Technology and Human Values* 20, no. 4 (1995): 408-37, here 417-19.

121 Huré, "Les réseaux transnationaux du vélo," 117.

with high cycling levels like the Netherlands was adopted in local circumstances in Belgium or France.[122] What distinguishes expertise as a form of action in the Dutch context, is its genuinely user-generated nature. Dutch organizations operated in a context where many citizens still used, or had to use, the bicycle. Consequently, the type of knowledge these organizations created was not so much technical, about how to construct cycling infrastructure, or how to manage cycling governance. As I have shown, Dutch policymakers had access to this type of knowledge: the construction of cycling infrastructure had continued without interruption from the 1920s, albeit with ebbs and flows. The Cyclists' Union switched its attention to user-based knowledge: about the routes everyday cyclists took, the obstacles they faced on their journey, and which bottlenecks were a priority. In other words, the fact that so many people in the Netherlands still cycled also meant that these cyclists possessed a large and untapped pool of experiential knowledge.

The expert role the Cyclists' Union took up was inspired by the Dooievaar's pioneering work on designing cycling path plans. A few months before the founding of the Cyclists' Union, the advocates of participative processes, Eisse Kalk and the Dooievaar group, asserted that local cyclists should come together to design the main cycling routes through their cities.[123] Indeed in the second half of the 1970s, the Cyclists' Union branches worked with many Dutch cities on a "bottleneck memorandum" (*knelpuntennota*), mapping in detail the greatest problems for urban cyclists. One example is the city of Amersfoort, near Utrecht. This branch of the Cyclists' Union reported on its experience with the bottleneck memorandum in *De Ketting* (The Chain), an internal Cyclists' Union publication for sharing news on local activism.[124] The newly founded Amersfoort group announced its presence with a critique of the city's traffic policy in 1975, but because "it was not very elaborate, we never heard anything."[125] Council members demanded a cycling policy document from the city government. However, the Amersfoort branch objected to this "seen and written from behind a car window" memorandum.[126] In other words, the cyclist's perspective of traffic problems was nowhere to be seen. To compile a better counterproposal, the activists handed out 3,000 questionnaires to cyclists in Amersfoort.

122 Ibid., 131-68, esp. 58-59.

123 Elly van der Ster, "Hoe sterk is de eenzame fietser," *De Volkskrant*, March 29, 1975, 35.

124 "Afdeling Amersfoort: 'Bericht over het ontstaan van een fietsrapport'," *De Ketting* 1 (1976): 2-3.

125 Ibid.

126 Ibid.

The questionnaire was also printed in the free local paper and distributed at high schools.[127] This resulted in 300 responses.

Cyclists were asked to describe points where they felt unsafe, where road surfacing was poor, where waiting times at junctions and traffic lights were long, and similar daily annoyances, dangers, and risks. The group also managed to convince the city council to postpone debating its official memorandum to allow the activists time to finish their input in the final policymaking phase. Applying the detailed survey input, the activists indicated on city maps which streets were unsafe and which crossings had to be redesigned to suit cyclists. They also devised an ideal cycling path network throughout the city. Their report in The Chain detailed how to compile, duplicate, and distribute the memoranda, including the costs.[128] The Amersfoort branch stressed two points: good contact with (some) council members was indispensable. Otherwise, the group would not have known what went on behind the scenes and could not have delayed decision-making to include their input. Secondly, earlier attempts at a general policy critique had failed. The more robust counterproposals had more effect. The detailed reports made it easier for policymakers to do something for cyclists, but it also put them on the spot. Point-blank refusing these concrete suggestions was harder than rejecting vague calls for system changes. Nevertheless, activists found the implementation of measures painfully slow, and so it was in this Amersfoort case.[129] The first steps to address bottlenecks were easy, but seeing concrete results in the street took more time and effort.

The bottleneck memorandum became a popular and innovative tool for addressing everyday cyclists' issues.[130] When the Cyclists' Union branches met in May 1976, many were in the process of compiling a memorandum: Amsterdam, Arnhem, Amersfoort, Delft, Enschede, Haarlem, 's-Hertogenbosch, The Hague, Maastricht, Rotterdam, and Utrecht.[131] Within two years, the method had become quite popular

127 IISG, Archive Cyclists' Union, inv. no. 1, Christiaan la Poutré, 'Inventarisatie knelpunten', 5.

128 "Afdeling Amersfoort: 'Bericht over het ontstaan van een fietsrapport'."

129 Paulien Osse, "'Ieder van ons kan zich voor de geest halen hoe het fietsparadijs eruitziet'," *De Vogelvrije Fietser* 3, no. 4 (1978): 9.

130 Matthew Bruno, Henk-Jan Dekker, Letícia Lindenberg Lemos, "Mobility protests in the Netherlands of the 1970s: Activism, Innovation, and Transitions," *Environmental Innovations and Societal Transitions* (forthcoming).

131 IISG, Archive Cyclists' Union, inv. no. 1, "Verslag ledenraad dd. 1 mei 1976 E.N.W.B. te Utrecht." The Hague's plan was based on 175 cyclists' experiences (GADH, Archive Dooievaar (0772-01), inv. no. 10). The Amsterdam branch compiled a memorandum in 1977, opposing the city's Traffic Circulation Plan that proposed a city-wide network of separated cycling paths. The Amsterdam

within the organization.[132] Guidelines issued at the time reflected on branches' experiences and the aim of this action. The text included many proposals for cycling paths or facilities, but "lacked the data to elaborate these proposals, or provide evidence [*hard maken*]."[133] Guidelines instructed the local Cyclists' Union branch to list all local dangerous situations, barriers, and missing links "in order to indicate where and how the government has to provide facilities for cyclists."[134] The tactic worked in some municipalities: in the mid-1970s, the Helmond branch noted that its bottleneck survey was sent directly to the municipality "which usually acts." It recommended this "simple but effective method" for Cyclists' Union branches "which have good contact with the municipality."[135] Elsewhere it produced less concrete results, leading a minority of activists to call for more street action, while others dismissed this as "sixties romanticism."[136] In the long run, the Cyclists' Union consolidated and professionalized its expertise-based approach.

We can see this task that the Cyclists' Union undertook voluntarily as a form of government outsourcing. In the Dutch political system, where consultation was a common feature, the action group took advantage of the opportunity to launch into local politics. By developing user expertise that the government did not have, the group tried to justify its role in cycling governance. In certain ways, this parallels the way the tourist organization ANWB leveraged its engineering expertise in the early twentieth century to gain a role. In the Dutch polder model, these groups succeeded in positioning themselves not just as the representatives of cyclists, but also as expert groups.

The Cyclists' Union always felt that its key role in cycling governance was providing (local) authorities with the knowledge they needed to improve cycling policy. Evidence for this is the Cyclists' Union's request for national subsidies in 1979. Such an attempt, although unsuccessful, indicated the

Cyclists' Union argued that not only was this unrealistic, but also unnecessary, since many bottlenecks for cyclists could be solved with simple and cheap means. They also argued that cycling should be safe everywhere and not depend on the existence of separated cycling paths. The entire city's infrastructure had to be rethought from the cyclists' perspective, not the car's; see Feddes and De Lange, *Fietsstad Amsterdam*, 91-94.

132 Paulien Osse, "De lange mars door de instituties," *De Vogelvrije Fietser* 3, no. 5 (1978): 8-9.

133 IISG, Archive Cyclists' Union, inv. no. 1, Christiaan la Poutré, 'Inventarisatie knelpunten,' 1.

134 Ibid.

135 Ibid., 6.

136 Paulien Osse, "'Beroepskrachten vind ik een gevaarlijk element'," *De Vogelvrije Fietser* 6, no. 1 (1980): 4-5; Theo de Jong, "Knelpunten-nota's grootste knelpunt," *De Vogelvrije Fietser* 3, no.6 and 4, no. 1 (1978): 14-15.

Cyclists' Union's more conciliatory stance towards government and is not unique: most action groups were – and are – subsidized in the Netherlands.[137] For example, the young activist organization considered its six (part-time) staff in the national office inadequate for the job at hand: "the ENFB is not sufficiently able to fulfil a part of its aspirational goals."[138] The activities it listed included "systematically processing relevant literature," "learning about activities and research of many organizations and institutions in the transport field," "establishing our own research, encouraging research in other organizations, attracting interns, and formulating study assignments."[139] The organization believed it needed to develop this expertise "for a mature functioning of our organization and for our role in the overall decision-making."[140] Remarkably, given its initial opposition to an expert role, the Cyclists' Union implied that this power – the ability to play a role in the Dutch decision-making process – depended on the knowledge activists could bring to the table. It believed the more playful happenings of the early 1970s, inspired by the anarchist Provos and the like, were no longer needed or effective. By shifting towards more expertise-based activism, producing reports, and discussing them with governmental actors behind closed doors, the Cyclists' Union became more embedded in policy circles and soon started to resemble the ANWB. How then did these two organizations differ in the 1970s?

The Cyclists' Union's commitment to bottom-up democratic policymaking for cycling contrasted starkly with the ANWB's technocratic approach. Moreover, the public increasingly considered the ANWB a car lobby group. In response, the organization claimed it merely defended everyone's mobility interests, while leaving the choice of mobility mode to the individual. If most Dutch people chose the car, the ANWB would lobby for better roads and anti-congestion measures. This stance obscured how groups like the ANWB actively promoted automobility. Its focus on individual mobility also meant it left public transit to the state and to labor parties, while the organization conducted its lobbying activities behind the scenes. It led to critique in the 1970s that the organization was too car-minded and ignored

137 R. Koopmans and J.W. Duyvendak, "Sociale bewegingen en het primaat van de politiek," in *Tussen verbeelding en macht: 25 jaar nieuwe sociale bewegingen in Nederland* (Amsterdam: Sua, 1992), 29-30; R. Koopmans, "Van provo tot RARA. Golfbewegingen in het politiek protest in Nederland," in *Tussen verbeelding en macht: 25 jaar nieuwe sociale bewegingen in Nederland*, ed. J.W. Duyvendak, et al. (Amsterdam: Sua, 1992), 65.

138 NA 2.02.28 (Archief Tweede Kamer), inv. no. 6272, letter ENWB to Minister of Public Works, May 18, 1979.

139 Ibid.

140 Ibid.

cyclists' interests. Letters to newspapers frequently complained that the ANWB failed to live up to its name.[141] A long critical article in opinion magazine *Vrij Nederland* aptly summarized public sentiment.[142] Journalists Hugo Arlman and Rudie van Meurs interviewed ANWB director Ijntema and press officer Kroon in 1980, who expressed a non-interventionist small-government ideology: if people wanted to drive, the ANWB would support that, even harmonize it with other interests.[143] This harmonization was an illusion, since the organization was rife with internal contradictions. Trying to give drivers, cyclists, and pedestrians their due, both in utilitarian and recreational mobility, while also protecting the Dutch natural and cultural landscape, was a tall order. The Cyclists' Union criticized the ANWB because its advocacy work for cycling was inspired by the goal to separate traffic and make driving faster.[144]

The well-established tourist and traffic organization ANWB might have agreed with those statements on paper, but the Cyclists' Union leveled implicit critique at its rival's more technocratic approach. As far as the Cyclists' Union was concerned, the aim to improve the situation for cyclists should be achieved "not by preaching from above or providing even better experts."[145] Road users' complaints should reach local policymakers as directly as possible. The Cyclists' Union saw its role as helping organize those groups and making their voices heard to put cycling on the local political agenda. The Cyclists' Union also criticized the ANWB by stating that a policy truly favoring cyclists should include restrictions on automobility. Many political parties were "willing to do something for the bicycle, without simultaneously curbing car traffic."[146] In other words: "the double moral standard of choosing the bicycle and the car has to be broken."[147]

In time, relations between the two organizations improved. Cyclists' Union member Jos Vernooy claimed that by 1980, the ANWB had started paying more attention to cycling.[148] ANWB chairman Frederik Hendrik van

141 D. de Vries, "Over fietsen gesproken," *De Kampioen* 94, no. 12 (1979): 1158. Technically, the ANWB called itself the Royal Dutch Tourist Organization ANWB; the letters ANWB no longer stood for anything but were retained for the sake of recognizability.

142 In the same year, newspaper *Het Parool* published a similarly critical piece: Bommels, "ANWB, kampioen op snelwegen en zijpaden," 25-26.

143 Arlman and Meurs, "'De adem van de ANWB'," 4.

144 GADH, Archive Dooievaar (0772-01), inv. no. 26, ENWB document "Waarom de 'W' van de E.N.W.B.," c. 1975.

145 Ibid.

146 Ibid.

147 Ibid.

148 Bommels, "ANWB, kampioen op snelwegen en zijpaden."

der Linde van Sprankhuizen was willing to admit publicly that the "ENFB youngsters" had had "a useful effect" on the ANWB.[149] The urban-based Cyclists' Union was also widening its scope to include recreational cyclists – the ANWB's traditional domain for activism. It embraced the idea that broadening its membership beyond the protesters would generate influence and visibility. After all, recreational cycling remained popular in the Netherlands.[150] This also brought the organization in closer contact with the ANWB. The rapprochement resulted, for example, in their collaboration in the 1990s on mapping recreational cycling routes and designing road signage.

The ANWB also experienced internal tensions in trying to reconcile nature conservation with promoting driving. These were exposed in a late 1960s proposal for a road through the scenic dunes between the coastal towns of Scheveningen and Zandvoort. Representing the ANWB roads and traffic interest, August Boost claimed future traffic demanded its construction. The organization's director Anton Blankert and chairman Van der Linde van Sprankhuizen, however, opposed the road because it would damage nature and make the dunes less attractive for (cycling) recreation.[151] Similarly, the ANWB supported protests against a highway between Amsterdam and Rotterdam right through the Green Heart area near Amelisweerd.[152] It was never constructed.[153] This critical stance did not represent opportunistic environmental activism. Nature conservation had been a core element of the tourist organization ANWB for decades. Within the organization, nature conservation interests continued to compete with the road lobby for dominance. The ANWB always sought to somehow reconcile its identity as traffic organization with that as recreation organization.[154] Promoting motorized mobility never had free rein within the ANWB, even though it dominated at times. Elsewhere, this tension between utility and aesthetics also played out in discussions about American parkways or how the German Autobahn and Italian autostrade should be embedded in the national

149 Ibid.

150 IISG, Archive Cyclists' Union, inv. no. 22, *Kruipolie* 3 (1984): 15-17.

151 Linders-Rooijendijk, *Gebaande wegen II*, 853-55.

152 Amelisweerd is a forest near Utrecht that engineers planned to demolish to make way for a highway. Fierce protests ensued – environmental activists occupied the forest and managed to delay and eventually prevent any trees being cut down. See Odette van de Riet and Bert Toussaint, "Learning from a Contested Project in the Netherlands: The Clash over the Amelisweerd Forest, 1957-1982," *Transfers* 4, no. 1 (2014): 63-82. The Green Heart is a large agricultural area in the West of the Netherlands; it has been a central principle of Dutch spatial planning to keep this area free of large-scale development. See Mom, "The West and the Rest," 143-99.

153 Buiter and Staal, *Het avontuur van de ANWB*, 133.

154 Linders-Rooijendijk, *Gebaande wegen II*, 728.

landscape.[155] The main discussion in the Netherlands did not center on how to embed a road into nature. The question was whether a road was necessary at all, especially if that nature could also be enjoyed by bicycle – a vehicle that the Dutch still widely owned and used, unlike in Germany and certainly the US, even in the 1960s.

In the critical weekly *Vrij Nederland*, many (anonymous) critical voices accused the ANWB of being a conservative, elitist, and power-hungry organization. Stop the Child Murder founder Langenhoff caustically characterized the ANWB's remaining involvement with the bicycle: "While children in their neighborhood and on the way to school were run over and killed, the ANWB still did something for the bicycle: they opened a touristic cycling path here and there in remote pine forests with the corny display of a Royal Commissioner on a bicycle like a circus bear. Meanwhile, it neglected the jungle of commuter traffic in cities and villages."[156] In its defense, the ANWB stated it supported cycling in the background, not seen by the public. While the ANWB had once sent hundreds of letters each year to many different authorities in its defense of cycling, that stream had slowed down to a trickle by the 1970s.[157]

The ANWB had historically served – and continued to do so – as an expert organization for traffic policy. Through a series of local government surveys, the ANWB attempted to gain an overview of current policymaking, critique it, and enable more coordinated traffic policy. Tellingly, in the same year the Cyclists' Union was established, it first surveyed the country's local cycling policies.[158] The exceedingly high response rate (81 percent) demonstrates that, despite the critique, municipalities still considered the ANWB a relevant player in the mobility governance network. Each municipality was asked about its cycling infrastructure type (separate or mixed), crossings designs, parking possibilities, a cycling path plan, and an overarching vision for cycling network. Larger cities often reported on a cycling plan incorporated in their Traffic Circulation Plan (VCP), smaller ones did not.[159] The ANWB

155 Zeller and Mauch: *Driving Germany*; Mauch and Zeller, *The World beyond the Windshield*.

156 Arlman and Meurs, "'De adem van de ANWB'," 20.

157 In 1975, ANWB director Blankert wrote to the minister of Public Works complaining of the low budget for cycling, arguing that the bicycle would solve congestion, parking, unprofitable public transit, and environmental problems. See NA 2.02.28, inv. no. 6245, letter ANWB/Blankert to minister of Public Works, February 18, 1975.

158 "(Brom)fietser niet vergeten," *Verkeerskunde* 26, no. 6 (1975): 285; "Voorzieningen voor fietsers en bromfietsers," *Verkeerskunde* 26, no. 6 (1975): 286-95.

159 Arnhem planned a 275 km cycling path network as part of its VCP. See "Arnhems plan voor fietspaden," *Verkeerskunde* 26, no. 6 (1975): 296-97. Berkers and Oldenziel, *Cycling Cities: The Arnhem and Nijmegen Experience*.

urged the smaller communities to create a plan for identifying missing links in the cycling network and creating a coordinated cycling policy.[160] The ANWB also published papers from a 1977 congress discussing the governance process around demonstration cycling routes in The Hague and Tilburg and learning best practices.[161] A wide array of local engineers and planners attended this congress.[162] Traffic Engineering (*Verkeerskunde*), the main Dutch traffic engineering journal run by the ANWB, regularly featured articles on cycling. The rising appreciation of cycling, boosted by the Cyclists' Union, pressured the ANWB to revisit its cycling roots.

By the late 1980s, as cities focused on plans to boost cycling levels, the ANWB agreed that the "city centers are choked by car traffic and would absolutely benefit from more bicycle traffic."[163] The ANWB remained committed to a national network of separated cycling paths, a policy it had proposed half a century earlier. It believed in traffic separation models: Dutch cyclists should have access to a nation-wide network of cycling paths separated from roads as much as possible. Otherwise, it recommended on-road paths, including on quiet streets.[164] The ANWB claimed to follow the wishes of its members, most of whom were car drivers, reinforcing the ANWB's public image as a car organization.[165] Nationally, its behind-the-scenes activities did not make for a promising candidate to build better cycling cities. Absent locally, it concentrated on promoting uniform and national engineering norms and rules. And while the organization did sustain a system of local consuls catering to suburban and rural cycling, it had less footing in urban areas where the Cyclists' Union was strong. Re-appropriating urban space for cyclists in a bottom-up process of activism and user expertise was not the ANWB's modus operandi.

160 "Voorzieningen voor fietsers en bromfietsers," 295.

161 These demonstration projects were "reviewed' in detail, including extensive photographs, in *Verkeerskunde*. The ANWB did more to spread knowledge about these important experiments than the (national) government, which largely funded the projects. See F. Stoovelaar and W. van Tilburg, "De fietsroute Den Haag (1) – verkeerstechnische aspecten," *Verkeerskunde* 30, no. 11 (1979): 544-48; F. Stoovelaar and W. van Tilburg, "De fietsroute Den Haag (2) – verkeerstechnische aspecten," *Verkeerskunde* 30, no. 12 (1979): 597-601; W. van Tilburg and F. Stoovelaar, "De fietsroute Tilburg: verkeerstechnische aspecten," *Verkeerskunde* 28, no. 4 (1977): 163-69.

162 Koninklijke Nederlandsche Toeristenbond ANWB, *Congres fiets- en bromfietsvoorzieningen Tilburg 1977* (Den Haag: Koninklijke Nederlandsche Toeristenbond ANWB, 1978).

163 Niek Schenk, "De fiets als redding," *Algemeen Dagblad*, April 15, 1989.

164 "Fietser in de berm van het verkeersbeleid. Fietsen is een volwassen verkeersvorm," *De Kampioen* 89, no. 4 (1974): 246-48, here 248.

165 Bommels, "ANWB, kampioen op snelwegen en zijpaden," 26.

The local and urban approach made the Cyclists' Union successful. Though its core of activists was relatively small, they acted as intermediaries between cyclists and the (local and often urban) authorities – a role the ANWB no longer fulfilled. Their quick and successful integration in governmental circles resembled the relationship between the early ANWB and government. After a decade of activism, the Cyclists' Union cycling governance became more participatory and collaborative. Now, cyclists could convey their concerns and complaints to local policymakers.

The Cyclists' Union's role as a non-state actor in the governance network around cycling was both old and new. Compared to the ANWB, the Cyclists' Union gave ordinary and urban cyclists a clear voice, despite a similar expertise-centered approach to the political process. Historian Popkema has aptly said of the ANWB that it functioned as a "free consultancy agency" for state agencies rather than an antagonistic action group.[166] Since the 1980s, the Cyclists' Union has functioned in a very similar way. There are also marked differences. While the ANWB in the first half of the twentieth century and the Cyclists' Union since the 1970s mediated between users and the government, the Union did so based on direct input from cyclists through surveys and local meetings. Much more so than the well-staffed and nationally funded ANWB, the Cyclists' Union, as a federation of local branches, remained an action group with a strong local basis, but was less effective nationally as lobby organization compared to the ANWB.

6.3 Working with the Government: Activists and Cycling Governance

In the late 1970s, local Cyclists' Union activists sought contact with policymakers in many cities and towns. This often led to a working relationship, but certainly not everywhere. How do we explain how the Union developed such good contacts with local policymakers in many cities and towns in such an extraordinarily short period, considering it had started out as oppositional organization?

Both the circumstances of the time and the wider Dutch political culture made this possible – and both need a more in-depth discussion. First, the Dutch political circumstances of the 1970s were favorable for cycling activism. It gathered momentum under the national government coalition of Labor

166 Popkema, "Tussen techniek en planning," 79.

leader Den Uyl. The electoral shift to the left helped translate many of the activists' demands nationally and offered ample opportunities locally. The political opportunity structure was much better than ever before or would be in the later 1980s.[167] Government stakeholders and activists considered a cooperative rather than an antagonistic strategy was more beneficial to both parties. Still, the Dutch political tradition was very favorable to action groups from the start. According to political scientists Koopmans and Duyvendak, in the Netherlands, "almost independently of political color, movement organizations are often and strongly professionalized due to the relatively strongly developed subsidy and consultation culture."[168] Both the larger political culture as well as the specific circumstances of the 1970s gave cycling activists a chance to succeed.

Early on, Dutch (urban) policymakers wanted to cooperate with activists for various reasons: these groups' expertise, and the way they communicated citizens' demands and preferences, were valuable input to policymakers. Action groups could credibly claim to represent a large share of Dutch society; consequently, policymakers felt unable to ignore many of their demands. Urban planners also faced the impact of the car on public space and urban livability. Cities like Delft and Groningen were already experimenting with ways to curb this effect. Those planners, in other words, did not need much convincing by cycling activists.

Local governments found cooperation with action groups useful because city officials were not trained in dealing with urban mobility conflicts facing cyclists and pedestrians. City engineers and traffic experts claimed to be open to activists' suggestions if they were helpful – that is to say, solid and specific enough. They were interested in citizen input. According to G.H.A. Hoogenboom, an engineer at Amsterdam's Traffic Office (*Verkeersbureau*), in 1968 many citizens wrote to suggest improving traffic conditions, sometimes even submitting comprehensive plans for the entire city.[169] Because reviewing and replying to them was time-consuming, he preferred to "receive suggestions with regard to concrete facts, concrete situations in the city, where something is wrong from the standpoint of traffic engineering."[170] Again in 1972, the Amsterdam traffic engineer called

167 Hanspeter Kriesi, "The Political Opportunity Structure of the Dutch Peace Movement," *West European Politics* 12, no. 3 (1989): 295-312, here 298.

168 Koopmans and Duyvendak, "Sociale bewegingen en het primaat van de politiek," 30.

169 Hoogenboom went on to become a high official in the Ministry of Public Works, bringing his experience as a city planner into national engineering circles.

170 "Directeur Amsterdams verkeerbureau: Geen doorbraken in binnenstad," *Algemeen Handelsblad*, April 25, 1968, 2.

for detailed information: "It is important for action groups not to aim their actions at city-wide issues, but at issues they have to deal with in their street."[171] Hoogenboom believed activism was important. He also warned activists should be both patient and persistent if they wanted to get anywhere with the government. By and large, engineers believed that it was impossible "to turn back the clock" by introducing car-limiting measures.[172] His article in *De Tijd* (the newspaper promoting Stop the Child Murder) advised activists to develop expertise and do as much groundwork as possible for cycling path construction: by talking to property owners, activists could prepare time-consuming expropriation procedures and design alternative routes for cycling paths. "If you take up a part of the preparatory work and remove obstacles, you can accelerate the preparation."[173] In other words, he actively encouraged the "outsourcing" of the government's cycling policies to citizens and activists. And the Cyclists' Union developed in this direction to such an extent that in 1985 new chairman Jaap Rijnsburger pushed back on the sometimes-heard claim that activists functioned as "unpaid civil servants." He repeated the idea that the Cyclists' Union had as much expertise as municipal engineers, but also that some form of external pressure was often needed to make their ideas a reality.[174]

Local policymakers felt compelled by citizen pressure, admitting that urgent calls for better cycling facilities and safer traffic led them to action. The Association of Dutch Municipalities (VNG) concluded in an illuminating study on traffic safety policy in the late 1970s that many local authorities were finding it hard to cope with mounting citizen pressure. Citizens were becoming more assertive when it came to traffic safety. Eindhoven reported that many citizens insisted the city create speed bumps in their neighborhoods. The city was "extremely hesitant" to do so but "the pressure from local society – one street constructed speed bumps without authorization – has contributed to a reassessment" and the city complied with citizens' demands.[175] Jan de Ruiter, council member for a confessional party in the well-to-do town of Heemstede, noted in 1972 that it was "no surprise that precisely the topic of traffic had established such an interesting dialogue with sections of the population." Because it affected everyone directly, the

171 "Zo wil Amsterdam zijn verkeersknooppunten doorhakken," *Het Parool*, February 23, 1974, 3.

172 G.H.A. Hoogenboom, "Actie is zinvol maar vereist veel geduld," *De Tijd*, October 24, 1972, 3.

173 Ibid.

174 Henk Dikker Hupkes, "ENFB heeft inhoudelijk net zoveel in te brengen als ambtenaren," *De Vogelvrije Fietser* 10, no. 2 (1984/85): 8.

175 NA 2.19.185 (Archive VNG), inv. no. 2428, Concept-eindrapport Gemeentelijk Verkeersveiligheidsbeleid: Een inventariserende case-study van tien gemeenten, 11.

traffic issue generated a lot of interest among citizens. For this council member it was "of paramount importance" to listen to local action groups like The Troublesome Cyclist (*De Lastige Fietser*).[176] The head of the ANWB's traffic department De Regt also thought that democratization had led to a greater focus on cycling. He believed the growing interest in cycling "has been stimulated by the greater attention to the opinions that road users *themselves* express through public surveys ... Democratization has had a clearly positive impact here."[177] According to Albert de la Bruhèze, in the old textile town Enschede, citizens' vocal demands in the media created pressure that was impossible for policymakers to ignore.[178]

Dutch urban planners too felt ambivalent about the car already in the 1960s. According to historian Tim Verlaan, "the shift from car-centered to traffic-calming policies was already well underway when urban social movements came to the fore."[179] The 1963 publication of Colin Buchanan's report *Traffic in Towns*, addressing the deteriorating quality of life in British cities due to automobility, generated concern in the Netherlands as well.[180] In the late 1960s and early 1970s, prominent Dutch experts like traffic engineer Hendrik Goudappel, transport economist Geurt Hupkes, and urban planner Niek de Boer advocated the pedestrianization of urban areas by banning the car from city centers.[181] Livability as a concept became popular: the threat of unrestricted economic growth to living conditions led policymakers to commission numerous reports on the topic.[182] The input from the Cyclists' Union in Amersfoort was due to dissatisfaction with the city's cycling memorandum. This might not have been a good memorandum – written by car-driving officials with little understanding of cyclists' needs – but a plan existed. At a minimum, it implied that officials, too, had begun to take cycling a bit more seriously.

176 Krantenviewer NHA, Raadsnotulen Heemstede, December 14, 1972, 390. https://nha.courant.nu/issue/RNH/1972-12-14/edition/Notulen/page/19?query=%22lastige%20fietser%22&sort=relevance

177 "Verkeerstechnische leergangen 1976: Samenvattingen lezingen," *Verkeerskunde* 26, no. 12 (1975): 614-15. Italics in original.

178 Albert de la Bruhèze, "Enschede: An Experiment in Cycling," 46-47.

179 Verlaan, "Mobilization of the Masses," 9.

180 Colin Buchanan, *Traffic in Towns: A Study of the Long Term Problems of Traffic in Urban Areas* (London: HM Stationery Office, 1963).

181 Mom and Filarski, *De mobiliteitsexplosie*, 347-48. On *Traffic in Towns*, see Marco Bianconi and Mark Tewdwr-Jones, "The Form and Organisation of Urban Areas: Colin Buchanan and Traffic in Towns 50 Years On," *Town Planning* Review 84, no. 3 (2013): 313-36; Gunn Simon, *Introduction to Traffic in Towns* (2015).

182 Kaal, "A conceptual history of livability," 540-42.

A younger generation of politicians in cities like Groningen also pioneered traffic-calming measures. A social democrat politician in Groningen, twenty-five-year-old Max van den Berg, became councilor for traffic in 1970. Strongly advocating democratization and politicization of urban planning, politicians tried to involve the public in designing a new traffic circulation plan, although planning scholar Tsubohara argues that ultimately a more technocratic, authoritarian approach pushed through traffic-calming measures in Groningen.[183] In 1974, a citizen action group published its survey on how Groningen citizens moved through the city. The largest group (37 percent) cycled while 60 percent wanted to see a car-free city marketplace. After a bicycle demonstration in September 1974, activists handed this report to Van den Berg. And in Amsterdam, the city's urban planner, Hoogenboom, went even further, arguing for restricted car traffic in the capital. In 1972, he had launched a new traffic plan to concentrate car use on certain roads and implement traffic-calming measures on others.[184] Car drivers "will now say that they have not been considered." His message was loud and clear: Amsterdam offered good public transit and if drivers "don't want to use it, then that is up to you, but I have built rows of stationary cars into my plan."[185] Hoogenboom was saying no to unlimited automobility – drivers had alternatives and if they chose not to use them, so be it. He hoped that slowing down drivers would ultimately make them so fed up that they would choose the tram or no longer try to reach the city center by car.[186] This convergence of a new generation of progressive planners with a wave of activism generated momentum for new traffic policies in Dutch cities.[187]

In the prevailing positive political climate, a pragmatic approach made more sense than antagonistic strategies. Looking for cooperation, one early

183 S. Tsubohara, "Democracy through political parties and public participation: the case of the planning history of Groningen, The Netherlands," PhD diss., University of Groningen, 2010, 65.

184 The car lobby disliked these measures: car industry lobby group RAI asserted that Hoogenboom's plan would cause drivers to fall into a trap (*fuik*) when they tried to reach the city center: "Geen gefundeerde visie op verkeer: Secretaris RAI kraakt verkeerscirculatieplan," *NRC Handelsblad*, April 21, 1972, 2.

185 "Ir. Hoogenboom: Bloed, zweet en tranen," *NRC Handelsblad*, January 19, 1972.

186 Hoogenboom's experience with urban traffic in Amsterdam (and before that in Nijmegen) matters because he then became an important figure within the national Public Works administration as secretary for the *Stuurgroep Verkeer en Vervoer*. There he continued to spread his critical perspective on unbridled car use and support for better public transport and cycling, informed by first-hand experience of governing urban mobility. See Willem de Bruin, "Nationaal verkeersveiligheidscongres: Belangen fietsers en voetgangers voorop," *De Waarheid*, April 23, 1980, 4.

187 In Enschede, 1974 elections returned a left-wing city council working with local action groups: Albert de la Bruhèze, "Enschede: An Experiment in Cycling," 44.

activist in the Amsterdam Cyclists' Union, André Guit, soon started to think strategically. In 1978, he helped compile memoranda listing all the major and minor cycling bottlenecks in the capital. Social democrat councilor Michael van der Vlis, a trained econometrist, supported cycling. His party had prepared a similar memorandum, creating synergy between activists and policymakers.[188] The activists were divided: when invited to join the municipal Working Group Cycling, some members wanted to remain a more activist, outsider group, leading to "principled and fierce" debates.[189] Looking back on these years, Guit remarked that while the activists were "very much anti-car," for strategic reasons they "prudently did not spread this message."[190] Dooievaar activist Van Beek also opposed radical and swift change, believing that "through education and discussion the mentality has to be adjusted" before implementing sweeping measures.[191] Being pragmatic in order to achieve as much as possible was how many activists operated in Dutch cities.

Most but not all Cyclists' Union figures preferred to work with the government, strengthening the Dutch polder model of governance. Initially they disagreed on action strategies. In 1976, the Maastricht branch was disappointed with how "we are being completely taken in [*ingepakt*] by the cycling path plans in many cities."[192] It was one thing getting issues solved for cyclists, but to what extent should the group be taking action against cars? This Dutch mobility strategy of trying to do it all, avoided making hard choices. Car levels in Maastricht were 180 per 1,000 residents. According to the local Cyclists' Union, the city considered it progressive policy to limit that number to 340 by 1990.[193] Similarly, in traffic action groups outside the Netherlands, such as Transportation Alternatives in the US, there were debates about the ideology between "those who wanted to continue with anti-automobile advocacy and those who wanted TA to pursue a more positive direction that emphasized the benefits of cycling and a less contestatory anti-car rhetoric."[194] While many branches of the Cyclists' Union chose to work with local government by participating in working groups and committees, some rejected this close

188 Anne-Mariken Raukema, "Meter voor meter fietspaden bevechten," *Op eigen kracht: Blad van de afdeling Amsterdam van de Fietsersbond*, no. 77 (2009): 12-13, here 12.

189 Ibid.

190 Ibid., 13.

191 "Dooievaar wil Hagenaar weer laten fietsen," 39.

192 "ENWB afd. Maastricht," *De Ketting* 1 (1976): 5. Eventually some members who preferred action became disillusioned and left the organization: Gert van Engelen, "Overleggen is niet fout, ingepakt worden wel," *De Vogelvrije Fietser* 16, no. 1 (1990): 24-25.

193 "ENWB afd. Maastricht."

194 Furness, *One Less Car*, 63.

relationship. The Nijmegen branch for instance, in 1980, preferred to stay outside the system so that it could still organize street protests. Participating in committees was time-consuming and, as one activist said: "the municipality doesn't expect tough action because you sit on the committee."[195] Over time, the more oppositional Cyclists' Union voices lost out. As a result, protest activities declined markedly over the group's first decade.

The Cyclists' Union activists did establish a good working relationship with local policymakers, in part because there was a lot of social mobility between activism and government. Within the wider world of environmental activism in the Netherlands, this was not uncommon.[196] Although young, the activists were from a similar socioeconomic background as their interlocutors within the professional bureaucracy. Cities like Delft hired young traffic engineers with new ideas about urban planning (*Woonerf*), traffic calming, and cycling facilities. Activist Michael van der Vlis co-authored a memorandum on Cycling in Amsterdam in 1978, the same year he became the alderman responsible for traffic on behalf of the social democratic party.[197] Pax Kroon, a civil engineer for Rotterdam's traffic department, was also a Cyclists' Union board member in the 1970s.[198] Eisse Kalk, an important inspiration for cycling action groups, joined Amsterdam's city government in 1981 to make policymaking more participatory from the inside out. He reasoned that citizens were willing to take part, thanks to groups like Werkgroep 2000 raising more political awareness. By working inside the government bureaucratic apparatus, Kalk hoped to achieve more.[199] Steven Schepel, a prominent member of Stop the Child Murder in the 1970s, became a government official in traffic (safety) in the 1980s, first in Rotterdam, later in Amsterdam. He ended his career at the Ministry of Public Works, working on traffic safety programs.[200] Again, the Danish case was remarkably similar. Martin Emanuel argues that Copenhagen city lacked planning expertise and turned to "former bicycle activists" who got jobs in the city's planning office.[201] Bringing their focus

195 Paulien Osse and Theo de Jong, "'Als je in de commissie zit, verwacht de gemeente dat je geen harde acties meer voert'," *Vogelvrije Fietser* 5, no. 4 (1980): 18-19, here 19.

196 Cramer, *De groene golf*, 100.

197 Feddes and De Lange, *Fietsstad Amsterdam*, 86

198 Other examples in the early 1980s: Olaf Ostendorf was a traffic civil servant in Deurne and Cyclists' Union activist in nearby Venray; prominent board member Jos Vernooy became an Amsterdam civil servant in 1982. "Ambtenaar in Deurne, actievoerder in Venray," *De Vogelvrije Fietser* 11, no. 1 (1985): 12-14.

199 "'Burger moet echt kunnen meespreken'," *Het Parool*, March 24, 1981, 7; "Eisse Kalk: 'In dit werk is een brede rug nodig'," *Het Parool*, April 13, 1983, 6.

200 Interview Author with Steven Schepel, November 13, 2019.

201 Emanuel, "Making a Bicycle City," 515.

and expertise on cycling within city halls created a more cycling-friendly planning culture at every traffic department level.

This mutual willingness to cooperate in transforming Dutch cities helped establish many (informal) contacts. Bureaucracy did not always move as fast as activists wished and implementing their vision was a time-consuming and gradual process, as the next two chapters will discuss.[202] In addition, despite the widely circulating ideas about democratization and participation at the time, citizens and cyclists' real influence over planning procedures remained limited. Still, in countries like the Netherlands and Denmark, the lines between citizens and urban policymakers were relatively short, cycling's status was rising, and more got done for cyclists as a result than elsewhere.

Participation did have its limitations. Good contacts with policymakers were relatively easy to come by. It did not immediately lead to concrete results. Activists wanted results fast, and civil servants' slow response frustrated them. Despite many statements of support for cycling policies in city councils, the actual implementation of policies lagged behind. At a Cyclists' Union study day in 1980, the pro-cycling The Hague council member R.W. Heus (a VVD conservative liberal and engineer by training) and his social democrat colleague Burgers-Molendijk both acknowledged the gap between political ambitions and implementation: "The plans that civil servants at city hall design do not connect enough to the ideas in the city council." The true "promotion of slow traffic" by public works engineers did not conform with the wishes of legislators and activists.[203] By 1982, the Amsterdam cycling action group Amsterdam Fietst, supported by the Cyclists' Union, organized its first mass demonstration in four years: Their ideal bicycle city was still a (distant) ideal because a "lack of political will" and the inertia of the municipal bureaucracy stood in the way of change. The (young) activists had to learn how (local) politics worked and not let themselves "be brushed off with pseudo commitments [*pseudo-toezeggingen*]."[204] The choice to cooperate with city government had been deliberate because the Cyclists' Union wanted "to get out of the phase of hollow phrases" to achieve something. But change was slow to come.[205]

Results were not just slow; citizens' roles in shaping policy did not go very deep either. Kalk held a lecture on "Communication between

202 Oldenziel and Albert de la Bruhèze, "Amsterdam: World Bicycle Capital, By Chance," 24.

203 "Fietsers hebben weinig aan de politiek," *Trouw*, June 10, 1980.

204 Paul Godefrooij and Toni Niël, "'Rijkswegen, dat is de grootste prioriteit in Brabant'," *De Vogelvrije Fietser* 9, no. 3 (1984): 18-19, here 19.

205 "Fietsvriendelijke stad is nog toekomstmuziek," *Trouw*, May 14, 1982; Jan Ploeger, "De komende tien jaar zouden wel eens oogstjaren kunnen worden," *De Vogelvrije Fietser* 11, no. 1 (1985): 6-7.

Citizen and Government" at the 1978 National Traffic Safety Congress.[206] Here he criticized the government's idea of participation, distinguishing "participation-for-the-government" from "participation-for-residents." According to Kalk, the government used the former, seeing participation as a "social technique" to gain insight into public opinion, a process that took place on the government's terms and meant citizens joining in the conversation but not the decision-making. Ideally, in Kalk's view, experts should service citizens who design their own plans with experts' help. In this way, "the weakest groups of society" in particular are not only heard but allowed a direct role in decision-making. Through the Cyclists' Union surveys, Dutch traffic planners gained valuable insight and data without having to use their manpower or hire expensive consultants. But was this more than tweaking a system, whereby the basic tenets were only indirectly open to participation through elections?

The Cyclists' Union was also worried about its own efficacy. It identified four sources of power: position, expertise, numbers, and publicity. Position meant having a place in the "circuit of organizations and committees involved with traffic policy in the Netherlands."[207] A seat at the policymakers' table allowed them direct influence, a timely awareness of new policies, and the possibility of harmonizing activities with other actors involved in cycling. Lack of manpower made it difficult for the Cyclists' Union to fully utilize these possibilities. It had no complaints about access: within a decade, members had gained access to the relevant policy circles.[208] For expertise power, the process was similar. The Cyclists' Union had built up a lot of (user) expertise and was regularly asked to join local working groups or committees. Finding knowledgeable volunteers with time on their hands who also could function well in governmental circles was harder. Regarding power in terms of numbers, the Cyclists' Union membership at 14,000 around 1985 was modest and declining. The organization faced the limited power of membership support (which the ANWB could claim) as well as revenues from membership fees.[209] The massive street protests and many local actions

206 E. Kalk, "Communicatie tussen burgerij en overheid," in *Congresverslag: de veiligheid van voetgangers, fietsers en bromfietsers binnen de bebouwde kom. Verslag van het Eerste Nationale Verkeersveiligheidscongres, gehouden op 19 en 20 april 1978 in het RAI-congrescentrum te Amsterdam*, ed. Stichting Wetenschappelijk Onderzoek Verkeersveiligheid SWOV ('s-Gravenhage: Koninklijke Nederlandse Toeristenbond ANWB, 1978), 40-41.

207 IISG, Archive Fietsersbond, inv. no. 22, *Kruipolie* magazine, October 1985, 'Concept beleidsplan 1896-1988', 8.

208 Ibid.

209 Ibid., 9.

throughout the 1970s demonstrated the huge public support – as did the stable cyclists' modal share of 20-30 percent (in cities). The Cyclists' Union believed it could credibly claim to represent a much larger share of the population than its membership numbers suggest. Publicity came from street protests, though in recent years "the willingness to carry out (street) actions within the ENFB is declining," an internal document acknowledged in the late 1980s.[210] Media attention for this type of activism, moreover, was harder to generate: the novelty had worn off.

After a successful decade of public protests, the Cyclists' Union concluded there was no longer need for street happenings and protests by 1985. Drawing attention to the positive aspects of cycling and the negative impact of cars had been achieved as government policy documents started to resemble the Cyclists' Union view – at least on paper.[211] Certainly, implementation of solutions for the bottlenecks it identified, or the cycling paths it requested, was slower than expected. But the Cyclists' Union's Tom Godefrooij believed that his organization had changed policymakers' willingness to consider cycling in traffic planning.[212] Other members, looking back at this first decade, concurred: a change in mentality was taking place among traffic engineers, a new generation of more bicycle-friendly traffic engineers came into office, and some activists became civil servants.[213] From the start, the organization had been "an assembly and breeding ground for (cyclist) traffic expertise [*(fiets)verkeersdeskundigheid*]."[214] And in time, the Union expected the demand for such expertise to increase. Politicians no longer needed much convincing but (local) traffic engineers often tried to block change by arguing that the Cyclists' Union preferred solutions were technically impossible. Activists believed they needed to provide detailed traffic expertise rather than organize street actions to combat these arguments.[215] While this expertise was important for solving bottlenecks and, importantly, traffic deaths were decreasing, by 1985, the Cyclists' Union was dissatisfied with its failure to truly convince both the authorities and drivers of the need to limit car use. Facilitating both car and bicycle use was, to an extent, contradictory. If car and bicycle interests clashed, the

210 Ibid.

211 Ibid., 5.

212 Cees van den Berg, "Op de barricaden voor de fietser," *Algemeen Dagblad*, April 17, 1986.

213 Jenny Schoemaker, "Uitzicht op succes is geen voorwaarde om door te zetten," *De Vogelvrije Fietser* 11, no. 1 (1885): 9-11.

214 IISG, Archive Cyclists' Union, inv. no. 22, *Kruipolie* magazine, October 1985, 'Concept beleidsplan 1896-1988', 5.

215 Ibid.

car still usually won.[216] Politicians also saw "spending 3 million on the car as a bargain, but 1 million for the bicycle as outrageously expensive." As a result, cycling in Dutch cities in the late 1980s was still a matter of hopping on and off the sidewalk, maneuvering past bollards and parked cars, and struggling to find a parking space.[217] Transforming streets and mobility patterns was a long-term process. In short, the Union may have moved relatively quickly from its oppositional position into policy circles, but it meant giving up more militant anti-car protesting.

6.4 Conclusion

In a political system where non-governmental organizations could be part of the policymaking – and in a decade when the car came under heavy fire – the Cyclists' Union quickly carved out a major role in cycling governance. As a key intermediary between state officials, users, and other interest groups locally, the federated Cyclists' Union became an important player in the local Dutch cycling policy network.[218] It successfully led the social questioning of the distribution of (urban) road space (and traffic risks), and the reappraisal of urban cycling as a public good.

Activists made several claims. First and foremost, they denounced the state's negligence in failing to protect vulnerable road users, and cyclists were particularly at risk. Concerning the causes of these risks, activists pointed to the lack of safe cycling infrastructure in cities and villages – a consequence of prioritizing automobility when allocating road space and infrastructure budgets. The activists also claimed that more cycling would have major benefits for society as a whole. Bicycles took up much less space than cars and did not produce noise or exhaust fumes. The bicycle was a primary vehicle for transforming the car-dominated city into a livable cycling city.

In terms of chronology, different waves of activism influenced and complemented each other. Groups like the anarchist Provo and the Kabouters succeeded in alerting policymakers to the fact that citizens had their own voice and demanded to be heard – and also introduced the bicycle as symbol to imagine an alternative future. The many local protests in

216 Niek Schenk, "Fietser wil meetellen," *Algemeen Dagblad*, May 6, 1989.

217 Schenk, "De fiets als redding."

218 Similarly, environmental interest group *Stichting Natuur en Milieu* was a spider in the environmental actors' web. H.A. van der Heijden, "Zandzakken, notenbomen en een taart. Nederlandse sociale bewegingen en mondiale klimaatverandering," *Jaarboek voor Ecologische Geschiedenis* 2008 (2009): 117-143, here 137-38.

the 1960s and 1970s lent the parent-teacher organization Stop the Child Murder considerable political leverage: the parents and teachers could not be dismissed as a bunch of extremists. It became the fountainhead for widely supported Dutch anger over traffic conditions for their children. The young Dooievaar activist-architects pioneered a counter-expertise-based method of translating this anger into practical and user-based designs for improving the urban space. All these currents came together in the Cyclists' Union's activism after 1975.

Some aspects of Dutch cycling activism can also be found in Denmark and Switzerland. In particular, an expertise role for cycling activists also emerged in other high-cycling-level countries with consensual political traditions. According to Emanuel the Danish cyclist federation DCF, like its Dutch counterpart, developed a bicycle plan for Copenhagen, and while not officially adopted, "planners did use the proposal as a blueprint to develop the bicycle network in Copenhagen in the following years."[219] The capital's pro-bicycle mayor helped the cause; historically, the Danish case also shows how good relationships between action groups and (local) government have played a crucial role in creating bicycle cities. In Switzerland's similarly consensual political model, cycling activists also enjoyed good relationships with local government.[220] The Zurich cycling action group IG Velo (*Interessengemeinschaft Velo*) participated in the "Velopicknick": regular meetings between city representatives, public works, and police departments. Initiator of these meetings in 1980 was Ruedi Aeschbacher, a politician who took an interest in cycling.[221] Looking back on this cooperation, Aeschbacher described IG Velo as a "decent but persistent" lobby which, like the Dutch Cyclists' Union, contributed to the governance network by providing expertise local governments lacked (*Facharbeit*).[222] Relationships between the cyclists and authorities professionalized quickly like in the Dutch case.[223]

In many countries, such as the US, cycling activism was (and remains) much more confrontational and antagonistic.[224] Dutch activists were well-

219 Emanuel, "Making a Bicycle City," 512.

220 The Netherlands and Switzerland differ in other aspects: Kriesi, "Federalism and Pillarization: The Netherlands and Switzerland Compared," 539-56.

221 Mijnssen-Hemmi and Boller, *30 Jahre Veloförderung in der Stadt Zürich, 1975 bis 2005*, 7-11, 21.

222 Ibid., 21.

223 Ibid., 29.

224 In a telling statistic cited by cycling historian Evan Friss, New York City spent twice as much on policing Critical Mass bicycle activists in the early 2000s than on actual cycling path

informed about cycling and worked with state officials and engineers to spread engineering knowledge on cycling, for instance by writing a design manual. There was less cycling knowledge in the US and New York city's Department of Transportation only started working on this around 2009.[225]

From an international perspective, it is also worth noting that the different cycling conditions from where Dutch activists started also influenced their strategies and discourse. American cycling advocates, whom Zack Furness analyzed, mainly focused on environmental issues. In Dutch activism, the safety argument was much more powerful because there was a much larger group of active (and endangered) cyclists. This is not to say that ecological arguments were absent from the Dutch action scene; cycling advocates constantly used those arguments to make a case for cycling. The safety argument made Dutch cycling activism simultaneously more urgent and political: the image of endangered children and concerned parents was explosive in ways that no politician could ignore. In another sense, it was also less political because almost everybody could get behind it. Pragmatic cooperation with the government paid off for Dutch activists. In lower-cycling-level contexts, protest seems a much more apt strategy.[226] American cycling activists of the 1970s recall being firmly outside the system or even being marginalized.[227] We cannot say the same about Dutch cycling activists in the same period.

In the 1970s, the consensus in the Netherlands about cycling was that it was far from disappearing as travel mode. Instead, it was valuable and should be promoted. This chapter has shown that the political process of making a claim for cycling, then convincing policymakers and engineers to take cyclists more seriously, was crucial. However, turning convictions into reality – and changing the distribution of road space – took decades and required an arduous governance process. We turn to this topic in the next chapter.

construction (Friss, *On Bicycles*, 153). Policies redistributing space away from cars to bicycles are extremely controversial. When in 2010 Janet Sadik-Khan's Department of Transportation proposed a bike lane in New York City to replace a car lane and parking spots, a powerful coalition of opponents sued the department (ibid., 164). This coalition consisted of Brooklyn borough president Marty Markowitz, Iris Weinshall, Sadik-Khan's predecessor at the DOT, former deputy mayor Norman Steisel, congressman and aspiring mayor Anthony Weiner, and prestigious law firm Gibson Dunn LLP, who worked pro bono to sue Sadik-Khan.

225 Ibid., 159.

226 Furness repeats this argument for the Montréal-based action group Le Monde à Bicyclette, which tried to work the government by letter-writing and other official means, but whose protests and street actions were much more successful: Furness, *One Less Car*, 64.

227 Furness, "Biketivism and Technology," 67.

7 Catching Up: The State Acknowledges Urban Cycling as Public Good, 1975-1990

"Citizens couldn't care less whether the national or provincial government provides a solution, but a solution has to be found."[1] With these words, Christian democrat MP Pam Cornelissen drew attention in 1975 to the urgent need for new cycling infrastructure – and the lack of understanding citizens had for governments looking at each other to take up this task. Ten years later, in 1985, at another parliamentary meeting, social democrat MP Hessel Rienks sympathized with parents in rural areas where school children had to use dangerous and poor roads without cycling paths. Meanwhile, in these same regions there were new, comfortable asphalt paths for recreation, which only attracted cyclists in the summer. Unaware of the governance system producing this outcome, locals could only conclude that for the government, "our daily existence is less important than attracting money-spending recreational cyclists."[2] What governance struggles were Cornelissen and Rienks highlighting? And why were more pressing utilitarian cycling paths sometimes built later than these recreational ones?

From the mid-1970s onward, policymakers at the national, provincial, and urban level tried to address cyclists' demands. Giving voice to these demands in parliament, the national Public Works department invested more in cycling. Wresting these concessions from the national government was a tough battle for national politicians. As a result, cyclists saw that the changes to the streets they used daily were slow to come. Behind the scenes, civil servants and Public Works engineers were arguing about the division of cycling governance responsibilities. Which role should each of the three government levels take? Who would govern recreational cycling and utilitarian cycling? Finetuning was a long process, and sometimes led to puzzling outcomes, when more urgent issues were addressed later or not at all.

Historians have shown how in the 1970s and 1980s urban policymakers sought to reverse car-centered policymaking. The implementation of measures to make cycling safer and more attractive, however, was a gradual and challenging process. Dutch urban policymakers responded to citizens'

1 Permanent Committee for Public Works (*Vaste Commissie Verkeer en Waterstaat*), meeting minutes February 3, 1975, 523. All parliament records are available at https://zoek.officielebekendmakingen.nl/uitgebreidzoeken/historisch.

2 Vaste Commissie voor Verkeer en Waterstaat, meeting November 11, 1985, 5.

demands for cycling policies by experimenting with new traffic measures and cooperating with activists. In contrast, we have little insight into what role, if any, provincial and national governments played in supporting or obstructing these urban policy goals. This chapter places our existing knowledge of urban cycling governance in the 1970s and 1980s in a wider context of provincial and national authorities as well as interest groups. It does this by using new material, including records from parliamentary committees, the Ministry of Public Works, and provincial and municipal agencies.

The chapter demonstrates the problematic relationship between government levels in cycling governance. As politicians demanded engineers step up their investment in and engagement with cycling, national Public Works officials remained reluctant to take up a larger role. The records show that politicians put pressure on the civil service to take cycling more seriously. Civil servants found translating that into actual policies more problematic. While the Ministry did subsidize cycling for some time – and with concrete results – it never firmly believed that cycling fell under the mandate of the central government and that funding should not be made permanent. Investments in automobility were of a different order of magnitude and remained a larger concern for state engineers than cycling. Deeply rooted Dutch policy beliefs about decentralized and local governance influenced this. When neoliberal ideas about a smaller government became internationally dominant in the 1980s, the modest national role in cycling governance temporarily disappeared altogether. I also show that the cooperation between various ministries was difficult because of the distinct funding streams and governance procedures engineers insisted on making between utilitarian and recreational cycling. The chapter ends in the late 1980s, when the national government decided to withdraw from cycling policy and leave it to lower government.

7.1 Expanding National Cycling Governance, 1975-1985

By the mid-1970s, social movements from Stop the Child Murder and Cyclists' Union to environmental groups, particularly in cities, had created broad public support for prioritizing cycling over driving. The pressure from activists, with broad backing from society, reached parliament, where many MPs pushed the Ministry of Public Works to invest more in cycling. In these debates and the wider public discourse, no one opposed doing something about the lack of traffic safety. What that something should be, and more importantly, who should do it, were less clear. Having reduced their role in cycling governance mid-century, the national government's engineers and civil servants faced a challenge. Historically, national engineers had little involvement in infrastructure policies within city limits. The national cycling policies engineers created in the 1920s only applied to national and provincial authorities' roads, not in cities. Now, society and politicians demanded national engineers also adopted a supportive role in urban cycling policies. It was a new role. Creating new governance links proved demanding in this multi-level governmental setting where traditionally few ties existed. Additionally, the competing frames of recreational and utilitarian cycling complicated governance arrangements in practice.

The calls for cycling funding, which were eventually heeded, took place in a worsening economic context. After the fall of the progressive Den Uyl coalition in 1977, the following governments were more conservative in orientation. Both the ideological and the economical background explain why the national government increasingly favored austerity measures in the 1980s. During the first Van Agt government (1977-1981) and second one (1981-1982), national investments in cycling put in place during the Den Uyl government continued. It was only in the mid-1980s under the Lubbers I and II governments (1982-1989) that ideas about small government and privatization led to the national government's (temporary) withdrawal from the cycling governance coalition.[3] In this period, the responsible ministers of Public Works were Tjerk Westerterp (1973-1977), Danny Tuijnman (1977-1981), Henk Zeevalking (1981-1982), and Neelie Smit-Kroes (1982-1989). They had to grapple with a parliament which often denounced the ministry's minor role in stimulating cycling.

MPs took up the defence of cycling infrastructure as a public good in the 1970s. They had come to acknowledge utilitarian cycling facilities as a public good in the 1970s and demanded more state funding and cycling policies. In some respects, this was reminiscent of the 1920s, when many MPs extolled the virtues of cycling for the nation in protest against the bicycle tax. This time policymakers were motivated by the negative effects of driving, not just on cyclists but also pedestrians, urban livability, and the environment. The start of this check on the almost unlimited growth of automobility became visible in 1976 when, in a highly symbolic move, parliament voted to forbid car parking in its own historic Binnenhof (Courtyard) at The Hague's parliament buildings.[4]

3 Remieg Aerts et al., *Land van kleine gebaren: Een politieke geschiedenis van Nederland, 1780-1990* (Nijmegen: SUN, 1999), 315-24.

4 *Handelingen II* 1975/76, 89 (June 9, 1976), 4537.

Politicians argued that cycling should be part of the state's mobility plans. In 1976, under the left-wing ruling coalition of Den Uyl, with Tjerk Westerp as Minister of Public Works, the national government increased its focus on traffic safety. The Permanent Committee for Public Works – a committee of MPs who had mobility in their portfolio – discussed the Multi-Year Plan for Personal Transport (*Meerjarenplan Personenvervoer*), a programmatic policy document discussing the future of automobility. According to MP Koningh, of the small left-wing DS'70 party, the Plan ignored the bicycle. A former dockworker and union organizer, Koningh asserted that his own trajectory through transport modes was quite typical: when he started with the union he rode a bicycle, which he exchanged for a moped, and ended up driving a car; this was "about common practice."[5] Despite using a car, Koningh advocated limiting the car's access to the city. The car threatened to make all other forms of transport impossible and "the government has to put a check on [*paal en perk stellen*] car use in the city."[6] According to Koningh, the car "takes up a disproportional amount of space for the number of people it moves," while in the form of the bicycle, the solution was already at hand. Sales of bicycles were on the rise and cycling was still very much common practice, he said.[7] This opinion was widely shared in the Public Works committee – left-wing and centrist parties talked about cycling in a very positive light. Tjerk Westerterp, the Minister responsible for a series of safety measures to stem traffic accidents at the time such as the moped helmet and seatbelt rule, rightly concluded in 1975 that "the general opinion emerging from this committee is that we have to urge lower government to prioritize cycling paths and other facilities for the weaker road users."[8] This meant a novel role for national government.

Partly from personal experience, MPs insisted the Ministry of Public Works do more to help local government make cycling a better option. MPs on the Permanent Committee for Public Works, who consistently lobbied for the bicycle, included the social democrats Van der Doef, Rienks, and Castricum (PvdA), the progressive liberals Imkamp, Zeevalking, and Tommel (D'66), as well as members of smaller left-wing parties like Koningh (DS'70) and Van der Heem-Wagemakers (PPR).[9] This advocacy is illustrated by one

5 Vaste Commissie voor Verkeer en Waterstaat, meeting May 10, 1976, 968.

6 Ibid.

7 Ibid.

8 Vaste Commissie voor Verkeer en Waterstaat, meeting February 3, 1975, 521.

9 Meeting November 11, 1973, 1-2, 8; Meeting February 2, 1975, 507, 520-21; Meeting October 27, 1975, 177-84; Meeting May 10, 1976, 968; Meeting May 17, 1976, 989, 1001-1002; Meeting February 2, 1978, 120; Meeting October 16, 1978, 73, 78-80.

of the most vocal supporters of cycling, MP Pam Cornelissen of the Catholic KVP party (which merged with other Christian parties to form CDA in 1980). A civil engineer, Cornelissen was a project manager before entering the Lower House (Tweede Kamer) in 1967. He also served as councilor for the municipality of Best (1966-1974) and on the regional Eindhoven board (1966-1970).[10] Cornelissen's allegiance to cycling has something to do with the fact he went everywhere by bicycle.[11] He had also cycled 800 km recreationally throughout the country and "experienced that all is not well with the Dutch cycling paths."[12] Cornelissen was very concerned about traffic safety. He was from Noord-Brabant, the founding location of Stop the Child Murder and a province where traffic casualties were relatively high during the peak of traffic-related deaths in the Netherlands. Cornelissen's local research showed that 20 percent of the fatalities among cyclists and moped riders were in his province. Among the victims was a considerable number of schoolchildren because many of them had to cross busy roads.[13] All this made him one of the most persistent cycling advocates in Dutch parliament.

Cornelissen often advocated more national cycling funding in the Permanent Committee for Public Works.[14] In the late 1970s, he initiated several parliamentary resolutions demanding a larger budget for cycling paths. As the important 1980 national policy document Structural Scheme for Traffic and Transport (*Structuurschema Verkeer en Vervoer*) failed to mention bicycles, Cornelissen proposed adding a section on cycling.[15] The Lower House accepted this proposal and it became officially part of the final document.[16] In 1980, he called for a national "policy plan for slow traffic." Arguing that the Netherlands "is still a cycling country and should remain so," he emphasized the importance of different types of cycling. Besides recreational, he mentioned cycling as the dominant school transport, and

10 Data kindly provided by Parlement.com: (https://www.parlement.com/id/vg09lkzd26zc/p_a_m_pam_cornelissen)

11 Bijzondere Commissie voor het Structuurschema Verkeer en Vervoer (14 390) en het Meerjarenplan Personenvervoer 1980-1984 (15 885), meeting March 10, 1980, 1208.

12 Vaste Commissie voor Verkeer en Waterstaat, meeting October 27, 1975, 177.

13 "'Brabant is onveiligste provincie'," *Het Vrije Volk* April 9, 1975, 7; "In het kort," *Verkeerskunde* 26, no. 6 (1975): 318.

14 See minutes of following committee meetings: November 26, 1973, 1-2; February 3, 1975, 520; October 27, 1975, 177-78, 182; February 27, 1978, 120; October 16, 1978, 78.

15 *Kamerstukken II*, no. 14390, sub-no. 67. See Bijzondere Commissie voor het Structuurschema Verkeer en Vervoer (14 390) en het Meerjarenplan Personenvervoer 1980-1984 (15 885), meeting two, 1001.

16 *Handelingen II* 1980/81, 23 (November 18, 1980), 1332. Government's proposal of SVV was part D, while Parliament's approved final version was part E.

as before-and-after transport for railway passengers.[17] An integral policy plan to support these activities – which Cornelissen called "friendly to the environment, requiring little space or oil, and healthy" – was lacking. The Cyclists' Union had already written a document laying out such a policy plan, which Cornelissen recommended the national government used.[18] The majority support in Parliament for many of these proposals shows how cycling's status as a public good increased in the 1970s.

The lack of formal institutions and rules on cycling governance meant the national government tried to devolve the responsibility to local authorities. Cyclists were not aware of these internal debates. On multiple occasions, Cornelissen confronted Public Works Minister Westerterp with the lack of cycling paths in his province.[19] Westerterp's response was that he did not have authority over provincial roads. Lower government autonomy had to be respected at all costs. Cornelissen did not accept this, arguing that the 25-million-guilder subsidy for 1976 was not nearly enough to address "the enormous backlog we have in the Netherlands" in cycling path construction. It was "insufficient if the minister says he is willing to consult with the province about spending funds. I believe as Minister of Public Works, he has a clear responsibility and should be willing to stimulate the construction of cycling paths."[20]

The parliamentarians repeated pleas for the national government's more proactive role were met with reluctance from the Public Works ministry. Nevertheless, politicians voted for national funds during this period that helped lower government's cycling policy with subsidies paid from the national Public Works roads budget, the National Road Fund (*Rijkswegenfonds*). The progressive Den Uyl government (1973-1977) jumpstarted the process. Between 1975 and 1985, Parliament facilitated local investments in cycling, prompting Minister Westerterp at the start of the period to remark that "we [national Public Works] are starting an entirely new project."[21] Compared to the 1950s and 1960s, the minister believed this national involvement constituted a significant break in cycling governance. Apparently he was

17 Bijzondere Commissie voor het Structuurschema Verkeer en Vervoer (14 390) en het Meerjarenplan Personenvervoer 1980-1984 (15 885), meeting February 18, 1980, 998. Cornelissen cited numbers showing 30 per cent of train passengers traveled to the station by bicycle. In the city of Gouda, it was even more than 50 per cent.

18 Ibid.

19 Vaste Commissie voor Verkeer en Waterstaat, meeting February 3, 1975, 523; Vaste Commissie voor Verkeer en Waterstaat, meeting October 27, 1975, 182.

20 Vaste Commissie voor Verkeer en Waterstaat, meeting October 27, 1975, 182.

21 Ibid., 181.

unaware of the government's history or his own agency's role during the 1920s and 1930s, when national politicians and engineers had already engaged with cycling as a result of the bicycle tax controversy and political compromise. Since then, the national Public Works departments had increasingly left cycling to lower authorities and were now facing the problem that the national government lacked the knowledge to create policies. Westerterp admitted that it had taken a long time for the ministry to gain insight in each province's specific situation and problems: decentralization created a problematic knowledge gap at national level, particularly regarding cycling.[22] Similarly, when MPs requested data on cycling safety or progress with cycling path construction, the ministry often did not have these numbers, and had to turn to the provincial public works departments.[23]

The national government employed cycling subsidies for lower governments as its main policy tool in response to pressure from the country's activists and politicians. Local government had always been able to construct cycling paths along new or reconstructed highways because they had control over regional and local road design and engineers considered cycling paths next to these roads an integral part of the road. The new subsidy indicated that lower authorities should prioritize constructing cycling paths where these were still missing but urgently needed.[24] In the late 1960s and early 1970s, cities could request a 50 percent subsidy for traffic projects, including cycling. In 1974, under Terpstra's direction, this level increased to 80 percent for cycling paths and a new subsidy of 50 percent for paths along roads outside built-up areas. The funding scheme reinforced the Dutch decentral model: the national government had limited options to steer the provinces and municipalities' policies, but did control almost all revenue and its distribution. By creating favorable subsidy conditions for certain policy objectives like cycling infrastructure, national policymakers could shape local priorities.

As figure 27 shows, from 1975 to 1985, the national government spent between 25 and 50 million guilders annually on local cycling facilities. In 1976, the Ministry approved 30 projects for cycling paths and a few cycling tunnels and bridges, averaging from 50,000 to 2.5 million guilders per project. Many local cycling projects were realized thanks to this national

22 Vaste Commissie voor Verkeer en Waterstaat, meeting February 3, 1975, 520.

23 See Tweede Kamer, *Aanhangsel van de Handelingen*, 1974-1975 no. 607 and 1975-1976 no. 728.

24 NA 2.27.5215 (Archief Ministerie CRM, Beleidsterrein Recreatie), inv. no. 2167, document "Aantekeningen naar aanleiding van een gesprek bij het Ministerie van Verkeer en Waterstaat op 10 september 1974 over rijksbijdragen in aanlegkosten rijwielpaden," 1-2.

subsidy (see 7.3 and Appendices 2 and 3).[25] In 1977, the national government subsidized local communities for 30 million guilders and gave provincial engineers another 25 million to construct separate cycling paths alongside provincial roads.[26] Compared to the highway budget (figure 28), however, these numbers were still negligible.

Under pressure from politicians, who in turn reacted to social and activist pressure, national state involvement came about but remained limited to modest national subsidies and centralizing expertise. National politicians strongly supported more (national) cycling policies within the Public Works department, but regarded civil servants and engineers' policy response as half-hearted. Despite the efforts of people like Cornelissen, the Ministry of Public Works never fully committed to a central role in cycling governance. Labor MP Henk de Hamer (PvdA), a former councilor for the successful cycling city Groningen, criticized the ministry in 1981 for failing to address "policy" for cyclists and pedestrians: "There is coordination, regionalization, research, and awareness, but still no policy."[27] In 1980, De Hamer, like Cornelissen, had demanded that the ministry compile a Multi-Year Plan for Slow Traffic (*Meerjarenplan Langzaam Verkeer*) while debating the Structural Scheme for Transport (*Structuurschema Verkeer en Vervoer*) and the Multi-Year Plan for Personal Transport (*Meerjarenplan Personenvervoer*) of 1980-1984. When Parliament demanded cycling policy, De Hamer argued, the Public Works minister tended to produce "beautiful phrases" but "no real policy."[28] In practice, cyclists were still discriminated against, for instance by giving car drivers plenty of parking spaces at shopping centers but providing no bicycle parking. De Hamer's plan for a separate cycling (and walking) policy document won majority support from the political left and the middle (PvdA, CDA, D'66, and PPR), and was adopted in November 1980. However, these resolutions were not binding. Under prime minister Van Agt's new and more right-leaning government coalition, Public Works minister Danny

25 *Kamerstukken,* no. 13711, sub-no. 6, 16-17.

26 "In het kort," *Verkeerskunde* 29, no. 6 (1978): 274-75. The distribution over the provinces (in guilders) was:

– Groningen	1,262,000	– Utrecht	1,840,000
– Friesland	933,000	– NH	2,544,000
– Drenthe	2,710,000	– ZH	7,113,000
– Overijssel	1,000,000	– Zeeland	769,000
– Gelderland	2,329,000	– Noord-Br.	5,105,000

This distribution was related to the provinces' population density and road network.

27 *Handelingen II* 1980/81, 24 (November 19, 1980), 1374.

28 Bijzondere Commissie voor het Structuurschema Verkeer en Vervoer (14 390) en het Meerjarenplan Personenvervoer 1980-1984 (15 885), meeting February 18, 1980, 1017-18.

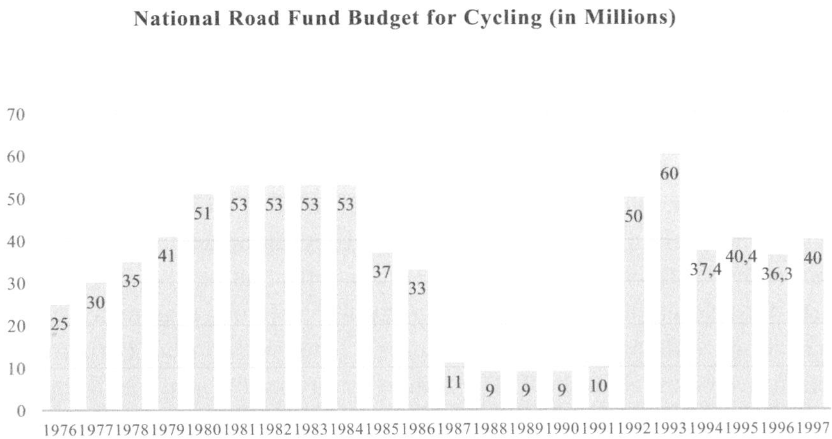

Figure 27 The annual amount reserved for cycling in the National Road Fund budget. Clearly visible is the decentralization in the mid-1980s. Small sums were invested due to prior agreements, but in the late 1980s, national government contributed very little to cycling infrastructure. Source: Annual budgets National Road Fund (Rijkswegenfonds) until 1993, when its name changed to Infrastructure Fund (Infrastructuurfonds).

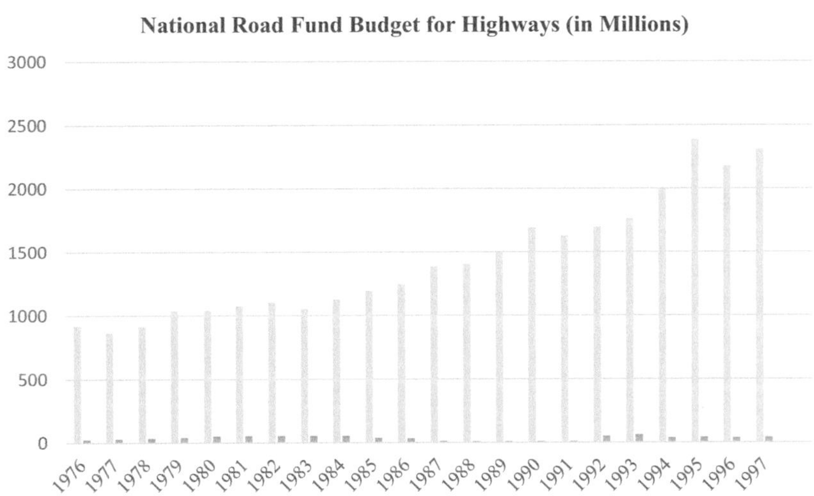

Figure 28 Investments in cycling fluctuated, but the national highway budget remained stable in the 1980s. The cycling budget is shown here in dark, illustrating the very minor investments in cycling compared to highway budgets. Source: Annual budgets National Road Fund (Rijkswegenfonds) until 1993, when its name changed to Infrastructure Fund (Infrastructuurfonds).

Tuijnman (VVD) objected, stating his officials were already overburdened. His remark revealed that cycling and walking were not that much of a priority after all for national engineers – and as minister of a pro-car liberal party, he was not going to push either.[29]

Some politicians opposed a national role in cycling governance. Few if any, however, openly contested the social value of either utilitarian or recreational cycling in the Netherlands. Vocal opposition to bicycle planning was negligible. Occasionally members of the conservative liberal VVD, the party most closely associated with pro-car policies, argued that money spent on cycling would be better spent on cars.[30] However, another Permanent Committee member for the VVD, Herman Lauxtermann, a builder from the Catholic south, made it known he was a keen recreational cyclist. He may not have used the bicycle for other purposes, but his extensive recreational cycling made him question how the various ministries coordinated their cycling subsidy programs.[31] Lauxtermann demonstrates that (recreational) cycling survived the 1960s in the Netherlands much better than in surrounding countries and made a difference to garnering a wider support base for pro-cycling policies in the 1970s, even among the most car-minded politicians.

The new subsidies concerned traffic needs, and what engineers and politicians considered utilitarian cycling paths. Older funding schemes for recreational cycling infrastructure continued to be in place to boost provincial tourism and recreation. The Ministry of Culture and Recreation (CRM) program funded recreational cycling infrastructure. The Ministry of Agriculture – responsible for the heavy agricultural machinery traffic on narrow country roads – funded a small subsidy program for rural (quaternary) roads.[32] According to the ministry in 1978, the situation on rural roads had received "heightened attention" recently "especially with regard to the safety of school children."[33] In all likelihood, the many protests by parents of school-going children, described in the previous chapter, contributed to this sense of urgency. The Land Use Service (*Landinrichtingsdienst*),

29 *Kamerstukken II* 1979/80, no. 14390/15885, sub-no. 90; Vaste Commissie voor Verkeer en Waterstaat, meeting November 3, 1980, 2; *Handelingen II* 1980/81, 25 (November 20, 1980), 1489.

30 Smit-Kroes (later minister of Public Works) at meeting May 10, 1976, 954; De Beer at meeting October 18, 1976, 2.

31 Bijzondere Commissie voor het Structuurschema Verkeer en Vervoer (14 390) en het Meerjarenplan Personenvervoer 1980-1984 (15 885), meeting February 18, 1980, 1004.

32 NA 2.27.5215, inv. no. 2167, letter Minister of Agriculture to Minister of Public Works, June 5, 1978.

33 Ibid.

the agency dealing with land consolidation and spatial planning in the countryside, awarded these subsidies. In the late 1970s and early 1980s, the Ministry of Agriculture spent 5 million guilders on constructing roughly 20 km of cycle tracks every year.[34]

The varying subsidies also tell us something about priorities. Politicians and officials presumed they could classify cycling infrastructure based on whether its main function was recreational or utilitarian. The higher level of subsidy for recreational cycling infrastructure shows that national policymakers attached great importance to providing recreation facilities. The underlying idea was that the facilities would attract tourists – and as government spending, the investment in cycling paths would provide a good return. Some politicians disapproved, arguing that citizens were justified in criticizing this logic. That is why social democrat MP Hessel Rienks concluded, "our daily existence is less important than attracting money-spending recreational cyclists."[35] The subsidy program for recreational cycling was rarely contested politically and never really disappeared. In contrast, the focus on utilitarian cycling infrastructure only returned after the heavily contested political process when city activists put it on the agenda. The diverse subsidies created conflicts between national ministries as well as confusion between national and local authorities.

These subsidies were not integrated, despite pleas by politicians and activists. The Ministry of Public Works rejected activists' calls for it to function as national cycling policy coordinator. Action group Stop the Child Murder wrote to the Ministry of Culture and Recreation (CRM) proposing a national cycling path plan to connect all the existing subsidy arrangements.[36] They argued that the subsidy system "led municipalities to construct cycling paths in the strangest places, depending on their inventiveness."[37] Indeed, it was almost a game for local policymakers to hunt down funding – not necessarily connected to cyclists' actual needs. The activists were particularly appalled by the numerous recreational cycling paths constructed in forests and

34 *Kamerstukken II* 1982/83, no. 17745, sub-no. 2 (*Nota Fietsverkeer*), 39.

35 Vaste Commissie van Verkeer en Waterstaat, meeting November 11, 1985, 5.

36 MP Cornelissen asked the minister whether he had responded. Minister Westerterp had addressed the group in person but told them he considered cycling paths a task for lower government. The National Road Fund was meant for highways, and a national cycling path plan would also need funding, which he did not want to provide (Vaste Commissie van Verkeer en Waterstaat, meeting November 26, 1973, 1-2, 8).

37 NA 2.19.198 (Archief Stop de Kindermoord), inv. no. 233, press communiqué "Staatssecretaris Meijer wil nationaal fietspadenplan."

national parks, ignoring the much more urgent school routes for children. In response, the Culture and Recreation Ministry (CRM) supported this idea in principle and wrote to the Ministry of Public Works requesting "a certain form of coordination and synchronization." It also made clear that these "coordinating measures would have to be taken primarily by you, the member of government responsible for traffic."[38] In other words, responsibility for cycling governance should lie with the Ministry of Public Works. This overture resulted in a meeting in September 1974 between the three Ministries – Public Works, Culture and Recreation, and Agriculture – to "explore the mutual fields of activity regarding the construction of cycling paths."[39] Prior to 1974, at a national level, each department had worked in relative isolation. This illustrates the prevailing lack of central guidance and coordination in Dutch cycling governance as well as the lack of communication between Dutch ministries.[40]

CRM Ministry officials continued to be critical of Public Works engineers' refusal to work more closely throughout the late 1970s and early 1980s: separating utilitarian and recreational cycling was misplaced, as each cycling path should combine these features. "Concretely this would mean, from the perspective of outdoor recreation, that a cycling path constructed for a 'utilitarian' function like commuter or school traffic would also be made as attractive as possible, both for specific recreational use as for the recreational experience during 'utilitarian' use."[41] After specifying how to do this (route choice, embedding in the landscape, and attractive greenery), the Ministry concluded that "a cyclist, while cycling, should not be able to tell whether they were on a utilitarian or recreational path."[42] According to the officials, cycling paths should not have a predetermined function and cycling commuters would also benefit from pleasant routes. As the agency responsible for traffic, the Ministry of Public Works would have to take the lead on this. The procedures and organization of the subsidy arrangements had to be synchronized.[43]

38 NA 2.27.5215, inv. no. 2167, letter State Secretary CRM to Minister of Public Works, January 8, 1974; letter State Secretary CRM to Stop de Kindermoord, January 8, 1974.

39 NA 2.27.5215, inv. no. 2167, document "Aantekeningen naar aanleiding van een gesprek bij het Ministerie van Verkeer en Waterstaat op 10 september 1974 over rijksbijdragen in aanlegkosten rijwielpaden," 1.

40 Andeweg and Irwin, *Governance and Politics of the Netherlands*, 149, 59.

41 NA 2.27.5215 (Archief Ministerie CRM, Beleidsterrein Recreatie), inv. no. 1971, document "Notitie van het secretariaat n.a.v. de nota fietsverkeer van Verkeer en Waterstaat."

42 Ibid.

43 Ibid.

Other agencies also attempted to convince the Ministry of Public Works to take on a larger role in cycling governance – and failed. In 1977, for example, the Dutch provinces lobby organization (IPO) argued that it was strange that both Public Works and Agriculture ministries subsidized cycling infrastructure for traffic safety reasons. According to IPO, improving traffic safety was the Ministry of Public Works' mandate. It should find more flexible ways to subsidize cycling infrastructure. The provinces proposed closer cooperation between ministries without the non-sensical boundaries between subsidies.[44] The Minister of Public Works refused to budge; while admitting that the logic of the subsidies overlapped, he also argued that the well-funded Ministry of Agriculture should deal with municipalities' funding requests: there was no need for Public Works to cooperate with this Ministry.[45] Parliament questioned the lack of central coordination in 1980, during discussions about the Multi-Year Plan for Personal Traffic. Three MPs remarked that since the proposal was an integrated, holistic approach to transport, no other ministries could subsidize transport facilities.[46] The Minister of Public Works persisted: cycling subsidies had "clearly demarcated areas" and where there was an overlap, "consultation takes place on an ad-hoc basis" so that "a separate coordination agency is deemed unnecessary."[47]

Consequently, recreational and utilitarian cycling governance remained separate spheres within national government – an artificial institutional arrangement that clashed with the reality of cycling. Cyclists had little patience for distinguishing between utilitarian and recreational paths. The two emerging forms of cycling path networks somehow had to be interconnected. Policymakers observed that safe cycling facilities leading out of urban centers to recreational areas were particularly lacking.[48] The government official responsible for outdoor recreation, engineer K. Kuyken, realized his ministry, CRM, had to come up with a strategy to convince Public Works to fund the construction of these mixed-use paths. One problem was that CRM subsidized 75 percent of the costs but Public Works only 50 percent. Potentially, municipalities and provinces would all frame their cycling paths

44 NA 2.27.5215, inv. no. 2167, letter IPO to Ministry of Public Works, December 23, 1977.

45 NA 2.27.5215, inv. no. 2167, letter Ministry of Public Works to IPO, September 20, 1978.

46 *Kamerstukken II* 1979/80, no. 15885, sub-no. 53.

47 *Kamerstukken II* 1979/80, no. 15885, sub-no. 201, 5.

48 NA 2.27.5215, inv. no. 2167, document "Aantekeningen naar aanleiding van een gesprek bij het Ministerie van Verkeer en Waterstaat op 10 september 1974 over rijksbijdragen in aanlegkosten rijwielpaden," 2; NA 2.27.5215, inv. no. 2167, letter Minister of Public Works to Minister of Culture and Recreation, June 23, 1975.

as recreational in order to obtain the higher CRM subsidy.[49] This was the unequal fight for subsidies that Stop the Child Murder activists already observed at local levels in the early 1970s. In short, national politicians, responding to the electorate, demanded the government coordinate cycling policy, but the state engineers at the Public Works administration did not heed this call.

7.2 Frictions and Distrust: Struggles with Multi-Level Cycling Governance

By the mid-1970s, cycling's social status as a public good was high. So was political support on the highest levels. Yet controversy ensued about (multi-level) cycling governance: who was to be responsible for policy implementation? In the 1930s, national politicians had decided cycling should be governed locally – and because the Dutch administration was strongly decentralized, national policymakers resisted national governance of issues like cycling. Since the state controlled funding, and its allocation of funds could indirectly steer policy priorities, it still made sense for politicians to demand a national role.

How workable was this system for provincial and local engineers? It turned out those willing to improve cycling facilities but in need of funding locally found the government's rules, regulations, and funding schemes confusing and laborious. At a meeting between national government and the provinces in 1976, a member of the Utrecht Provincial Executive pointed out "the artificiality of the proposed distinction between different categories of cycling path, the various overlapping functions of cycling paths, and the fact that there are different subsidizers each with different subsidy arrangements."[50] Cities and provinces often struggled with the complexity and laboriousness of the subsidy system. The provinces, interceding between municipalities and the national government, played a crucial role in coordinating cycling governance.

Provinces expanded their role in cycling governance in response to the new financial opportunities created in the 1970s combined with the national

49 NA 2.27.5215, inv. no. 2167, letter Minister of Public Works to Minister of Culture and Recreation, June 23, 1975.

50 NA 2.27.5215, inv. no. 2167, document "Verslag gesprek tussen de minister van Verkeer en Waterstata en de colleges van Gedeputeerde Staten inzake de aanleg van fietspaden, op 19 februari 1976," 2.

government's lack of coordination. They assisted municipalities in raising awareness of subsidies as well as in administrative procedures. The confusing mix of subsidies, spread over no fewer than four ministries, required the provinces to take the lead in coordinating municipal funding requests as well as building plans, to ensure a connected provincial network.[51] The chief-engineer-director of the Utrecht Provincial Public Works department proposed that municipalities could be informed of new subsidies. They could pass on their requests to the province, which would collect and submit them at the national level.[52] In 1975, only 3 municipalities utilized the subsidy, rising to 18 in 1976, and peaking with 35 in 1979. Up to 1980, between 80 and 90 of the country's 811 municipalities had made use of the subsidies.[53] By the time local policymakers found their way to the funding, it was already abolished (see 7.4).

The central government's functional and financial separation of cycling types clashed with the municipalities' more integrated approach. Provincial policymakers and politicians repeatedly criticized the unworkability of the system at provincial lobby IPO meetings. They complained that the central government's system was "obscure because of the different state support arrangements concerning the subsidizing of cycling paths."[54] Provinces and cities had to translate their local needs into the various ministries' terms to comply with subsidy applications. To maintain their negotiation position with the national government, the provinces had to come up with a unanimous counterproposal. That proved difficult. Some provinces favored further decentralization of cycling policy and wanted to confront the national Public Works Ministry. Others were hesitant to stir up trouble given the considerable efforts to secure the current subsidy system.[55]

The Cyclists' Union activists backed the provinces and municipalities in their critique of the ponderous procedures. In a letter to parliament in 1978, the activists formulated three criticisms. First, the Ministry wrongly undertook an unnecessary and time-consuming technical check of municipal plans.[56] The subsidy's only condition should be whether the proposed

51 HUA 1205, inv. no. 5065, letter GS Utrecht to municipalities, January 14, 1977.

52 HUA 1205, inv. no. 5065, letter PWS Utrecht to GS Utrecht, February 2, 1976.

53 *Kamerstukken II* 1982/83, no. 17745, sub-no. 2 (*Nota Fietsverkeer*), 40; Trudy Lisci-Wessels, "Ontwikkeling van het aantal gemeenten sinds 1900," *Bevolkingstrends. Statistisch kwartaalblad over de demografie van Nederland* 52, no. 1 (2004): 56-57, here 56.

54 NA 2.19.063, inv. no. 1826, meeting minutes, IPO committee for Traffic August 28, 1980.

55 NA 2.19.063, inv. no. 1826, minutes meetings IPO committee for Traffic October 29, 1981.

56 Criticism from the provincial cooperative body IPO in 1976/1977 resulted in discussions with the Ministry of Public Works when engineers acknowledged that the strict Rijkswaterstaat

cycling facility fit into a larger program of cycling-promoting and car-limiting measures. The implication was that municipalities had enough technical expertise to design their own cycling facilities to a standard agreeable to the cycling activists, maybe even more so than to the national government. Second, the exact conditions and timeline for subsidies were unclear. This made it hard for local government to plan ahead: potential funding was unreliable and the national government did not actively explain the subsidy. Third, the subsidy was too one-sidedly geared to cycling infrastructure (paths), "when cycling traffic is better served with a reorganization of the total traffic space."[57] It was not feasible to construct cycling paths everywhere. Cycling conditions could often be improved with much cheaper, car-limiting measures, they argued. The national government's cycling policy did not make these radical decisions.

The ANWB likewise complained about the distinctions between utilitarian, recreational, and mixed cycling paths. Depending on the time, place, and season, cyclists used paths for commuting and going to school, as well as recreation. The related division of subsidies over different ministries was "illogical" and "forms an obstacle to an energetic policy."[58] In addition, the crucial lanes leading out of the city did not fit this categorization. Because of the paths' mixed use (people coming into the city for school or work, and those leaving it for recreation), it was hard to know where to get subsidies; as a result, ministries ended up pointing at each other.

Local government and action groups recommended a solution: give the provinces a sum of money to spend on cycling in cooperation with municipalities and activists. Eliminating national bureaucracy would speed up and simplify the process. A 1982 letter from all outdoor recreation stakeholders complained about the slow implementation of cycling path plans due to the technicalities of the subsidy arrangements. "By omitting the distinction in types of cycling path, the differently earmarked ministries' payments can be combined into one subsidy fund." Then, local actors "do not have to rack their brains over unwieldy 'dominantly recreational' and 'utilitarian' concepts."[59] The large cycling governance role the national government

technical test would be cancelled and provinces would be more independent (NA 2.19.063, inv. no. 1826, meeting minutes, IPO committee for Traffic, February 27, 1976, October 14, 1976, June 9, 1977 and December 8, 1977).

57 NA 2.02.28, inv. no. 6267, letter ENWB/R.B. van Putten to Parliament, November 7, 1978.

58 F. van Schelven and G. Bakker, "Overzicht belangrijkste fietspadenplannen in Nederland," *Recreatievoorzieningen* 12, no. 8 (1980): 373-77.

59 NA 2.27.5215, inv. no. 2167, letter Stichting Recreatie to Ministry of Culture and Recreation, June 9, 1982.

envisaged for provinces would be much easier if the provinces received a lump sum that they could divide over lower authorities as they saw fit. This decentralization of cycling funding (and mobility funds in general) eventually happened in the late 1990s.

During the progressive Den Uyl government, the Minster tried to end the impasse. In order to streamline the new forms of cycling governance, many meetings took place between national and local policymakers in the 1970s and 1980s. In July 1975, a high-level delegation from the Ministry of Public Works – Minister Westerterp and high-ranking *Rijkswaterstaat officials* – met with representatives of the provinces to discuss cycling policy.[60] Most provinces delegated a politician from the Provincial Executive, an engineer from the Provincial Public Works department, and a public administration expert from the Provincial Registry (*griffie*). The Public Works Ministry initiated the meeting as it started to develop an interest in (cycling) traffic safety and the promotion of cycling and public transit as alternatives to automobility in the mid-1970s.

National officials were surprised to discover that the provinces were already creating cycling policies. G. Ph. Brokx, a provincial politician in Noord-Brabant as well as the IPO, pointed to his province's plan to construct 100 km of cycling paths over a five-year period along major, so-called secondary roads. Joop Borgman, the labor politician who had promoted cycling in Zuid-Holland since 1974, explained that 320 of the 350 km of major secondary roads in his province already had cycling paths. He and other provincial representatives also pointed out that they struggled particularly with making crossings safer. To also address problems on smaller roads and in cities and villages required more money. A follow-up meeting in 1976 discussed "whether and to what extent an improvement in the coordination of the activities of the national government and the provinces is possible."[61] In short, it was in the mid-1970s that the national government became aware that for decades, cycling policy had been carried out largely locally without much support from or contact with the national government.

This opening up of plan-making had to do with the increased call for more democratic and participatory politics, discussed more fully in the next chapter. It led to participation procedures (*inspraakprocedures*) for

60 NA 2.27.5215, inv. no. 2167, document "Verslag oriënterend gesprek tussen de minister van Verkeer en Waterstaat en de Colleges van Gedeputeerde Staten inzake de aanleg van fietspaden, July 23, 1975."
61 Ibid., 5.

the public's input in local transport and spatial planning proposals. The scope for participation since the 1970s included informative consultations and written reactions. And while citizens complained that they could only comment on completed plans, the reality was that their input did bring about changes: the 1976 Urbanization Report was revised in 130 places after public consultation.[62] In the same vein, the provincial government published the revised concept of the Zuid-Holland cycling path plan in newspapers and invited comments from the public. The almost 300 respondents ranged from municipalities to interest groups and individuals.[63] In the context of new cycling activism emerging in that period, local cycling activists also critiqued the plan, resulting in alterations.[64] Some municipalities also reacted critically. Dordrecht was "disappointed" with the program's lack of ambition and wanted "more attention for environmentally-friendly traffic solutions like cycling paths for inter-local and recreational use."[65] Another community also hoped for more cycling paths, expecting "that the bicycle will fulfil a larger role in the near future."[66] Opening up plans to participation (albeit not in the design phase) was in sharp contrast to the 1950s plan.

Still, the national government refused to trust local officials with cycling policy and decentralize funding entirely. Instead, it tried to steer policy through subsidy rules. This became clear at a second meeting of all the (governmental) stakeholders on February 19, 1976.[67] In addition to the Ministry of Public Works and provincial delegations, this time there were also officials from the other subsidizing ministries (Culture and Recreation; Agriculture). On behalf of the provinces, the Noord-Brabant representative Brokx welcomed a rise in national cycling subsidies but also demanded that the provinces should take on the coordination of cycling policy and the distribution of the money. In addition, he demanded more than a 50 percent subsidy: provinces did not have the funds to come up with the other half.[68]

62 Arjen van der Burg, Odette van de Riet, and Bert Toussaint, "Contested Spatial Planning: Ca. 1965-1985," in *Builders and Planners: A History of Land-Use and Infrastructure Planning in the Netherlands*, eds. Jos Arts et al. (Delft: Uitgeverij Eburon, 2016), 222.

63 NA 3.02.46, inv. no. 2915-2917. Letters from citizens in no. 2916, e.g., M. de Visser-Ameling (July 23, 1975), C. Skov Hansen (July 26, 1975), and A.J. Wtenweerde (August 19, 1975).

64 Ien Deltrap, "Er is nog hoop voor Zuid-Holland," *De Vogelvrije Fietser* 3, no. 1 (1977): 9.

65 NA 3.02.46, inv. no. 2915, letter Dordrecht municipality to Zuid-Holland Provincial Executive, September 3, 1975.

66 Ibid., letter Hazerswoude municipality to Zuid-Holland Provincial Executive, August 28, 1975.

67 See HUA 1205, inv. no. 5065 for list of attendees.

68 NA 2.27.5215, inv. no. 2167, document "Verslag gesprek tussen de minister van Verkeer en Waterstaat en de colleges van Gedeputeerde Staten inzake de aanleg van fietspaden, op 19 februari 1976," 1.

Minister Westerterp rejected the provinces' attempt to gain more influence. While acknowledging that cycling facilities were a local and provincial policy matter, he also argued that the allocation of funds over the provinces fell under the Public Works department based on a justification of what the provinces intended to do with the money and why.[69] The state's strategy of controlling policy indirectly rather than with top-down directives was also known as the "golden strings" approach. This was not unique to cycling policies. According to political historian Harm Kaal, the Ministry of Public Housing and Spatial Planning funded urban renewal projects in the 1970s and 1980s, "the national government was prepared to invest heavily in urban renewal projects, but in exchange demanded that projects were subject to governmental approval."[70] By keeping control over subsidies, the national government could to some extent influence the shape of cycling infrastructure, even in a decentralized system where local government had considerable autonomy.

In the case of cycling, this meant that national engineers developed quantifiable, seemingly neutral, and technical guidelines for where cycling infrastructure was most needed. National Public Works engineers developed the RONA norms (*Richtlijnen Ontwerp Niet-Autsnelwegen*, "Design Guidelines for Non-Highways"). RONA included a working group tasked with formulating basic design criteria. This in turn had a Cycling Paths subcommittee, which attempted to formulate a more scientific basis for cycling path construction in three stages. First, the engineers would formulate preliminary norms based on a literature study. Secondly, an external engineering consultancy (DHV) would investigate what types of cycling infrastructure correlated with the lowest accident rates. In the long run, thirdly, the Scientific Organization for Traffic Safety (SWOV) would investigate the results of the measures.[71] A concept version of the first stage – determining which roads were most in need of cycling paths – was sent to the provinces in January 1976. Ideally, prioritizing these roads would be based on a wide range of data: the intensity of car traffic, the types and speed of motorized traffic, the intensity of bicycle traffic, the types of cyclists (schoolchildren, older people), the width of the existing road, the level of obstruction of the view, and the type of road surfacing. Since most of the data was not available, a more practical measure was to

69 Ibid., 4.

70 Kaal, "A Conceptual History of Livability," 541.

71 A. Wilmink, "Onderzoek betere richtlijnen fietsvoorzieningen," *Verkeerskunde* 28, no. 2 (1976): 76-77.

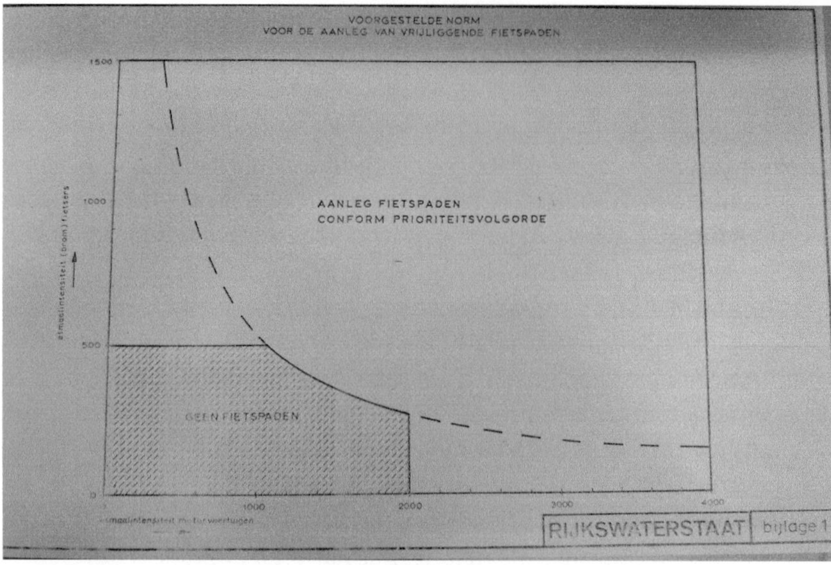

Figure 29 Policy guideline for the cycling path priorities set by Dutch engineers around 1975. The y-axis shows the number of cyclists per 24 hours, the x-axis the number of car drivers. Engineers would consider a cycling path if the numbers exceeded 500 cyclists and 2,000 drivers. [Source: Kamerstukken II 1982/83, no. 17745, sub-no. 2 (Nota Fietsverkeer), 23.]

combine the intensity of car and bicycle traffic and the width of the road. The formula for this was "$Pi = I_{EMI} \times I_{EFI} \times f_{Bi}$" where Pi stands for Priority of road section I, I_{EMI} stands for Intensity motorized traffic, I_{EFI} stands for Intensity bicycle traffic, and f_{Bi} for the Factor width (see figure 29).[72] The development of these engineering norms marked a new stage in the standardization of cycling planning in the Netherlands. It also ushered in a governance phase that led the national government to consider its role in cycling governance to be only producing and disseminating engineering knowledge.

In practice, local conditions and cyclists' subjective experience turned out to be as important in planning cycling facilities as these engineering norms. One example comes from the conurbation board (*agglomeratieraad*) for the Eindhoven region – a regional cooperation between municipalities below the provincial level. In 1980, the board subsidized a cycling path that policymakers scored relatively low on RONA's "danger index" – fifteen

72 NA 2.27.5215, inv. no. 2167, letter Ministry of Public Works to Provincial Executive provinces, January 26, 1976, attached concept, 4-5. The formula could be finetuned by giving truck traffic a higher weight in motorized traffic counts or giving very young or very old cyclists a higher weight in cyclist counts.

projects were supposedly more dangerous. Cycling advocates had little patience with this kind of ranking. School directors had pressed for the construction of this path since children used it intensively. One board member speculated that there might be more dangerous situations in the conurbation than RONA's norms showed, "but everything is more or less subjective after all." Another member mentioned the hard numbers but also asserted that "you should not only rely on such numbers." The responsible official acknowledged that it was the "exerted pressure ... from the side of the users" that had led them to not strictly follow the engineering norms in this case.[73] The fact that so many school children used this route and that school directors demanded it trumped the path's low "objective" score. On a local level, the Cyclists' Union succeeded in countering these norms by gathering input from cyclists who influenced policymakers' decisions. Here, the activists' belief that cyclists' everyday (qualitative) experience should be the foremost consideration in planning cycling infrastructure could and did take precedence over more impersonal (quantitative) norms.

At the same time, national engineers formulated guidelines for cycling path design with less input from the Cyclists' Union. Municipalities dutifully followed suit to ensure they would receive the subsidies. In 1979, theoretical physicist Hendrik van Ouwerkerk, active in several nature conservation groups, wrote a critical article about the RONA norms in the action group's journal *Natuur en Milieu* (Nature and Environment). The RONA framework led to wide road profiles and straightened curvy roads. Both interventions were detrimental to the attractive quality of rural roads.[74] Van Ouwerkerk found this deplorable, but recognized that provincial engineers were not keen to depart from the norms out of "subsidy-fear."[75] The official rules did not stipulate RONA as a condition for receiving subsidies, but in practice functioned this way: "the designers prefer not to take any risks, so the RONA norms are being followed in practice as much as possible."[76] He criticized the conflict between the road users and nature and recreation users' interests, but also how the RONA norms delayed cycling path construction: the norms made (re) constructing roads and cycling paths so expensive that a large province

73 RHCE 10884 (Agglomeratieraad Eindhoven), inv. no. 151, excerpt from meeting minutes, conurbation board June 19, 1980.

74 Hendrik J. van Ouwerkerk, "Fietspaden of RONA-banen? Norm over de schreef," *Natuur en milieu* 1979, no. 7-8 (1979): 18-23, here, 20.

75 Ibid., 18.

76 Ibid.

like Noord-Brabant could only construct about 20 km yearly.[77] Likewise, in a meeting between Public Works minister Zeevalking and the Association of Dutch Municipalities (VNG) in 1982, the social democratic mayor and VNG member Henk de Wilde also thought the requirements regarding width and quality of cycling paths were too high, implying that municipalities were only willing to invest in cycling if the subsidy conditions were lowered.[78] This problem was compounded by the fact that legally, municipal governments were not allowed to run budget deficits.[79] On paper, the Dutch government was strongly decentralized, giving lower government considerable autonomy in designing traffic facilities. In practice, the national government exercised significant control over lower government through its financial power.

A measure of distrust characterized the relationship between local policymakers and the national government in the 1970s and 1980s. Local policymakers experienced national intervention in cycling governance in the 1970s as both heavy-handed and insufficient. The national government distributed cycling funds sparsely yet required an arduous bureaucratic procedure that local policymakers found hard to master. According to the Public Works ministry, municipalities showed a lack of initiative and stamina in applying for subsidies and constructing cycling infrastructure. In 1978, Public Works minister Tuijnman said that he could not spend all the money he had reserved for cycling because local authorities failed to submit enough projects.[80] MPs faced the problem they could not judge who was right: what went on behind the policymaking scenes at the Ministry and in municipalities was hard to gauge.[81] All the same, the national subsidies formed a major share of the money provinces and municipalities invested in cycling, as tables 6 and 7 show. The differences are also visible though: some cities relied for a large majority of their cycling funding on national money. Others, like Eindhoven, received almost none. This might have had to do with different knowledge of the subsidy system or adroitness at manipulating it.

77 Ibid., 20.

78 NA 2.19.185 (Archive VNG), inv. no. 2431, meeting VNG with Minister Zeevalking, January 13, 1982.

79 Kickert, *The History of Governance in the Netherlands*, 28.

80 Vaste Commissie van Verkeer en Waterstaat, meeting October 16, 1978, 80; "Onzeker of provincie aanleg paden met geld kan steunen," *Leeuwarder Courant*, November 6, 1978.

81 Vaste Commissie van Verkeer en Waterstaat, meeting October 16, 1978, 102.

Table 6 City investments in cycling infrastructure in built-up areas, 1975-1980, in millions of gulders

Municipality	Total Investments	National Public Works Subsidies
Rotterdam	15.0	11.0
Eindhoven	7.0	0.3
Groningen	9.0	8.0
Haarlem	–	–
Nijmegen	4.0	0.4
Arnhem	0.5	0.3
Maastricht	0.2	–
Leiden	11.0	4.0
's-Hertogenbosch	4.0	1.0
Leeuwarden	4.0	3.0
Delft	9.0	4.0
Ede	–	–
Zwolle	13.0	10.0
Assen	5.0	4.0
Middelburg	2.0	1.0
Geldrop	1.0	0.4
Voorburg	0.4	0.1

Source: Kamerstukken II 1982/83, no. 17745, sub-no. 2 (Nota Fietsverkeer), Table 3.8, 41

Table 7 Total amount that provinces spent on cycling paths along secondary and tertiary roads between 1975 and 1985, in millions of guilders

Province	Total Investments	National Public Works Subsidies
Groningen	8	4
Friesland	8	4
Drenthe	32	4
Overijssel	19	6
Gelderland	15	6
Utrecht	5	2
Noord-Holland	21	6
Zuid-Holland	36	10
Zeeland	7	2
Noord-Brabant	107	23
Limburg	8	4
Total	**266**	**70**

Source: Nota Fietsverkeer, 37

7.3 Governing the Redistribution of Urban Road Space

Most innovations came from cities, however. In the mid-1970s, a combination of factors created critical momentum for a change in urban mobility policy. The wide social unrest over traffic safety, growing environmental awareness, increasing knowledge about the health benefits of cycling, and general unease about the car's dominance of public space, triggered a strong wave of activism. A new generation of urban planners experimented with ways of limiting the car's access to living areas and public space. The historical development of cycling in the Netherlands as a practice, in infrastructure, and embedded in personal networks and engineering norms, provided a great starting point. Rather than having to create a cycling culture from scratch, activists could build on and upscale an existing practice that made it easier than elsewhere. The provincial recreational cycling path plans, for instance, could be extended to include utilitarian cycling. The design and engineering knowledge as well as the bureaucratic procedures were already in place. Drawing up province-wide plans including both types of cycling was possible – getting national funding for these required addressing different national governmental bodies. Provincial planning experience facilitated cycling planning, but funding structures stalled progress.

Local engineers and planners undertook many experiments. They worked with local cycling activists to find the best ways to redesign the city for non-motorized transport. Having allowed cars an ever-larger place in city centers, planners now had to pioneer ways to undo this situation. Partly with national government subsidies (see tables 6 and 7), urban planners created new traffic circulation plans which included more measures for cyclists and the first tentative traffic-calming measures.[82] Making it impossible for cars to traverse city centers but forcing them to go around is one example. The wider application of the *Woonerf* concept, where the speed of pedestrians rather than cars or cyclists is the norm, is another. Gradually policymakers realized that cycling could solve parking and congestion issues caused by the car. According to Drenthe public works engineer and cycling advocate Van den Muyzenberg in 1973, this could only happen if the quality of cycling facilities improved: once "there are perfect cycling facilities, it will be easier for the government to responsibly limit car use."[83] In his own province, the

82 Oldenziel and Albert de la Bruhèze, "Amsterdam: World Bicycle Capital, By Chance," 23-24; Buiter, "Utrecht: Bicycles Rule – Again," 35-36.

83 G.A.M. van den Muyzenberg, "De fiets en het verkeer: Pleidooi voor het maken van voorzieningen voor de fietser," *Tijdschrift voor Verkeerstechniek* 24, no. 5 (1973): 239-42, here 242.

town of Assen had already started this: by 1973, the city had built nine cycling tunnels to eliminate the barriers and safety risks that cyclists faced when crossing major roads or railway tracks. In other words, at least some urban policymakers and engineers were already working on improving cycling conditions when activists were still getting organized.

Famously, the government sponsored experiments involving the construction of two highly visible and expensive main cycling routes in government city The Hague and textile town Tilburg in 1975. The government's large investments signaled it appreciated urban cycling – though in the end the pilots failed to attract more cyclists to these routes. The experiments did generate attention for cycling in the engineering community and helped engineers develop new expertise regarding urban cycling infrastructure.[84] Nevertheless, cycling activists in The Hague criticized these high-publicity pilots because they masked "structural underfunding of cyclists in the national Structural Plan for Traffic and Transport (*Structuurplan Verkeer en Vervoer*)."[85] Indeed, these incidental, expensive, and high-publicity policies did not solve cyclists' disadvantaged position in Dutch cities. As activist Eisse Kalk pointed out, the investments in cycling were still dwarfed by those in automobility and public transit – and were in no way proportional to cyclists' actual modal share.[86] Neither activists nor government evaluations considered these pilots cost-effective. Each recommended a more piecemeal, street-by-street approach. The Cyclists' Union complained that the projects in The Hague and Tilburg were so expensive that they would deter other cities from implementing cycling path plans.[87] The cycling paths in The Hague would cost about 1 million guilders per kilometer. Thus the government's

84 A.A.J. ten Broecke and J. Hoekwater, "Kwalitatieve evaluatie van de demonstratie fietsroutes in Den Haag en Tilburg (1)," *Verkeerskunde* 30, no. 9 (1979): 417-21, and A.A.J. ten Broecke and J. Hoekwater, "Kwalitatieve evaluatie van de demonstratie fietsroutes in Den Haag en Tilburg (2)," *Verkeerskunde* 30, no. 10 (1979): 492-96; A. Dersjant and C. Wentink, "Het gebruik van de demonstratie fietsroutes in Den Haag en Tilburg," *Verkeerskunde* 30, no. 8 (1978): 358-61; P.B. van Gurp and R.A. Sangen, "Tilburgers op de fiets: een demonstratieproject," *Verkeerskunde* 26, no. 11 (1975): 550-56; P.B. van Gurp and R.A. Sangen, "De fietsroute Tilburg," *Verkeerskunde* 28, no. 4 (1977): 170-71, 191; Tilburg and Stoovelaar, "De fietsroute Tilburg: verkeerstechnische aspecten," 163-69; Dienst Gemeentewerken 's-Gravenhage, *Demonstratie-fietsroute Den Haag 1975-1979* (Den Haag: Dienst Gemeentewerken 's-Gravenhage, 1978).

85 Berkers, Botma, and Oldenziel, *Cycling Cities: The Hague Experience*, 39.

86 Metselaar, "De vuist van de gemangelde fietsers." According to Kalk that national government had reserved 13.5 billion up to 1990 for automobility, 3 billion for public transit, and 350 million for cycling, a proportion of 40:10:1 despite cycling's much larger share of commuter transport. Interview Author with Eisse Kalk, May 18, 2021.

87 IISG, Archive Fietsersbond (ARCH01969), inv. no. 24, letter ENWB to Ministry Public Works, November 20, 1975.

extra budget for cycling would cover at most 30 km of cycling paths. Instead, the Cyclists' Union preferred to do more with less. The organization pointed to its local chapters in The Hague, Leiden, Groningen, Amersfoort, and Utrecht, who had made cycling path plans.[88] These cities combined their high-cost separated cycling paths on major streets with traffic calming, cycling and pedestrian underpasses and bridges, along with other solutions, to create a fine-grained network of safer cycling routes. In this mix, the national subsidies for building expensive traffic separation infrastructure were crucial. By contrast, traffic calming was cheap. Yet, this meant reducing car space, not always the case with cycling facilities. Cyclists' Union activists sometimes deplored the lack of political will to take this approach.

The government's policy goal was not just to make cycling safer – something engineers knew how to do: it had always aimed to reduce the potential for conflict on busy streets or at junctions. Now politicians and activists also wanted to make cycling inherently more appealing. Their next goal was convincing people to hop on a bicycle rather than drive their car, by providing shorter connections between home, work, and recreation. Civil engineer Rodenburg observed in 1975 that engineers previously only built cycling infrastructure in places where potential conflicts could arise between drivers and cyclists. Increasingly, cycling infrastructure like tunnels was built in places with low cycling levels, but where this infrastructure could offer more comfortable cycling and create more direct downtown connections.[89] This shift from a more reactive to a more proactive stance illustrates the changing public status of utilitarian cycling: no longer a necessity but the better mobility option.

Cyclists' Union activists and volunteers introduced a new strategy. Identifying the most urgent blackspots and bottlenecks in the city, they often worked closely with traffic planners to create a list of priorities. Before the 1979s, urban cycling policy – if it existed at all – was largely ad hoc. Now local planners and engineers had to pay more sustained attention to cycling, activists thought, preferably working with them. Urban planner Poelstra noted in 1977 that the planning procedures for car infrastructure or public transit were "professional" while cycling had to make do "with wishy-washy planning guidance [*halfzachte planmatige begeleiding*]."[90]

88 Ibid.

89 R.H.J. Rodenburg, "Enkele voorbeelden van fiets- en voetgangersvoorzieningen in en rond woonwijken," *Verkeerskunde* 26, no. 5 (1975): 256-68.

90 H. Poelstra "De (brom)fietser: zijn plaats in de plannen en planvorming. 4e preadvies: Plannen maken voor de fiets; enkele kanttekeningen," in *Congres fiets- en bromfietsvoorzieningen Tilburg 1977* (Den Haag: ANWB, 1978), 28.

Figure 30 Action groups like the Cyclists' Union, Stop the Child Murder, and the Pedestrian Association collaborated to convince politicians and engineers to introduce traffic calming measures like lowering the speed limit in urban areas. At this 1980 press conference they argued about reducing the 50 km/h maximum speed in built-up areas (the banner reads "50 is too much"). Within a few years, the national government allowed municipalities to create zones with a 30 km/h speed limit. Action groups and politicians converging on these issues from 1975 to 1985 made this an important decade for reducing the car's place (and speed) in the city [Source: Nationaal Archief, Rob Bogaerts/Anefo, no. 931-1415]

Poelstra's point was illustrated by fellow engineer M. Slop at Utrecht's Traffic Department. He acknowledged the many enthusiastic attempts by "private persons" (*particulieren*) to place cycling higher on the agenda: "The political pressure on the traffic engineer to do a bit more for cyclists/moped riders, is undeniable."[91] Cyclists and urban citizens' everyday experiences formed the input for policy. As a result of public-private cooperation, almost every community developed a cycling policy document in the 1970s or 1980s. Two Rotterdam Traffic Service (*Verkeersdienst*) officials stated bluntly in 1974, that "every self-respecting municipality is currently involved in making plans to improve cycling facilities" citing the energy crisis as to why awareness of cycling's potential was increasing.[92]

91 M. Slop, "De (brom)fietser: zijn plaats in de plannen en planvorming. 5e preadvies: Waarom niet meer groene golven voor (brom)fietsers?," in *Congres fiets- en bromfietsvoorzieningen Tilburg 1977* (Den Haag: ANWB, 1978), 30.
92 H. Been and G. van der Kolk, "Fietsverkeer in steden," *Tijdschrift voor Verkeerstechniek* 25, no. 4 (1974): 164.

In the 1980s, the Cyclists' Union consolidated and extended its user-based expertise.[93] Apart from local activism, national study and working groups were created around a specific topic, including stimulating bicycle parking (at train stations and shopping centers), combating bicycle theft, cycling in shopping areas, and collecting and recording cycling engineering expertise in design manuals for policymakers. In lobbying, the Cyclists' Union strove for lowering maximum speeds in built-up areas and more focus by local government on the maintenance of cycling paths.[94] Regarding traffic rules, the Cyclists' Union also wanted to allow cyclists to turn right when traffic lights were red. The Cyclists' Union strongly supported improving the bicycle-train combination in multiple ways. They worked with Dutch Railways (NS) in 1980 to assess problems with cycling routes to stations, parking facilities, and the possibility of taking the bicycle on the train.[95] Bicycle parking in general was a major talking point.[96] Cyclists' Union figures like Pax Kroon thought about the bicycle's role in before-and-after transport.[97] The Union also considered a national registration system to prevent bicycle theft, and a better cycling signposting system.[98] Much of this pioneering work returned in Bicycle Masterplan experiments and pilot projects in the 1990s (see 8.1).

Cyclists' Union activists contributed much to the production of urban planning design knowledge for cyclists. True to their belief that cyclists were the true experts, they came up with the essential qualities of a good cycling network: cohesion, directness, safety, comfort, and attractiveness. Directness, for instance, could be improved by building bicycle bridges or tunnels where cyclists previously had to take detours. Reducing traffic distances would make cycling routes shorter than competing car routes. The Groningen Cyclists' Union observed this in the early 1980s and contacted the Twente Technical High School, where students developed ideas for

93 Bram Duizer, *"In het nut van actie moet je geloven": Dertig jaar actievoeren door de Fietsersbond* (Utrecht: Wetenschapswinkel Letteren Universiteit Utrecht, 2005), 15-22; Interview Tom Godefrooij and Jaap Rijnsburger, June 2, 2021.

94 Dick Ringlever, "De deskundige is de fietser zélf," *Trouw*, Oct 18, 1980.

95 "NS inventariseert knelpunten voor fiets bij stations," *Nederlands Dagblad*, March 12, 1980; "Fietsverkeer trein te weinig gestimuleerd," *Leeuwarder Courant*, November 1, 1980; "ENFB organiseert actie 'Fiets en trein'," *Het Vrije Volk*, October 21, 1982.

96 "ENFB wil meer stallingsgelegenheid bij leeszaal," *Nieuwsblad van het Noorden*, April 10, 1982; "In hart Rotterdam tóch bewaakte fietsenstalling," *Algemeen Dagblad*, August 25, 1982.

97 Wim Bont, "Fiets met bus en trein mee," *Het Vrije Volk*, December 24, 1980.

98 "Plan om fietsendieven te remmen," *Algemeen Dagblad*, January 31, 1981; "Politie lost maar 1pct fietsendiefstallen op," *Het Vrije Volk*, November 9, 1984; Jacques de Jong, "'Aangepaste wegwijzers nodig voor fietsers'," *De Volkskrant*, September 27, 1982; "ENFB wil speciale wegzijzers voor fietsers," *Nieuwsblad van het Noorden*, October 6, 1983.

cheaper bicycle bridges. Groningen alderman Jacques Wallage used these in the city's subsidy applications to the national government.[99] Comfort included providing maintenance: the Cyclists' Union also lobbied for this often overlooked task.[100] A 1984 memorandum noted that tiles were often used as paving (for the city's easier access to underground pipes rather than for cyclists' comfort) and that budgets for the maintenance of these paths were insufficient and not likely to be increased.[101] These road-users' perspectives of the built environment were novel.

The Cyclists' Union also formalized their cycling engineering documentation. The experience gathered in Amsterdam on the width of cycling lanes, paving types, junctions, and so on, was too valuable to lose. Amsterdam subsidized the Cyclists' Union in 1987 to document design knowledge in a manual. As cycling policy was about to be decentralized from the municipality to neighborhood councils, cycling activists feared the knowledge would be lost. The design manual would help safeguard the quality of cycling infrastructure.[102] Amsterdam eventually published the manual in 1991.[103] It was well-received and – with a national subsidy – became a national manual.[104] Published in 1993, Drawing for the Bicycle: Design Manual for Bicycle-Friendly Infrastructure (*Tekenen voor de fiets: ontwerpwijzer voor fietsvriendelijke infrastructuur*) contained the knowledge garnered by engineers and local activists.[105] Not every member of the Cyclists' Union supported the move away from street actions. Duizer has shown how some members did not believe in writing memoranda and working with authorities. They wanted to be physically present in the streets. As one activist explained the rationale: "You have to believe in the usefulness of activism ... you cannot prove it ... an action forces a choice, it polarizes."[106] But it was precisely this polarization that was not welcomed

99 "Groningen ziet wel iets in fietsbruggen," *Nieuwsblad van het Noorden*, January 16, 1981.
100 "'Fietspad langs Stadsweg vertoont tal van gebreken'," *Nieuwsblad van het Noorden*, October 20, 1983; "ENFB in actie voor onderhoud fietspaden," *Nederlands Dagblad*, May 8, 1984.
101 Marc van den Broek, "Pad voor fietser is hobbelig en slecht aangegeven," *De Volkskrant*, August 28, 1984.
102 Marnix Bruggeman, "Handboek laat wegbeheerder door een fietsbril kijken," *Trouw*, October 9, 1987.
103 "Amsterdam publiceert handboek fietsbeleid," *NRC Handelsblad*, July 23, 1991.
104 Duizer, *"In het nut van actie moet je geloven"*, 25. See Jan Ploeger, *Wegwijzer fietsvoorzieningen: Uitgangspunten en planvorming* (Ede: Stichting CROW, 1990).
105 Jan Ploeger, "Tekenen voor de fiets," *Verkeerskunde* 44, no. 11 (1993): 22-25. Updated versions appeared in 2006 and 2016, in Dutch and English.
106 *Vogelvrije Fietser* 7, no. 1 (1981): 6. Cited in Duizer, *"In het nut van actie moet je geloven"*, 16. Duizer took the title of his study from this quoted passage.

by the majority of Cyclists' Union members: the Amsterdam branch, for instance, had identified no fewer than 400 bottlenecks in the city, and wanted to focus on working with city officials to solve these.[107] With the growth of occasional project subsidies from the government, the Cyclists' Union's national bureau could professionalize and hire staff members. At the same time, the local chapters remained entirely voluntary and their performance and character differed from place to place. The gap between these local chapters' activities and the national organization widened steadily.[108]

The Cyclists' Union 1980s focus on improving cycling facilities did not directly address the issue of rising automobility. Not all cycling-facilitating projects reduced car space or the conflict between drivers and cyclists. According to social democrat MP De Hamer – a former council member in the pioneering cycling city of Groningen – the state could not solve the fundamental power imbalance between car drivers and vulnerable cyclists with subsidies for cycling paths. Politically, the government could construct cycling paths because drivers appreciated how removing cyclists from the road improved the flow of cars. But the funding structure neither allowed for, nor was it applied to the task of making junctions safer, as shown in figure 31. Consequently, the conflict between driver and cyclist returned at crossings. More effective would be cycling tunnels or facilities that shorten travel time, which the cycling paths alongside major arteries did not necessarily achieve. Despite investments in such infrastructural solutions, safety was still an issue.[109]

Local policymakers repeated throughout the 1970s that they, more so than national authorities, were willing to make the more fundamental political choice: restrict automobility. When applying for national government funds, local engineers found it hard to convince state engineers of the need to facilitate "slow traffic." According to the municipality of Sassenheim in a 1979 Association of Dutch Municipalities (VNG) report: "there is a significant difference in point of view between an agency like the Public Works Department ... and the municipality. These agencies primarily consider the interests of motorized traffic flow (*doorstroming*) and the like. You could say that the weaker the road user, the harder it is for municipalities to get a facility for them from national or provincial

107 Ibid.

108 Interview Tom Godefrooij and Jaap Rijnsburger, June 2, 2021.

109 Bijzondere Commissie voor het Structuurschema Verkeer en Vervoer (14 390) en het Meerjarenplan Personenvervoer 1980-1984 (15 885), meeting February 18, 1980, 1018-1019.

Figure 31 Cycling tunnels like these near Vreeland, Utrecht, enabled cyclists to cross larger roads safely and prevent detours, making cycling a more attractive option. Since the 1970s, cities increasingly invested in such tunnels in response to broad public demand for cycling and safety. Part of the budget came from national governmental subsidies. [Source: Collectie HUA, fotograaf Provincie Utrecht, nr. 847441]

Public Works."[110] In the same report, the cities of Amersfoort and Eindhoven mentioned how the national government prioritized flow over traffic safety. At points where nationally or provincially owned roads crossed local roads, this could lead to conflicts with Public Works engineers, like the funding request from the municipality of Etten-Leur for a cycling tunnel.[111] The nationally administered road, also a major traffic artery for cyclists, ran right through the center of the built-up area. Plans to construct a bypass had been postponed indefinitely. According to the local policymakers "the safe and fast crossing of traffic on the national road onto local roads in our municipality is a primary responsibility for the national government."[112] In other words, these local policymakers believed the town was responsible for traffic safety, but if national roads created the problems, the state should provide the funds to mitigate them.

The conflict between investments in infrastructures or making political choices only increased. As the urban cycling landscape changed, the investments in separate infrastructure along suburban and rural roads that had started back in the 1920s and 1930s, continued gradually even in the 1950s and 1960s. At the same time, the national and provincial government investments in car infrastructure kept on rising after the 1970s. And despite the change of direction in urban mobility policy, urban policymakers still supported driving. The state built more and wider roads. So cyclists alongside these roads were now only safe on separate cycling paths. Table 7 shows that the provinces, partly with national subsidies, but predominantly using their own funds, spent 266 million guilders on paths between 1975 and 1985. They generally paid much more for cycling paths than their nominal 50 percent subsidy entitlement. The part the province paid indirectly still came from national government, in the form of the general roads budget that provinces received under state funding (*Wet Uitkering Wegen*).[113] This money could have been spent on roads rather than cycling paths: the fact that provinces also invested in cycling infrastructure shows engineers still thought they were necessary. However, in the context of the mid-1980s, where decentralization and small government were key ideas, the national engineers' old idea that cycling was a local concern returned with a vengeance and ended the subsidy program.

110 NA 2.19.185 (Archive VNG), inv. no. 2428, Concept-eindrapport Gemeentelijk Verkeersveiligheidsbeleid: Een inventariserende case-study van tien gemeenten, 20.

111 NA 2.02.28, inv. no. 6268, letter municipality Etten-Leur to Rijkswaterstaat, November 24, 1978.

112 Ibid.

113 *Kamerstukken II* 1982/83, no. 17745, sub-no. 2 (*Nota Fietsverkeer*), 37.

7.4 Decentralizing Cycling Governance (Once Again), 1985-1990

The 1980s changing political climate led to decentralization of the 1970s experimental national cycling governance and funding. In 1982, a new center-right government came into office. Facing economic problems and influenced by neoliberal ideas, it introduced austerity measures and privatized some state agencies. Decentralization also became popular again.[114] The national cycling subsidies, hard fought by activists and MPs, were prime candidates for cuts. National government officials had argued many times over the years that cycling governance belonged at the local level. As a result, in the mid-1980s, conservative liberal Public Works minister Neelie Smit-Kroes (VVD) ended the cycling subsidy program.[115] Although the agency did not question cycling's status as a public good, it reduced funding and slowed down the implementation of cycling policies. The agency now devoted more attention and budgets to rising car congestion issues.[116]

The break in policy came in 1983, when the government announced its intention to decentralize governance. The official document Cycling Traffic Memorandum (*Nota Fietsverkeer*), dedicated entirely to cycling, was written when Smit-Kroes headed the Ministry of Public Works (1982-1990). This first national cycling policy document represented the culmination of fifteen years of activists and politicians working hard to put (urban) cycling higher on the agenda. The memorandum also emphasized that, going forward, the national government considered cycling governance exclusively a matter for local government – and significantly – in cooperation with action groups.[117] National policymakers still envisioned playing a large part in cycling governance, even when Smit-Kroes refused a coordinating role: the provinces should solve any friction themselves without national intervention.[118] The only remaining national tasks would be knowledge-centered like sponsoring research on cycling safety. Citing a commitment to decentralization, by stressing in her report to Parliament the national

114 Michiel S. de Vries, "The Rise and Fall of Decentralization: A Comparative Analysis of Arguments and Practices in European Countries," *European Journal of Political Research* 38, no. 2 (2000): 193-224, here 206. Subsidies for urban renewal projects were also scaled back in the 1980s: Kaal, "A Conceptual History of Livability," 541; De Liagre Böhl, *Amsterdam op de helling*, 403.

115 Paul Pestman, "Dutch Infrastructure Policies: Changing and Contradictory Policy Arrangements," in *Political Modernisation and the Environment: The Renewal of Environmental Policy Arrangements*, eds. Jan van Tatenhove, Bas Arts, and Pieter Leroy (Dordrecht: Kluwer, 2000), 75.

116 Van der Burg, Van de Riet, and Toussaint, "Contested Spatial Planning," 256.

117 *Kamerstukken II* 1982/83, no. 17745, sub-no. 2 (*Nota Fietsverkeer*), 15.

118 Vaste Commissie van Verkeer en Waterstaat, meeting Dec 5, 1983, 15.

government's limited role, Smit-Kroes was indicating how little involvement the Ministry wanted in cycling in the 1980s.[119]

Arguing that national cycling subsidies were no longer necessary, the Ministry of Public Works replaced them with a general deposit in the Provincial and Municipal Funds. Without being explicitly earmarked, however, there was no guarantee the money would be spent on cycling facilities. In the meantime, MPs tried to continue the subsidies, but failed.[120] To defend its decision to discontinue the funding, Public Works agency argued that the subsidies had always been a temporary measure to stimulate local government to rapidly improve cycling safety. According to Minister Smit-Kroes: "The aim of this contributory arrangement for constructing cycling facilities was incentivizing lower government to implement an active policy that would promote safe cycling traffic and bicycle use." She believed that "we have achieved this policy goal." Stimulus was no longer necessary as "bicycle traffic has an important place in local government traffic policy." And as a result of the subsidies, "improvements to infrastructure have strongly increased the potential for using bicycles safely and comfortably." She summed up: "The need for national government to provide a financial incentive for constructing cycling facilities has diminished, which was the background to dismantling the contribution arrangement."[121] That was not entirely correct. The early 1970s policy documents did not say that goading local engineers and planners into cycling policy was the purpose of the cycling subsidies. It seems Smit-Kroes presented this narrative to boost the case for decentralization – the subsidies had always been intended as "an injection to address the municipalities politically," to point out "their own responsibility" for cycling.[122]

MPs and policymakers denounced the decentralization of cycling funding as thinly veiled austerity cuts. The small left-wing PSP party leader, André van Es, considered the end of the subsidies was "shifting poverty from the national government to the provinces and municipalities."[123] Tommel, of the center liberal party D'66, criticized the claim that this policy fitted into decentralization and that cycling governance belonged at the local level, as a transparent lie: "This has nothing to do with decentralization but is all

119 *Kamerstukken II* 1982/83, no. 17745, sub-no. 1, letter Minister of Public Works Smit-Kroes to Lower House, January 19, 1983.

120 *Kamerstukken II* 1982/83, no. 17745, sub-no. 5, motion Castricum.

121 Vaste Commissie van Verkeer en Waterstaat, meeting November 2, 1983, 26. See also meeting Dec 3, 1984, 16.

122 Vaste Commissie van Verkeer en Waterstaat, meeting December 5, 1983, 13.

123 Ibid., 1.

about prioritization within the national government's financial policy. Or, to put it bluntly, with this government, the cyclist pulls the short straw [*het kind van de rekening*]. Children are often the victims."[124] The end to cycling subsidies was even harder to swallow because cities could still request national funding for large infrastructural projects. This in effect excluded cycling facilities that were too small to qualify.[125] According to Tommel, the relatively small amounts made it all the harder to understand why this subsidy was stopped in the first place, citing a total annual spend of 500 million guilders on new car infrastructure (figures 27 and 28). Under the guise of decentralization, the minister's austerity cuts and political decisions hit cyclists. According to labor MP Rienks (PvdA), traffic safety and pro-cycling subsidies had been "interred in the deregulation graveyard, without opposition from the governing parties."[126] These MPs did not necessarily oppose decentralization. Rienks trusted provinces and municipalities were interested in providing better facilities for cycling, but feared that without extra funds, this policy would stall.[127] He pointed out how many parties had repeatedly requested funds, implying that they should find some middle ground, as was common in Dutch political culture. Minister Smit-Kroes, backed by the right-wing VVD and CDA majority, refused any concessions.

Still, it was not an easy right-left wing divide. Though the VVD came to be a pro-car party, prominent members advocated funding for cycling. Smit-Kroes's fellow party member MP Annemarie Jorritsma-Lebbink criticized the decentralization of cycling policy.[128] She wondered "whether the catching up is really sufficient to justify the dismantling of the national contribution," adding that she wanted to know how the Ministry would continue to incentivize lower government because "the backlog is still large."[129] Another prominent VVD politician, Hans Wiegel, at the time Royal Commissioner in Friesland, echoed this criticism. At a regional meeting of entrepreneurs and roadbuilders, Wiegel called the abolition of the subsidies "a painful development." After a slow start, local authorities were now developing cycling path projects in greater numbers, so money was urgently needed. Worse, the minister lied about her true motives: according to Wiegel, she intended to transfer the cycling subsidies to the National Road Fund, where the money would go to roads in the Randstad conurbation rather than

124 Ibid., 7.
125 Ibid., 13.
126 Ibid., 2.
127 Vaste Commissie van Verkeer en Waterstaat, meeting December 3, 1984, 28.
128 Jorritsma-Lebbink, too, became minister of Public Works.
129 Vaste Commissie van Verkeer en Waterstaat, meeting December 5, 1983, 10.

other provinces.[130] Wiegel believed national policies marginalized both the province of Friesland and cyclists.

Cities and provinces also objected to the notion that cycling facilities were adequate, requiring no further state support. The provincial lobby organization IPO, for example, criticized the idea that the cycling path backlog had been solved. The provinces, having aimed to provide 90 percent of secondary and 70 percent of tertiary roads with cycling paths in 1987, did not reach this target. The subsidy had to continue.[131] The provincial lobby found the prospect of the provinces going from sharing 24 million for cycling infrastructure in 1984, to sharing only 5 million in 1985, "plainly alarming."[132]

The national government's refusal to coordinate cycling governance created a governance vacuum, once again. It allowed non-governmental actors to continue their central role. In the 1980s, many European countries privatized public sector domains or placed them under independent regulatory bodies. The neo-corporatist notion that advisory bodies of social group representatives could balance interests was discarded.[133] For cycling governance in France, as Maxime Huré shows, advertising companies like JC Decaux and Clear Channel jumped into this governance gap and played a large role in establishing urban bike-share systems.[134] Industry actors' role was less prominent in the Netherlands. Still, since the 1960s, Dutch bicycle producers had collaborated through Bicycle Foundation (*Stichting Fiets*). It published brochures advocating more cycling infrastructure in an attempt to make cycling more attractive and sell more bicycles.[135] In the late 1980s, it also established a working group for utilitarian cycling, with major interest and industry groups on board.[136]

The Cyclists' Union continued to lobby at a national and local level. Increasingly, professional local branches established permanent working

130 "Nieuw pleidooi Wiegel voor snelweg Midlum naar Zurich," *Leeuwarder Courant*, October 25, 1984, 19.

131 NA 2.19.063, inv. no. 1826, minutes meetings IPO committee for Traffic April 28, 1983.

132 NA 2.19.063, inv. no. 1826, minutes meetings IPO committee for Traffic February 23, 1984.

133 Van Veen, "From Neo-Corporatism to Regulatory Governance," 86.

134 Huré, "Les réseaux transnationaux du vélo."

135 Stichting: fiets!, *Meer fietsen ... meer fietspaden: Over de noodzaak van het aanleggen van meer fietspaden voor verkeersveiligheid en recreatie* (Amsterdam: Stichting: fiets!, 1967); Stichting: fiets!, *Leefruimte voor de fiets: Over de noodzaak van het treffen van voorzieningen met betrekking tot het stallen en parkeren van fietsen*. (Amsterdam: Stichting: fiets!, 1969); Stichting: fiets!, *Optimale aandacht voor fietsers: 27 stellingen over voorzieningen voor het fietsgebruik*. (Amsterdam: Stichting: fiets!, 1972); Stichting: fiets!, *Fit door fietsen*.

136 IISG, Archive Fietsersbond (ARCH01969), inv. no. 22, *Kruipolie* magazine, June 1987, 'Verslag werkgroep utilitair fietsverkeer," 18.

relations with local policymakers.[137] Leading members believed acquiring traffic expertise to become a legitimate and useful governance partner had been a successful strategy, because "the public and policy world recognize the ENWB [Cyclists' Union] as an expert advocate for cyclists."[138] Like the ANWB, the Cyclists' Union gained access to most policymakers and traffic committees. National and local government repeatedly invited them to advise on cycling issues.[139] In a different context from fifty years before, the Cyclists' Union nevertheless managed something similar to the ANWB: enter into the policy coalition by employing expertise as a source of power and legitimacy.

Looking back at the period between 1975 and 1985 and examining the policy landscape, the Union believed it had managed to raise cycling's status as a public good. More politicians considered the bicycle a valuable transport mode worthy of promotion and enhancement. Importantly, traffic engineers took bicycle plans seriously as an integral part of mobility and infrastructure planning.[140] The Cyclists' Union denounced the national government's refusal to decide once and for all whether to back the bicycle, however. The same corporatist governance model (*poldermodel*) that gave cycling activists the opportunity to influence policy also meant the state was unwilling to antagonize car drivers. Measures limiting car use were much rarer than those favoring bicycle use with no overly negative impact on drivers. Even as the national government invested more in cycling facilities, the highway budget continued to rise throughout the 1970s. The Cyclists' Union differentiated three government level approaches. The national government invested increasingly more in the maintenance and construction of national roads and cut unprofitable public transit lines. Provinces on the other hand "create cycling path plans, but they do not provide a network of low-traffic [*autoluwe*] roads, a much more fundamental way to create safe cycling connections."[141] The Union found the municipalities the most innovative in their choice of a different, more bicycle-friendly policy, "often forced by practical necessity."[142] As elsewhere, Dutch urban policymakers were at the vanguard of bicycle planning.

The state's retreat from cycling governance created alternative formats to fill the national governance gap. To ensure a national network of recreational

137 Duizer, *"In het nut van actie moet je geloven"*.

138 IISG, Archive Fietsersbond (ARCH01969), inv. no. 22, *Kruipolie* magazine, October 1985, "Concept beleidsplan 1896-1988," 5.

139 Ibid., 5-6.

140 Ibid., 2.

141 Ibid., 3.

142 Ibid., 3.

cycling infrastructure in the absence of strong state guidance, for example, stakeholders founded the National Cycling Platform (*Landelijk Fietsplatform*) in 1987.[143] Both the ANWB and the Cyclists' Union held prominent positions in this platform. Next, citizens played a national role: Cyclists' Union volunteers were involved in designing – and organizing road signage for – long-distance recreational routes (so-called *LF-routes*) across the country.[144] The Ministry of Agriculture, which had taken over the role from the CRM Ministry after that ministry was dismantled in 1982, subsidized these recreational routes.[145] In many of these instances, national politicians and engineers showed through their subsidies they considered these types of cycling a public good, but the initiative for such projects had to come from non-state actors or local policymakers. In short, at the prompting of a cycling-friendly electorate during the 1970s and 1980s the national government created complicated and laborious funding instruments. It left cycling governance and innovations to lower governments and non-state actors, however.

7.5 Conclusion

Cycling's rising status as a public good in the 1970s and 1980s led to a proliferation of policies and initiatives. Many national politicians addressed the social protesters' concerns about the unacceptable dangers and prioritization of car traffic in extensive discussions in Parliament and parliamentary committees. They accepted and promoted the framing of cycling as a solution to increasingly pressing issues caused partly by automobility, such as environmental pollution, congestion, and the deteriorating quality of urban public space. Cycling remained enduringly popular as a recreation mode. As a result, policymakers and activists worked together at a local level to unmake the car-centered city. Provincial policymakers coordinated these efforts and mediated between local and national government. At the national level, MPs put serious pressure on the Ministry of Public Works to not only invest more in cycling, but also take on a larger directing role.

In short, the division of tasks and responsibilities was the most contentious aspect of 1970s and 1980s cycling governance. Historically averse to

143 Ad Snelderwaard, Eric Nijland, and Anita Bakker, eds., *Landelijk Fietsplatform 1987-2012: 25 jaar in beeld* (Amersfoort: Stichting Landelijk Fietsplatform, 2012).
144 Ibid., 5.
145 Ibid., 28.

interfering with local transport and mobility planning, and preoccupied with automobility and highway planning, national engineers resisted the idea they should also steer cycling governance. Paradoxically, they insisted on retaining control over awarding individual cycling subsidies to municipalities. Local policymakers experienced this form of micromanagement as a sign of distrust in their pro-cycling intentions. This was hard to swallow because local policymakers were often proactive, innovative, and forerunners with their cycling policy initiatives. Alongside these multi-level conflicts, national agencies were involved in inter-departmental struggles. Because cycling was defined as utilitarian, recreational, or rural-agricultural, different ministries were involved. Relations between these ministries were fraught and attempts to assign each other responsibilities failed. In short, while national civil servants all paid lip service to the value of cycling, few were willing to take up a new or larger role in this policy domain.

Despite these problems on the national level, the large coalition of cycling stakeholders made cities signifcantly safer for urban cyclists in the 1970s and 1980s. For all the struggles over the "rules of the game," cycling grants existed at all levels. The national government spent some 500 million guilders on cycling subsidies between 1970 and 1990, while local governments reserved sizeable shares of their road budgets for cycling projects and utilized national subsidies. Perhaps made more creative by ever scarcer funds, local engineers and planners also implemented cheaper measures for traffic safety than building infrastructure. Lowering speed limits or closing streets to traffic cost little and were as effective, even though such measures required political courage and willingness as they limited car access to parts of the city.

8 Self-Evident: Mainstreaming Cycling Policy and Practice since 1990

"A swimming pool attracts more voters than a cycling path. Before you know it, the money has been spent on something else." Speaking to Dutch MPs in 1991, Public Works minister Hanja Maij-Weggen expressed her doubts that transferring funds to municipalities for cycling would work without strong guarantees: "Provinces and municipalities might then spend the money too easily on different matters. I, too, have worked in a municipality."[1] But she was also critical of how her predecessor Neelie Smit-Kroes had terminated the national cycling subsidies. On a provincial level, after 1985, "cycling path policy has quite often collapsed." The country needed a new, national cycling policy with a "booster function" (*aanjaagfunctie*).[2] Maij-Weggen was suspicious of lower government's commitment to cycling policies: "it would be a sign of overconfidence if I just as merrily [*jolig*] transferred the funds in question to the provinces and municipalities. After all, last time they did not do the things they were supposed to do with them," she thought.[3] A role for national governance was called for. It was a remarkable turn of events. Just a few years after the national government had decisively ended its funding role in cycling governance, Maij-Weggen put national cycling governance back on the agenda. After seventy years of government involvement in cycling policies, the policy coalition still had no stable form. It kept changing. How has this developed and stabilized in the thirty years since? Is there a governance equilibrium now?

Cycling's status, coupled as it is to the fight against climate change, has now reached a new peak. Cycling infrastructure has a higher status as public good than before. The traffic separation system, however, once instigated to give drivers more space, is reaching its limits. Urban policymakers – facing overcrowded cycle lanes – no longer want to prioritize drivers over cyclists in the city. Yet they are still working within this system established in the 1920s. The growing popularity of traditional bikes as well as new developments in mobility like the rise of the e-bike as well as the continuing presence of mopeds, are making cycling paths overcrowded. As cars are increasingly banned from city centers and automobility is declining, new questions

1 Vaste Commissie Verkeer en Waterstaat, meeting October 28, 1991, 48.

2 Ibid., 47.

3 Ibid., 48.

arise about equal space distribution. In cities, officials have focused on speeding up cycling infrastructure construction, raising car parking fees, and blocking car access to inner cities. These measures have reinforced the rise in cycling levels. In the suburbs, provincial funders have helped build cycling highways to boost commuter cycling and promote regional recreational cycling networks. A large policy coalition, increasingly also with a European dimension, now shapes cycling policy. Given the increasing amount of cycling policies and policy documents (and greater availability of sources), "Self-Evident: Mainstreaming Cycling Policy and Practice since 1990" takes a different direction. It discusses recent governance developments and puts them in a historical perspective. Though by no means an exhaustive overview of the last three decades of Dutch cycling policy, it takes stock of traffic separation ideology and cycling policy after a century of governance.

8.1 National Government Settles on Expert Role

The national government, having backed away from cycling governance in the mid-1980s, introduced new cycling policies already in the early 1990s under May-Weggen. Again, this intervention would be short-lived. We could describe the state's role in Dutch cycling governance since the 1970s as occasional bursts of high activity followed by little to none. National Public Works officials frame these interventions as attempts to stimulate and energize cycling policy at the local level, where it belongs in their view, but not as a structural commitment.

The Ministry of Public Works introduced a policy document in the early 1990s called Bicycle Masterplan (*Masterplan Fiets*) – a misleading title given that the plan entailed funding for many small cycling projects like pilots, and increasing bicycle expertise.[4] The plan exemplifies the national government's conviction that its role was to foster innovation and expertise rather than coordinate or implement cycling policies. The government sought to create new ways to improve cycling safety. For the first time, the national program also tried to convince drivers to switch to cycling or the bicycle-train combination by making cycling more attractive.

National concerns about growing automobility in the late 1980s inspired components of the Bicycle Master Plan. It formed part of the major policy document Second Structural Scheme for Traffic and Transport (SVV-II),

4 This also funded the 1999 study by Albert de la Bruhèze and Veraart, pioneering Dutch cycling historiography.

which the national government introduced in 1988 – and Parliament approved in 1991. The state committed to large investments in public transit (the Rail21 program) and put the brakes on constructing new roads. Despite the Netherlands' reputation as a nation of cyclists, it is also very much a nation of drivers. The Dutch Institute for Public Health and the Environment (RIVM) reported in 1988, that the Netherlands geographically had the highest car density in the world, with 128 cars per square kilometer.[5] And although local governments facilitated cyclists to some extent, these same authorities did not hesitate to build more roads for cars, with the number of roads outside built-up areas increasing by 15 percent between 1970 and 1988, and within them even by 40 percent.[6] As long as separate cycling facilities were present alongside these roads, policymakers firmly believed that high cycling levels were compatible with high car levels. This avoided having to make a clear choice between either mode.

Nevertheless, there was a limit to growth. The 1991 national mobility plan SVV-II sounded the alarm bells: welfare, wellbeing, and prosperity were at risk from ever-rising automobility levels.[7] The policy document represents a period (the late 1980s and early 1990s) when the Dutch national government was most willing to take anti-car measures. Both before and after, this willingness was at a much lower ebb nationally. By contrast, cities have been consistently more committed to curbing car access since the 1970s.[8] Congestion threatened the accessibility of Dutch cities and annoyed car commuters; air and noise pollution plus the fragmentation of the Dutch landscape threatened the quality of life; and traffic deaths were on the rise again after a period of decline. To address these issues, the Dutch government embraced the definition of sustainable development in the 1987 Brundtland commission report *A Common Feature,* as "development that meets the needs of the present without compromising the ability of future generations to meet their own needs."[9] Employing a railway metaphor, policymakers formulated a number of "tracks" aiming to reduce pollution as well as mobility, through spatial planning – and "Track 15" of SVV-II was

5 Rijksinstituut voor Volksgezondheid en Milieuhygiene, *Zorgen voor morgen: Nationale Milieuverkenning 1985-2010*, ed. F. Langeweg (Alphen aan den Rijn: Samson H.D. Tjeenk Willink, 1988), 44.

6 Ibid., 44-45.

7 *Kamerstukken II* 1989/90, no. 20922 (Structuurschema Verkeer en Vervoer), sub-no. 16 (Deel D: Regeringsbeslissing), 5.

8 Interview Author with Koos Louwerse, August 12, 2019.

9 *Kamerstukken II* 1989/90, no. 20922 (Structuurschema Verkeer en Vervoer), sub-no. 16 (Deel D: Regeringsbeslissing), 8.

promoting bicycle use.[10] In these policy articulations, the status of cycling had fundamentally changed. It had turned into a vehicle to attain the government's overall policy goals.

Cycling could counter automobility's excess. According to the national policy designers, the bicycle was ideally suited to trips up to 5-10 km. The campaign's key aim was to convince drivers to switch to cycling for these short trips. National engineers and planners needed to take measures that made cycling more and driving less attractive. Track 9 of SVV-II focused on the reconstruction of urban space: citing successful experiments in the historic city of Groningen and the Utrecht commuter town of Houten, policymakers encouraged making city centers more or less car-free.[11] Bicycle policies revolved around more attractive cycling routes, more parking facilities, new measures to prevent bicycle theft, and ways to improve the bicycle's image. From a governance perspective, the national government defined its role as "supporting municipalities and provinces in creating attractive and safe cycling route networks" through increased funding. To stimulate recreational cycling, policymakers envisioned increased cooperation between the Ministries of Public Works, Economic Affairs, and Agriculture, as well as between provinces, municipalities, and citizen organizations.[12]

Bicycle Masterplan encompassed several projects, many of which included a component to increase cycling expertise and innovations. In Houten, a pilot with wind protection screens along a cycling path tested whether they could make cycling more attractive. Other projects encouraged bike-public transit connections with innovative bicycle parking facilities at bus stops. In total, the national Masterplan sponsored thirty-one research projects, forty-one pilots, and forty projects to develop and disseminate cycling expertise for instance with design manuals.[13] The national plan provided local policymakers with a more solid foundation for making cycling a safer and more appealing option – the national government's preferred knowledge role in cycling governance.

Non-governmental actors played a major role in implementing Bicycle Masterplan and cycling policy more widely in the 1990s. For an evaluation of the plan in 1997, researchers interviewed local policymakers as well as

10 Ibid., 57-59.

11 Ibid., 35.

12 Ibid., 57-59.

13 *Eindrapport Masterplan Fiets. Samenvatting, evaluatie en overzicht van de projecten in het kader van het Masterplan Fiets, 1990-1997*, (Den Haag: Ministerie van Verkeer en Waterstaat, Directoraat-Generaal Personenvervoer, 1998), 20-22.

Cyclists' Union representatives.[14] Its local branches possessed the knowledge and expertise to comment meaningfully. According to the municipalities, the Cyclists' Union was an "useful ad hoc contact" because of its long and active role in cycling governance.[15] Representatives of the cities Alkmaar and Vlissingen commented that the lack of an influential local Cyclists' Union branch had a negative impact on their ability to promote cycling.[16] Their own planners lacked the user expertise that cycling activists provided as policy input elsewhere. In contrast, Hengelo reported poor relations with the local Cyclists' Union, complaining that the organization's "details of cycling policy (like maintenance) are often lagging behind." A remarkable comment: apparently, the city expected detailed knowledge about the local cycling situation would come from activists rather than its very own traffic department.[17] The Flevoland municipality Noordoostpolder reported that its good relations with the local Cyclists' Union was a key factor in local cycling policy.[18] At least one journalist expressed surprise about this close relationship: after all, the national government's investments in cycling were a pittance compared to road budgets. So why not adopt a more critical perspective? The Cyclists' Union did not want to jeopardize this relationship, considering the state's early 1990s investments an improvement over the 1980s situation.[19]

Minister Maij-Weggen, cited in the opening of this chapter, expressed her ambivalence about national cycling subsidies. She believed municipalities had not done enough after subsidies ended in 1985, so national investments could serve as a kick starter. At the same time, she was suspicious if the money would actually be invested in cycling. National Public Works officials wanted to make sure local communities did this. In the eyes of Public Works officials, cycling should be governed locally. The national contributions to cycling funding reinstated in the early 1990s were meant to be just as temporary as those of the late 1970s and early 1980s. Still, Maij-Weggen, supportive as she was of cycling, believed that ultimate responsibility did lie with lower governments; the danger with a large role for national government would be "a certain laziness" and

14 Dirk Ligtermoet et al., *Gemeentelijk fietsbeleid: stand van zaken* (Rotterdam: Ministerie van Verkeer en Waterstaat, Directoraat Generaal Rijkswaterstaat, Adviesdienst Verkeer en Vervoer, Hoofdafdeling Vervoer en Maatschappelijke Ontwikkelingen, Afdeling Personenvervoer, 1997).

15 Ibid., 18.

16 Ibid., 34-35.

17 Ibid., 35.

18 Ibid.

19 Joop Bouma, "Fooi voor fiets, maar tòch een omslag," *Trouw*, September 27, 1991.

Figure 32 This (ultimately unsuccessful) experiment with a wind protection screen along a bridge was part of the Bicycle Masterplan, a policy document devised in the 1990s. It illustrates how with this plan and since then, the Dutch national government has encouraged and boosted innovation and knowledge production in cycling, leaving the implementation of pro-cycling and car-limiting measures mostly to local government. [Source: Collectie Het Utrechts Archief, Fotodienst GAU, no. 823903]

a "backing away" by lower government.[20] These words show that in the early 1990s, the national Public Works Ministry had faith in a strong national impetus for cycling, even as she continued to frame cycling – like her predecessors had done – as a local affair. It also highlights the sometimes-strained relationship between national and local government. We can see this as a result of the informal, unclear nature of the cycling governance arrangement. The rules of the game were constantly changing – and fiercely contested.

For Dutch policymakers and engineers, with their strong belief in decentralization, the reintroduction of cycling subsidies was controversial. According to MP Jan De Graaf in 1991, Maij-Weggen's colleague in the CDA party, decentralization was government policy – and the minister had to stick to that policy. He even commented that "eventually, as far as we are concerned already in 1992, no national funds will go to cycling paths."[21]

20 Vaste Commissie Verkeer en Waterstaat, meeting November 8, 1993, 22.

21 Ibid., meeting October 28, 1991, 62.

In response, the minister spoke out passionately against decentralizing cycling subsidies. Pleading for common sense, she said she was "against decentralizing for its own sake."[22] She correctly noted that compared to highway budgets, the cycling subsidies represented "a relatively small sum." Maij-Weggen insisted on the subsidies because they impacted traffic safety and human lives. What she considered failed decentralization in cycling governance had cost lives. Others agreed and attacked De Graaf for his proposal to cut cycling subsidies entirely. De Graaf defended this by blaming the provinces and municipalities for inactivity, framing his proposal as a punishment.

Many national politicians, however, believed that cycling subsidies worked. Cancelling them had caused cycling policies to stall in the late 1980s. According to left-wing MPs Wilbert Willems (green party Groenlinks) and Rob van Gijzel (labor party PvdA), no matter who was to blame, cyclists should not be the victim of this intergovernmental conflict.[23] Van Gijzel emphasized that under the Bicycle Masterplan, lower authorities had invested twice as much in cycling while the subsidies were in place, proving that "the money in the national budget is an incentive."[24] Van Gijzel also pointed out that many communities had submitted projects for subsidies: in his view, the national government could and should bear part of lower government's financial burden. In general, left-wing parties supported cycling, together with a sizeable contingent of the major conservative and even pro-car parties (liberal VVD and religious rural CDA).

Indeed, the government's decision to discontinue a national role for cycling policies after the Masterplan concluded in the late 1990s was controversial. The independent national advisory board for Public Works (*Raad voor Verkeer en Waterstaat*) protested vigorously.[25] Its report cited leading traffic experts who insisted the government set concrete targets for lower governments. These would have to include measurable quality improvements and the national government's commitment to enforce goals, contravening the Dutch tradition of decentralized autonomy. According to

22 Ibid., 48.

23 Ibid., 62.

24 Ibid., 25.

25 This advisory council included traffic scientists. See: Raad voor Verkeer en Waterstaat: Commissie Personenvervoer & Commissie Verkeersinfrastructuur, *Advies Rijk op de fiets: advies over de rol van de rijksoverheid in het fietsbeleid* (Den Haag: Raad voor Verkeer en Waterstaat, 1995). Cover letter from Committee to Minister Jorritsma-Lebbink, October 31, 1995. The policy goals were formulated in the Second Structural Scheme for Traffic and Transport around 1990. Bicycle Masterplan was a product of this policy document.

the board's report, Bicycle Masterplan had not achieved its policy goals – now left entirely to lower government. The independent experts considered this poor governance.[26] Achieving policy goals like lowering congestion and CO_2 emissions required promoting environmentally friendly transport modes.

Without strong national government direction, the sustainability goals were unattainable. The traffic experts agreed with the national government that constructing cycling paths was lower government's responsibility. At the same time, they argued that cycling was a means of achieving certain national policy goals. From a governance standpoint, it was problematic to formulate these goals then expect others to work towards them without a major national role.[27] Lower government, for example, could receive a subsidy for their regional traffic plans, without the national government strictly checking the amount of cycling facilities in the plans. Moreover, the national government spent perhaps 1 percent of its infrastructure budget on cycling.[28] In short, critics argued the national government made "paper policy" – it paid ample lip service to cycling, yet did not put its money where its mouth was.[29] The national ambitions became ensnarled in the decentralization politics of the 1990s.

The issue of decentralization occurred in a wider political context. In the late 1990s, the central government decentralized transport budgets entirely to local government under a key policy called the VERDI covenant. The Dutch acronym refers to the policy's aim: make traffic policy Regional, Decentral, and Integral. What made this covenant different to the earlier discussed transfer of responsibilities was that the national government now created a new funding structure whereby control over funds shifted to the provinces and city regions.[30] In 1996, the national government introduced GDU (*Gebundelde Doeluitkering*) to replace various national transport subsidies with one lump sum of un-earmarked money: provinces could divide this money as they saw fit between smaller infrastructural projects, including cycling. In 2004, this arrangement was extended. Now called BDU (*Brede Doeluitkering*), it included funding for larger projects.[31] With

26 Ibid., 1.

27 Ibid., 2-5.

28 Ibid., 14-16.

29 Ibid., 14-16.

30 In recent decades, the Dutch government like elsewhere has experimented with an intermediary level of government between the province and the municipality: urban region. Variously called *kaderwetgebieden, stadsregio's* and *plusregio's*, these were short-lived.

31 The GDU was meant for projects up to 25 million guilders (11.34 million euros), while the BDU was for projects up to 500 million guilders, or 225 million euros.

the introduction of these measures, the previous "golden strings" method, when national government could steer local policy priorities through funding allocation, disappeared. Provinces and municipalities became more autonomous, prompting a truly new phase in cycling and mobility governance. While this went some way to creating a clearer division of tasks in cycling governance, in practice, all layers of government remained involved in some capacity. Commentators feared the lack of specific cycling funds would put a halt to lower government investments in cycling. This fear proved unfounded, as I discuss in the next section.

Despite this funding decentralization, the national government has occasionally been involved in cycling policy since 2000 as more than a knowledge broker. As national involvement in automobile and public transit planning is vastly larger, national policies tackling congestion or promoting rail transport bring extra funding for cycling as well. The extra funding came about because the government embraced door-to-door mobility. It began to see cycling as ideal to bridge the first and last kilometers of people's journeys. This is part of a conceptual shift from transport to mobility as a leading planning principle. In this model, individuals' actual travel choices and routes instead of the infrastructure steer the provision of transport facilities.[32] As Jan Ploeger argues, it was decades-long activism by train users, who traveled to and from the station by bicycle, that brought this type of integrated travel to the attention of government and railway companies, who lacked an integral, bird's eye view of mobility.[33] Having divided up tasks and responsibilities, few policymakers or businesses had a systemic overview from the traveler's perspective of their door-to-door experience. It required user practice and expertise to point out the door-to-door potential of bicycle-train-bicycle journeys.

Since 2000, the role of cycling in chain mobility put the bicycle on the national radar in two main ways. First, the national government finances and coordinates the construction of more bicycle parking facilities at train stations to boost multi-modal bicycle-train journeys as a substitute for car journeys – responding to a long-standing social practice.[34] For decades, Dutch people have traveled to stations by bicycle. Engineers and planners discovered this only belatedly and have applied concepts like multi-modal or chain mobility to this old practice. The journey from the station to the

32 Ploeger and Oldenziel, "The Sociotechnical Roots of Smart Mobility," 15.

33 Jan Ploeger, "Waar laat ik mijn fiets?," Working Manuscript 2021.

34 For the insights to this issue and history, I thank Jan Ploeger whose forthcoming book discusses the mutual shaping of train-bike policy and practice.

destination (places of work, schools) is now also often made by bicycle as Dutch Railways have built a successful system of rental bicycles at train stations, the public transit bike (*OV fiets*). Second, recent investments in cycling highways between towns are increasing. These wide, high-quality cycling routes have priority crossings and cater for (e-)bike commuting as a solution for car congestion. What were the governance arrangements for these policy goals?

Bicycle parking is essential to encourage commuters to travel by train, with the bicycle as before and after transport. Policymakers consider the public-transit combination with cycling a premier alternative to car traffic. Since 2003, the highly successful introduction of shared bicycles at train stations, taken over in 2008 by Dutch Railways, has only reinforced people's habit of combining the bicycle with the train for their commute.[35] Where national policymakers often frame cycling as a local mode (to be governed locally), trains form a national network and fall within the national government's remit. To policymakers, the hundreds of haphazardly parked bicycles at major Dutch train stations presented a chaotic sight. In order to relieve overburdened facilities, the national government invests heavily in bicycle parking. Policy implementation is left to another national stakeholder, state-owned company ProRail, that is responsible for rail infrastructure. In 2011, the Ministry of Public Works launched Action Plan for Cycle Parking (*Actieplan Fietsparkeren*). As with previous plans, the policy document makes it very clear that national funding is only a temporary stimulus. Again, the state believed that long-term responsibility for planning and constructing bicycle parking facilities at stations should be fully decentralized.[36] Local government could sometimes count on national subsidies in order to speed up the construction of bicycle parking.[37] Utrecht, with its central position in the country's railroad network, completed the world's largest bicycle parking garage at its central train station in

35 Ploeger and Oldenziel, "The Sociotechnical Roots of Smart Mobility." For the more recent history of commercial bike-share operators, see Brett Petzer, Anna Wieczorek, and Geert Verbong, "Dockless Bikeshare in Amsterdam: A Mobility Justice Perspective on Niche Framing Struggles," *Applied Mobilities* 5, no. 3 (2020): 232-50, and "Cycling as a Service Assessed from a Combined Business-Model and Transitions Perspective," *Environmental Innovation and Social Transitions* 36 (2020): 255-69.

36 *Kamerstukken II* 2010/11, no. 32 404 (Programma Hoogfrequent Spoorvervoer: Actieplan Fietsparkeren [Program High Frequency Rail Traffic: Action Plan Bicycle Parking]), sub-no. 53, 3.

37 The ministry heavily subsidized Amsterdam's construction of extra bicycle parking facilities around its Central Station. While entirely local, the project was deemed worthy of investment in the 2010s.

Figure 33 Masses of parked bicycles at Utrecht Central Station in 2009 show the importance of the bicycle-train connection for Dutch commuters. Utrecht has since constructed the world's largest underground bicycle parking garage. The national government supports and invests in these facilities. [Source: Wikimedia Commons]

2019 (see figure 33). The nation's capital Amsterdam pledged to spend 15 million euros annually between 2012 and 2020 for the same purpose.[38] In 2020, ProRail announced its goal to increase the current 490,000 bicycle parking places at train stations to 600,000 by 2030, depending on local and national government's willingness to co-finance.[39] In 2016, another governmental agreement transferred national funds to local policymakers to increase bicycle parking places at train stations and strengthen the bicycle-train combination.[40] In recent years, the railroad authority has significantly upgraded and expanded many major Dutch bicycle stations.[41] We should note that these measures are in response to cyclists' demand and long-standing social practices. This is not necessarily the case with the second national policy.

The e-bike put cycling on the national governance radar as well. National policymakers have been promoting long-distance (e-)bike commuting as another solution to traffic congestion.[42] This forms an extension of the

38 Oldenziel et al., *Cycling Cities: The European Experience*, 26.

39 https://www.prorail.nl/reizen/stations/fietsen [accessed February 11, 2020].

40 Letter Ministry of Public Works to Lower House, November 26, 2020, 29. https://www.rijksoverheid.nl/documenten/kamerstukken/2020/11/26/bestuurlijke-overleggen-mirt-25-en-26-november-2020-voortgang-mirt-moties-en-toezeggingen [accessed November 27, 2020]

41 Folkert Piersma and Wout Ritzema, *Fietsparkeren bij stations. 20 jaar ontwikkeling, ontwerp en realisatie* (Bussum: THOTH, 2021).

42 I take these insights from the work of Matthew Bruno, George Liu and Sun Qi, undertaken within the Bicycle Challenges and Smart Cycling Futures research programs at Eindhoven University of Technology. Matthew Bruno's work on cycling governance and innovations like the bicycle highway or *fietsstraat*, George Liu's work on design norms and cycling infrastructure in

by now time-honored Dutch tradition of constructing separate cycling infrastructure. It presents new challenges. The national government coordinates and funds the construction of cycling highways, emerging as focal point of cycling policy in the Netherlands. Engineers design these cycling paths for longer-distance (and often higher-speed) trips, so the dimensions are more generous, with right-of-way at crossings.[43] While the bicycle is typically useful for 5-10 km distances, e-bikes and good cycling connections enable longer trips. Noteworthy is that bridging longer distances by bicycle represents a return to older mobility practices, common before the car made its breakthrough for commuting: until the 1950s, workers often commuted over an hour one way by bicycle.[44] The government's policy to fund bicycle highways aims to combat car congestion by offering cycling as an alternative to drivers – in other words not to help current cyclists, but, as Matthew Bruno argues, to create a new group by seducing car drivers to choose the bicycle for their commute.[45] In the multi-year program for infrastructure published in 2019, the Ministry of Public Works announced its aim to convince 200,000 car drivers to commute by bike (and train) before 2021.[46] In a push-and-pull model, the Dutch government prefers to make cycling more appealing rather than come up with politically sensitive car-limiting measures.

The Ministry of Public Works also formalized its role in cycling expertise by creating the central knowledge institution Cycle Council (*Fietsberaad*) in 2001. Combining representatives of municipalities, Dutch Railways, the Cyclists' Union, consultancy, and educational agencies, this national institution coordinates the production and exchange of cycling knowledge. It disseminates expertise through its magazine Bicycle Traffic (*Fietsverkeer*, 2002-2018, since then entirely online) and a website with a report database.[47]

the built environment, and Sun Qi's comparative work on the rise of e-bikes in the Netherlands and China have all benefited me enormously.

43 George Liu, Marco te Brömmelstroet, Sukanya Krishnamurthy, and Pieter van Wesemael, "Practitioners' perspective on user experience and design of cycle highways," *Transportation Research Interdisciplinary Perspectives* 1 (2019): 1-8.

44 Bek, "Governing Workers' Mobility," book manuscript, 2021.

45 Matthew Bruno and Anna Nikolaeva, "Towards a Maintenance-Based Approach to Mode Shift: Comparing Two Cases of Dutch Cycling Policy Using Social Practice Theory," *Journal of Transport Geography* 86 (2020): 1-11.

46 *MIRT Overzicht. Meerjarenprogramma Infrastructuur, Ruimte en Transport 2019*, (Den Haag: Ministerie van Infrastructuur en Waterstaat, mede namens het Ministerie van Economische Zaken en Klimaat, het Ministerie van Binnenlandse Zaken en Koninkrijksrelaties, 2019), 41. See https://www.mirtoverzicht.nl/.

47 https://fietsberaad.nl/

This reflects the government's core policy belief that cycling, as a local mode of transport, can best be governed locally. Investment in cycling expertise also reflects the government's belief that it is more efficient in supporting research and coordinating policy at a higher level. Over the course of the twentieth century, the cycling (engineering) expertise or bottom-up cycling user knowledge produced by the non-governmental tourist organization ANWB and the Cyclists' Union has always provided extensive expertise. This has informed the more evidence-based cycling policies in the Netherlands. In short, the national government has now also adopted that role.

At a national level, recent research points out, engineers are also interested in gathering expertise about cyclists' behavior and choices, along with their traditional interest in all kinds of pilot projects around infrastructural innovation. Most policies aim to turn drivers into cyclists, however, rather than facilitate cyclists already on the road. Only some national policy initiatives facilitate existing cyclists.[48] One example is the Keep on Pedaling (*Doortrappen*) program that finds ways to keep senior citizens cycling longer – a program financed by the Ministry of Public Works and supported by consultancy agencies and NGOs.[49] Another example is a 2020 best practices report identifying how to use large infrastructure projects for automobility to simultaneously improve cycling infrastructure in a cost-efficient way.[50] According to the report, "linkage opportunities" (*meekoppelkansen*), refer to the large (highway) projects requiring extensive groundwork that offer the opportunity to construct a bicycle tunnel or path in the process at much lower cost.[51] For instance, the agency plans to use the renovation of bridges on the A27 highway to create an extra river crossing for cyclists.[52] In a way, this is a continuation of older practices:

48 Bruno and Nikolaeva, "Towards a Maintenance-Based Approach." Between 2006 and 2009, the Ministry of Public Works launched the program FileProof. Like Bicycle Masterplan, its smaller projects aimed at reducing congestion. One was Fiets Filevrij ("Bicycle Congestion-Free"), encouraging substituting car journeys under 15 km with bicycle trips.

49 https://doortrappen.nl/default.aspx. The potential for higher cycling levels among senior citizens is significant: Lucas Harms, Luca Bertolini, and Marco te Brömmelstroet. "Spatial and Social Variations in Cycling Patterns in a Mature Cycling Country: Exploring Differences and Trends." *Journal of Transport & Health* 1, no. 4 (2014): 232-242.

50 Bike-minded Design Consulting, M. Lopes Cardozo, and M. Bielderman, *Voorbeeldenboek fiets: toewerken naar het structureler meenemen van de fiets in rijksprojecten* (s.l.: Ministerie van Infrastructuur & Milieu, Rijkswaterstaat, 2020).

51 Ibid., 11.

52 Ibid., 50. Other examples are ways to keep cycling routes open during major road works, including temporary cycling paths and bridges to prevent people from switching from cycling to driving.

twentieth-century engineers often saw the reconstruction of a road as an opportunity to simultaneously upgrade existing cycling infrastructure or construct new paths. It also exemplifies how Dutch engineers have historically tried to facilitate cycling and driving in tandem – with varying levels of success. When push came to shove, the bicycle also lost out often even in the Netherlands.

The national government is also facing legislative issues surrounding cycling paths: (e-bike) commuters, sport cyclists, and recreational cyclists are using the country's cycling paths in ever larger numbers, creating potential conflicts and prompting questions about who belongs where. The issue of overcrowding on the dedicated cycling infrastructures is a return to the 1950s debate surrounding the moped's place on cycling paths. Today, this debate has returned, prompting fierce debates about where moped riders (*brom-versus snorfiets*) belong. Cities like Amsterdam are now experimenting with banning these mopeds on paths and moving them to the roads.[53] In 2015, Parliament asked the minister of Public Works to investigate this issue. The ministry organized network meetings to discuss pilot projects and solutions with numerous stakeholders. These included governmental branches and agencies (provinces, municipalities and transport regions, traffic police, CROW central knowledge center), industry lobby organizations (RAI and BOVAG), interest groups (ANWB, Cyclists' Union, recreational cycling union NTFU), and traffic safety lobby group VVN and the scientific institute for traffic safety SWOV.[54] Cycling (at different speeds) is so popular that it is presenting congestion and safety issues in certain areas and on cycle paths inside and outside cities. The overcrowding of the country's existing cycle path system represents a new challenge for policymakers, but most of all highlights the limits of the traffic-separation model of engineering. After almost a century of following this model, urban planners are now exploring a return to the mixed spaces that were the norm before engineers adopted traffic separation in and outside cities.

Based on the analysis in previous chapters, national policymakers' involvement in the cycling policy coalition demonstrates a wave-like pattern. The national government's interest in cycling is currently on an upward trend as evidenced by rising investments. Until 2017, Dutch coalition agreements rarely mentioned cycling. But the 2017 center-right coalition agreement announced – again – a one-time sum for cycling: "Because cycling

53 E.g., at a traffic safety committee meeting with MPs on June 13, 2019 (*Kamerstukken II* 2018/19, no. 29398, sub-no. 735).

54 *Kamerstukken II* 2014/15, no. 29398, sub-no. 465.

can be a good alternative for public transit and the car, the government invests a once-only sum of 100 million euros for co-financing municipal and provincial investments in cycling infrastructure and bicycle parking at public transport junctions."[55] This funding went mainly to bicycle parking facilities at train stations and cycling highways.[56] As part of the country's Climate Agreement (*Klimaatakkoord*) to fulfil the Paris Agreement concluded in 2019, the government set aside an extra 75 million euros for bicycle parking.[57]

The national government's role in cycling remains contested: the different types of cycling are not all governed at the same level. In November 2020, two MPs called on the national government to take more "system responsibility" for recreational cycling and walking routes, to (economically) stimulate tourism. According to the MPs, "recreational cycling routes often coincide with utilitarian use and contribute to sustainable mobility goals."[58] It is fitting to close our final discussion on cycling governance with this example as it sums up the national government's historical role. In response to constant calls and political pressure from social organizations and politicians in parliament to play a greater part in cycling governance, the national government has been heeding these periodically. Permanent institutional embedding has not happened.

8.2 Provinces and Municipalities Double Down

The decentralization of transport funding to the provinces in the late 1990s strengthened their already significant intermediary role in mobility and cycling governance. The mandate allows provinces to coordinate cycling networks at the level just above the city and local communities. Still, a 1996 report on decentralization and cycling policy shows how the provinces were struggling to find their role in cycling governance. The experiments to establish so-called transport regions as regional governance networks between the municipality and the province were, in most places,

55 "Vertrouwen in de toekomst: Regeerakkoord 2017-2021, VVD, CDA, D66 en ChristenUnie," p. 40. [https://www.parlement.com/9291000/d/pdfs/Regeerakkoord20172021.pdf, accessed February 11, 2020]

56 74 of the 100 million euros are dedicated to bicycle parking facilities at train stations.

57 Klimaatakkoord, p. 77 https://www.klimaatakkoord.nl/klimaatakkoord/documenten/publicaties/2019/06/28/klimaatakkoord [accessed February 12, 2020]

58 *Kamerstukken II* 2020/21, no. 34 682 (Nationale Omgevingsvisie [National Environmental Vision]), sub-no. 59.

short-lived.[59] At first, the provinces thought they could withdraw from the cycling policymaking process and leave it to these regional administrative bodies. When this was not successful, the provinces had to step in again, if only to distribute funds over municipalities.[60]

National policymakers were not confident about how provinces would spend the funds, fearing that local policymakers would perhaps focus on large infrastructural or public transit projects, leaving little for cycling.[61] This fear proved unfounded. According to different assessments, local government spent 30-35 percent of the transport lump sum (GDU) funds on cycling facilities. Of the other 65-70 percent, half went to projects that indirectly benefited cyclists like implementing 30 km/h zones and constructing roundabouts and safer junctions.[62] Moreover, decentralization was a wider political trend. National policymakers have also increasingly made provinces responsible for traffic safety.[63] All in all, the suspicion some national politicians expressed about lower government's traffic policy priorities was unfounded.

Regional autonomy meant that the focus on cycling varied widely from place to place. In 2003, the Dutch cycling expertise center published a policy evaluation in its Cycling Traffic magazine. It found that eleven of the nineteen recipients of GDU funds used them to achieve their own traffic policy goals. The regional authorities in Amsterdam, Utrecht, Eindhoven, and the province of Brabant, spent more on cycling facilities than on other traffic measures. Some municipalities had guidelines reserving significant amounts of GDU funding for cycling.[64] The absence of national rules for spending GDU funds did not mean that municipalities neglected cycling – quite the contrary. This also shows the remarks by Ministers of Public

59 Thirty transport regions (*vervoerregio's*) were planned in the 1990s, but quickly disappeared again. Only Rotterdam-Den Haag and the region around Amsterdam still form a formal transport region.

60 Menno Post, *Decentralisatie van het fietsbeleid: Stuurloos naar de toekomst?* (Amsterdam: Adviesburo voor Mobiliteit, 1996), 45-46.

61 Lower government had an obligation to contribute financially to large infrastructural projects for which they received national funding. Contributions to costly projects might swallow up the GDU funds, which, however, does not seem to have been the case.

62 "Jaarlijkse GDU-bestedingen vanaf 1996: Aandacht voor fietsvoorzieningen houdt stand," *Fietsverkeer* 2, no. 5 (2003): 11. This was also concluded in an evaluation study from 2001: H.W. van Altena and B. Rakic, *GDU subsidies: Inventarisatie van probleemgebieden en maatregelcategorieën, nu en in de toekomst* (Veenendaal: Traffic Test, 2001).

63 Charlotte Bax, Pieter Leroy, and Marjan P. Hagenzieker, "Road Safety Knowledge and Policy: A Historical Institutional Analysis of the Netherlands," *Transportation Research Part F* 25 (2014): 131-33.

64 "Met GDU (fiets)verkeersbeleid regisseren," *Fietsverkeer* 2, no. 5 (2003): 10-11.

Works in the late 1980s and early 1990s in a different light: they argued that municipalities had spent little on cycling and that further subsidies would likely be misspent. The recent history does not bear out their misgivings.

In governance terms, the provinces opted for multi-disciplinary teams instead of making one person or provincial body responsible for cycling as most cities did. According to policymakers, from a governance perspective, civil servants in such teams should come from different departments and devote only part of their time to cycling, to integrate it with other aspects of their work like spatial planning and recreational policies.[65] In other words, cycling policy should not be cast in the narrowly defined traffic policies.[66] These provincial teams worked in informal ways with municipalities to coordinate cycling on a regional level. According to a Gelderland official in 2002, the province subsidized municipal cycling paths and solved blackspots, while its contact with municipalities was "indirect" and limited to the occasional nudge, with no binding agreements.[67] In 2003, Stan Lauret, a provincial coordinator for cycling in Zeeland, called its role in cycling governance a "search for the exact position of the bicycle in the larger whole of traffic and transport, spatial planning, tourism, and recreation."[68] He believed that a stimulating and coordinating role for a relatively small provincial government like Zeeland was important, if only because most officials lacked traffic design expertise for cyclists.[69] Provinces were committed to coordinating a provincial cycling route network with local cycling networks through subsidies for municipal infrastructure. However, the relationships with municipalities often consisted of non-binding agreements, mirroring the national attitude towards provincial and local policymakers. The rules of the game in this cycling governance coalition remained by and large informal.

At the urban level, cities have intensified their already active engagement in promoting urban cycling since the 1990s. According to a consultancy report, municipalities increasingly and innovatively shaped (intensive)

65 "Linda de Klein-Bevers, provincie Brabant: 'Wees realistisch: de wereld verandert niet in één keer'," *Fietsverkeer* 1, no 2 (2002): 9.

66 Around 2000 these provincial teams of civil servants existed, for example in Gelderland (Unit Fiets), Noord-Brabant (FietsAdviesTeam) and Zeeland (Zeeuws Coördinatiepunt Fiets).

67 "Johan Leferink, provincie Gelderland: 'Behoefte aan uitwisseling met andere provinces'," *Fietsverkeer* 1, no 2 (2002): 8.

68 "Coördineren en stimuleren: Het fietsbeleid van de provincie Zeeland," *Fietsverkeer* 2, no. 5 (2003): 8.

69 Lauret notes how not all officials were aware of CROW guideline publications, partly explained by the fact that only 25 percent of traffic policy officials had a traffic engineering education.

cycling policy.[70] Since the early 1990s, cities have continued to implement pro-cycling measures by addressing the blackspots activists mapped out, while simultaneously intensifying car-limiting measures in city centers.[71] The researchers concluded that history matters: communities and cities that started a cycling policy earlier, had a more mature cycling network than others.[72] Policymakers in cities like Den Bosch or towns like Emmen and Hengelo considered the longer local tradition of cycling policy a plus.[73] Civil servants confirmed the value of a stable, long-term commitment to cycling by building cycling cities. In a series of interviews with the cycling expert magazine Cycling Traffic, the experts believed the winning formula for creating cycling cities consisted of good design principles coupled with continuous financial and individual support. In 2005, Zwolle's government official for cycling policy, Willem Bosch, asserted that the principles and routes in the city's 1978 Cycling Traffic Plan still guided its policy. The city invested every year in cycling and its traffic officials spent half their time on cycling infrastructure: realizing something like the 1978 Zwolle plan was costly and its implementation was only gradual. It required a long-term commitment from policymakers. As a result, Zwolle boasts an extensive network of cycling paths and traffic-calmed streets, including many cycling tunnels.[74]

Local policymakers generally approve of their national counterparts' focus on producing cycling expertise. Civil servants, particularly in smaller municipalities, do not always have the resources or relevant knowledge to adequately design cycling facilities. Here the provinces and national government can play an important role in promoting the expertise of institutes like the cycling expertise center CROW.[75] Before the formal founding of such nationally funded expert centers, policymakers shared expertise and best practices informally. A case in point is the town of Veenendaal, which in the 1970s adopted the Delft-pioneered network design principle, providing cyclists with a fine-meshed network of safe cycling infrastructure and limiting detours as much as possible. Veenendaal's version – many other communities applied the Delft principle less strictly – included cycling tunnels at several hundred-meter intervals in a new railway embankment.

70 Ligtermoet et al., *Gemeentelijk fietsbeleid*, 7.

71 Ibid. in Utrecht: Buiter, "Utrecht: Bicycles Rule – Again," 36.

72 Ibid., 10.

73 Ibid., 34-35.

74 "De lange adem van Zwolle: Perfecte fietsroutes," *Fietsverkeer* 4, no. 11 (2005): 1-7.

75 "Coördineren en stimuleren." Lauret noted that 40 percent of Zeeland's civil servants had a background in tourism and recreation, a key sector in this coastal province.

Cyclists no longer had to make detours.[76] Cities also discovered that compact urban expansions resulted in shorter distances for cyclists: planners often designed new neighborhoods close to their town centers to keep distances to the city center cyclable.[77] Sharing and producing these insights centrally might create more uniformity in the future than the earlier history of Dutch cycling governance.

For now, local differences remain significant. And, as they have throughout the twentieth century, the right people at the right time can still make a major difference, especially at the provincial and city level. According to historian Hans Buiter, it was not until the 1990s that the city of Utrecht actually started constructing a city-wide network of cycling paths.[78] Pro-cycling policies were implemented more enthusiastically and speedily under left-wing traffic councilors like Hugo van der Steenhoven in the mid-1990s, he wrote.[79] According to Adri Albert de la Bruhèze, in the case of Enschede, the presence of an active social democratic councilor, Dick Buursink, made all the difference in curtailing car traffic and giving back road space to cyclists.[80] Other cities considered cycling so crucial that policymakers could implement policies for cyclists regardless of the ruling parties' political color.[81] The Cyclists' Union, on the other hand, has criticized such dependence on a local councilor's stance. The organization wants more long-term guarantees that local government will provide safe and attractive cycling facilities regardless of elections and who happens to be in power.[82] Importantly, it presents cycling as a non-partisan issue and public good. In the Union's view, a stable bureaucracy consisting of officials who remain in function when political coalitions change, guarantees some continuity to protect cycling.

Cities may have the freedom to promote cycling as they see fit. Urban policymakers occasionally articulate their frustration at the state's inconsistent attitude and the lack of firm backing. Echoing the national advisors' critique of decentralization in the late 1990s, they criticized the lack of national support in 2004. In a decentralized mobility governance structure without

76 "Het comfortabele netwerk van Veenendaal: Verwennerij per 300 meter," *Fietsverkeer* 4, no. 12 (2005): 4-7.

77 Dirk Ligtermoet, "Het fietsklimaat van Culemborg: waarom Nederland zo'n fietsland is," *Fietsverkeer* 10, no. 27 (2011): 10-14.

78 Oldenziel et al., *Cycling Cities: The European Experience*, 36-39.

79 Ibid. Van der Steenhoven later became chairman of the Cyclists' Union.

80 Ibid., 47.

81 E.g., Eindhoven: ibid., 60-61.

82 Post, *Decentralisatie van het fietsbeleid*, 75.

clear national support for anti-car measures, municipalities will conceivably end up competing in attracting business and residents by presenting their area as the most accessible by car. After all, the national government, citing economic motives, keeps on investing in Dutch highway infrastructure and encouraging high car use levels that impact cities. This leads to conflict between government levels. As of 2021, for example, the Dutch national government has committed to widening a highway through the Amelisweerd woodland near Utrecht – scene of a half-century-long protest starting in the 1970s (see 6.2) – over both the city and the province's objections. Urban authorities find it hard to achieve traffic calming outside city centers even after designing downtown areas for cyclists and pedestrians.[83] Increasing evidence shows that it is hard to promote cycling levels without simultaneously discouraging car use.[84] According to a civil servant in The Hague, like elsewhere, policy documents on cycling multiplied, for example with the Multi-year Bicycle Program 2003-2007. In the end, he argued, what truly made a difference for cyclists was the adoption of strict car restrictions in the city center in 2009 under the new Traffic Circulation Plan.[85] Has the Dutch mobility policy tradition of facilitating both driving and cycling reached its limits? Or can further shifts from driving to cycling only be achieved by making driving more expensive or otherwise less attractive?

Urban cycling governance has recently come to mean more than just infrastructure provision. For most of the twentieth century, cycling governance equated almost exclusively to the construction of cycling paths. Theft prevention and bicycle parking have been part of urban cycling policy since the 1950s. Fifty years on, local policymakers are working hard to make cycling more inclusive and accessible in order to improve the mobility of groups with few options (the so-called "traffic poor"). Around 2010, Amsterdam earmarked funds to boost cycling among groups of people who normally do not cycle much.[86] With low levels of cycling (for Dutch standards) and many immigrants, Rotterdam's policymakers in particular have committed to making cycling more inclusive and combat transport poverty.[87] The port city is particularly sensitive to how a lack of mobility options can hamper citizens' ability to access jobs or recreational possibilities and

83 "De plek van de fiets in gemeentelijk verkeersbeleid," *Fietsberaad* 3, no. 9 (2004): 15-17.

84 Lucas Harms, Luca Bertolini, and Marco Te Brömmelstroet. "Performance of Municipal Cycling Policies in Medium-Sized Cities in the Netherlands since 2000." *Transport Reviews* 36, no. 1 (2016): 134-62.

85 Berkers, Botma, and Oldenziel, *Cycling Cities: The Hague Experience*, 49.

86 Oldenziel et al., *Cycling Cities: The European Experience*, 25.

87 Berkers et al., *Cycling Cities: The Rotterdam Experience*, 55.

is working to solve this through cycling.[88] These initiatives highlight "a broader approach" to cycling policies instead of the almost exclusive focus on building cycling paths in earlier decades. Today, investments in aspects like bicycle parking and broadening user groups are part of cycling policy. This policy now comprises more than just safe infrastructure, showing that it has met cyclists' basic needs and that cycling has become a normal and mature activity in most Dutch cities.[89] It fits into a larger trend of policy attention for door-to-door mobility rather than a narrower focus on (physical) infrastructure.

A final governmental development to discuss here is the role of regional governance as an intermediate level between the city and the province. This is, in a sense, not new. The citizen cycling path organizations in the first half of the twentieth century also operated on a regional scale. When their operations were incorporated by provinces and municipalities in the 1950s, they often chose a regional scale of operation. This level fits quite well with providing primarily recreational cycling path networks. With the increasing popularity of commuter journeys by e-bike, policymakers recognize the regional scale as also appropriate for governing this type of cycling.

One outcome is that regional authorities jumped on the idea of cycling highways as their new mandate: wide cycling routes with priority at crossings, allowing (suburban) commuters on (e-)bikes to get to work quickly. The longer distance and cost of these routes require multiple municipalities to collaborate in the construction beyond their boundaries. Today, provinces play a key role in creating regional cycling routes, often in the form of cycling highways. The province Gelderland, for example, is very active.[90] In the Arnhem-Nijmegen region, policymakers have targeted drivers more explicitly than before, aiming to get more of them on bicycles. Facing rising automobility, the town of Arnhem, traditionally focused on public transit rather than a specific cycling policy, came up with a more dedicated cycling policy in 2013. This included regional cooperation to set up more cycling highways.[91]

Other policymakers invested in regional cycling networks. Policymakers in The Hague, which already has high rates of interurban cycling in the region, consider bicycle highways a solution to traffic congestion. According to Berkers, Botma, and Oldenziel, the investment represents a departure

88 Ibid., 55-56.

89 Ibid., 57.

90 https://www.snelfietsroutesgelderland.nl/ [accessed 07-02-2020].

91 Berkers and Oldenziel, *Cycling Cities: The Arnhem and Nijmegen Experience*, 45-46.

from older ideas about cycling as an urban mode associated with "traffic calming, urban livability, relaxation, and fun."[92] They suggest that these paths, and the increasing popularity of e-bikes and speed pedelecs, facilitate a new kind of fast cycling more akin to car driving than traditional cycling – in this region, interurban commuting by bicycle had a long history. Research in the busy Rotterdam-The Hague region showed that on the A20 highway, a third of all cyclists rode no more than 15 km each time. With the extended radius of the e-bike, these distances are bikeable, if the barrier of the highway can be solved with cycling bridges and tunnels.[93] In 2008, Enschede started a regional program for a 60-kilometer-long, 4-meter-wide cycling highway.[94]

From a governance perspective, these networks involved several municipalities that required regional cooperation. Arnhem, Nijmegen, and the surrounding communities started to collaborate regionally in 1988.[95] Such cooperation between city regions and the non-governmental Cyclists' Union succeeded in convincing the Ministry of Public Works to fund cycling highways.[96] Partly because of the many e-cyclists' longer range, Rotterdam is also working in the Metropolitan Region Rotterdam-The Hague to complete a regional cycling path network.[97] Rather than seeing these collaborations as a true innovation, they are the return of older but forgotten governance practices.

8.3 Cyclists' Union Professionalizes Further

In the 1970s and 1980s, cycling activists established themselves as important stakeholders in the Dutch cycling governance coalition. Increasingly throughout the 1980s, policymakers at various levels welcomed their input in cycling policymaking. Local branches, albeit to a varying extent, maintain good contacts with local policymakers and sometimes receive subsidies. The Cyclists' Union has become a part of the governance network that is taken seriously by other stakeholders. During parliamentary debates on the Bicycle Masterplan in 1992, MPs often invoked the Cyclists' Union

92 Berkers, Botma, and Oldenziel, *Cycling Cities: The Hague Experience*, 51.

93 Bike-minded Design Consulting, Lopes Cardozo and Bielderman, *Voorbeeldenboek fiets*, 98.

94 Oldenziel et al., *Cycling Cities: The European Experience*, 48. Total projected costs were 45 million.

95 Berkers and Oldenziel, *Cycling Cities: The Arnhem and Nijmegen Experience*, 44.

96 Ibid., 50.

97 Berkers et al., *Cycling Cities: The Rotterdam Experience*, 56.

opinions and comments on the proposed policy.[98] With a view to this plan, the national government awarded the Cyclists' Union a modest subsidy to "professionalize" its activities. According to the parliamentarians, these subsidies were valuable for improving the quality of regional and local cycling policies as "local branches of the ENFB usually have a good idea of how (local) interests of cyclists are best served."[99] Like the ANWB in the 1920s, the Cyclists' Union had legitimacy as a professional actor. Indeed, so valued was its input, that it became a semi-public institution – still nothing like the ANWB which employs several thousand people. The ANWB also has an independent source of income from its large member base as well as its commercial activities selling goods and services for tourism.

Feeling the competition from the Cyclists' Union and the general shift in public opinion, the tourist organization ANWB has increased its focus on cycling in the past thirty years. Due to the alternating evaluations of automobility and cycling since the 1970s, the ANWB has to some extent rediscovered its roots. In 2011, MPs noted that the ANWB had called for transferring money from car infrastructure budgets to cycling infrastructure. Socialist MP Paulus Jansen (SP) called this "revolutionary" because it was hard "to suspect the ANWB of leftist sympathies."[100] The organization has also undergone significant changes. Throughout the twentieth century, the ANWB developed more and more commercial activities. Unifying these business activities with lobbying and advocacy work in one organization was too difficult. In 1996, the organization split into a commercial branch and a foundation. Having jettisoned commercial activities, the foundation had to redefine its role. In the past twenty years, volunteer activities, including in cycling, have increased again.[101]

The Cyclists' Union continues to call on national government to direct and support lower government more robustly in providing better and safer cycling environments, and to formulate concrete policy framework and intervene if local government fails to achieve its goals.[102] Traffic safety and the quality of cycling infrastructure have advanced dramatically over the past fifty years, and the Cyclists' Union keeps pushing for further improvements. In particular, it would like to lower maximum speeds in built-up areas from 50 to 30 km/h. For drivers this is a controversial issue. However,

98 *Kamerstukken II* 1991/92, no. 20922 sub-no. 120.
99 *Kamerstukken I* 1991/92, no. 22300 XII/22 300 A/22 300 G, sub-no. 177b, 19.
100 *Kamerstukken II* 2010/11, no. 32660, sub-no. 15, 4.
101 Buiter and Staal, *Het avontuur van de ANWB*, 156-81.
102 Post, *Decentralisatie van het fietsbeleid*, 75.

the activists point to the data showing that many accidents happen on 50km/h roads. In 2020, the Dutch Parliament accepted a motion to make 30km/h the default speed in these areas and only allow 50 km/h on certain roads. Lobbying at the national level is combined with continued pressure from local branches.

While developing professional expertise, the Cyclists' Union still collects cyclists' local experiences as input for better cycling policy. First, an alternative means of stimulating municipalities to undertake pro-active cycling policies is the Cyclists' Union tool Cycling Balance (*Fietsbalans*, introduced in 2000). From the start in 1975, the Cyclists' Union defended user expertise as an indispensable source of good cycling policy. Second, through its innovative "bottleneck memoranda," it relayed this user expertise of where cyclists encountered difficulties in their journey to local engineers. With the publicly visible Cycling Balance tool, the Cyclists' Union ranks Dutch cities' cycling environments on a host of issues. Third, coupled to this is the election of the best cycling city in the Netherlands – the fact that this is a coveted annual title is further proof of the improved status of urban cycling over the past fifty years.

This initiative also shows that the Cyclists' Union upholds the value of gathering user-based expertise: via Cycling Balance, it collects a large amount of data, using among other things a specially equipped bicycle to measure the maintenance quality of cycling infrastructure. Instead of leaving this measuring to civil servants, the Cyclists' Union puts pressure on cities to raise their standards and make themselves invaluable as governance partner.[103] In 1990, Gelderland province, in a memorandum on a provincial cycling path plan, wrote that "there is a need for structural contact with the Cyclists' Union" because they had "local knowledge" and "know what problems occur in practice."[104] As early as 1994, Union volunteers did a quality check on cycling infrastructure in the province Flevoland. The province processed the results in policy documents.[105] And in 2006, the province Noord-Holland asked the Cyclists' Union to make an inventory of its problems, resulting in a report with 190 points of improvement.[106]

103 Veraart, Emanuel, and Oldenziel, "Eindhoven: Engineering a Path for Bikes?," 60.

104 Jan van den Broek, "Op de fiets de provincie in!," (Fietsersbond ENFB, 1990), 40.

105 "Fietsersbond ENFB keurt fietsroutes in Flevoland," *Fietsverkeer* no. 4 (1994): 3.

106 "Provincies willen meer grip op besteding gemeentelijke subsidies," *Fietsverkeer* 6, no. 17 (2007): 23. The reporters are Fietsersbond Landelijk Bureau and afdelingen Noord-Holland, *Top 10 knelpunten op de hoofdfietsroutes per regio in Noord-Holland* (Haarlem: Fietsersbond, 2008). An update appeared in 2016: Piet van der Linden, *Knelpunten in het hoofdfietsnetwerk in Noord-Holland 2016* (Utrecht: Fietsersbond, 2016). For online access, see respectively https://

In all these instances, provincial policymakers could have asked their own engineers to do this research or hired a consultancy agency. Instead, they looked to the Cyclists' Union for expert advice. In an evaluation of a Bicycle Masterplan study, many municipalities mentioned their relations with the local branch, proactive or otherwise, as a key factor in their extensive cycling policy success.[107] Because of the Cyclists' Union structure, its relationships with policymakers differ, depending on local branches' visibility and activity.[108] Increasingly, working with local policymakers is hampered by the fact that (smaller) municipalities possess ever less in-house expertise and outsource many of their tasks to consultancy agencies. This makes it harder for the Cyclists' Union to lobby local government: who to address in the absence of knowledgeable and influential civil servants? Where in the past these civil servants served for a long time in one place, allowing activists to build lasting personal relationships and gain influence, this has changed in recent decades.[109]

In short, the Cyclists' Union maintains its role as intermediary between users and government officials. The bottleneck memoranda idea still exists in digitized format. In 2005, the local Arnhem branch created a "digital hotline" to forward complaints about dangerous or annoying situations: by 2009, the city had solved 95 percent of these cases.[110] Local policymakers mostly take such citizen complaints seriously. Although one civil servant in the village of Hardenberg opined in 2010 that a new separated cycling path was not strictly necessary since no traffic accidents had occurred on this road, he believed "too many cyclists feel unsafe" and "keep insisting." It was sometimes "public pressure" that led to the reconstruction of a road junction or crossing.[111] This relationship between cyclists and local government through the Cyclists' Union, dating back to the 1970s, remains a workable governance arrangement at local level. As of 2020, the Cyclists' Union had 150 local branches across the country.

With the state's decentralized approach to cycling governance and municipalities spreading cycling policy over various departments, the Cyclists' Union believes it is the guardian of coordination. According to Fred Feddes and Marjolein de Lange, Amsterdam's Cyclists' Union

www.noord-holland.nl/Onderwerpen/Verkeer_vervoer/Fiets/Beleid/Knelpunten_in_het_hoofd-fietsnetwerk_in_Noord_Holland_2016.org [last accessed March 16, 2020].

107 Ligtermoet et al., *Gemeentelijk fietsbeleid*, 34-35.

108 Van den Broek, "Op de fiets de provincie in!" 53, 56, 70.

109 Interview Author with Tom Godefrooij and Jaap Rijnsburger, June 2, 2021.

110 Berkers and Oldenziel, *Cycling Cities: The Arnhem and Nijmegen Experience*, 45.

111 Dirk Ligtermoet, "De geleidelijke weg naar een prettig fietsklimaat," *Fietsverkeer* 9, no. 26 (2010): 24-28, here 25.

illustrates this role very well. In 1979, the city created a cycling working group with different department officials as well as local Cyclists' Union representatives.[112] When Amsterdam decentralized its civil service into district councils, Feddes and De Lange claim that the Cyclists' Union safeguarded the coherence of urban cycling policy. Just as the ANWB did at an earlier stage, civil society activists thus served as experts and intermediaries safeguarding policy uniformity. Amsterdam funding enabled the local Cyclists' Union to employ Clasien Slebos, a social geographer, to serve the city's Bicycle Working Group. One journalist remarked how the "once so radical action group has now been taken up by the official bureaucracy."[113] Reflecting on this experience in 1990, Slebos acknowledged that there were objections to this close relationship, but she thought, "we could achieve more by shaping policy from the inside," adding that there was nothing wrong with "going pragmatic if it is fruitful."[114] The slow workings of the official bureaucracy frustrated the Cyclists' Union at times, but as Slebos said, ultimately it improved life materially for Amsterdam cyclists.[115] In short, cycling activists are still happy to work with urban policymakers despite the drawbacks. There is enough common ground for a fruitful cooperation.

8.4 Conclusion

At present, cycling is thriving in the Netherlands – both as a practice and in policies. While the governance status of cycling has remained the same, there has been a subtle but marked shift in the status of cycling. No longer are policies only geared towards protecting cyclists; they are now also considered a tool for achieving sustainability goals. Some policies respond to the demands and needs of existing cycling practices like bike parking for train commuters, others serve the government goals of reducing automobility like cycle highways. The numerous governance coalitions that have grown historically and are still being created, mirror the many types of cycling and cyclists. The policies attempting to make Dutch city centers less accessible to cars and more accessible to cyclists and pedestrians

112 Feddes and De Lange, *Fietsstad Amsterdam*, 98.

113 Kees Beekmans, "Amsterdam werkt aan 'fietssnelwegen'," *NRC Handelsblad*, March 13, 1990, 7.

114 Peter Kee, "Amsterdam moet de fiets nog ontdekken'," *Het Parool*, February 20, 1990, 12.

115 Beekmans, "Amsterdam werkt aan 'fietssnelwegen'," 7.

– in part based on the continuing cooperation between policymakers and the Cyclists' Union over many decades – are successful. In the most urbanized areas of the Netherlands, car usage declined between 2005 and 2015, while cycling and walking are increasing.[116] Between 2004 and 2014, statistics show a 9 percent increase in the total kilometers cycled in the Netherlands.[117] Researchers have also found, however, that the increase in cycling kilometers is limited to urban areas.[118] Car trips in rural areas may also be in decline, yet the average trip length is increasing.[119] People in rural areas are cycling more kilometers, but much more moderately than in cities. Looking at the overall trends, mobility researchers conclude that the modal split direction is in line with Dutch cities' policy goals.[120] Bicycle sales number around one million a year; increasingly, consumers are leaning towards more expensive (e-)bikes.[121] Based on the entrenched principle of traffic separation, ever busier cycling paths, with speed differences between traditional bicycles and e-bikes or mopeds present new policy challenges. Can one type of cycle path, although wider than before, serve the needs of all these different cyclists? Fundamentally, cars still have priority and cyclists have to cope with the margins – though with cycling paths wider, more comfortable, and more ubiquitous than elsewhere. Further limiting maximum speeds is another way to make cities safer and more attractive for non-motorized traffic. Increasingly, this also means that more policymakers are considering abandoning the longstanding commitment to traffic separation for an approach centered around traffic calming and shared space. Can we turn back the clock on an engineering model that has become so literally ingrained in the Dutch landscape over the past century?

116 Kennisinstuut voor Mobiliteitsbeleid, *Mobiliteitsbeeld 2019* (Den Haag: Ministerie van Infrastructuur en Waterstaat, Kennisinstutuut voor Mobiliteitsbeleid, 2019), 39.

117 Kennisinstituut voor Mobiliteitsbeleid, *Fietsen en lopen: de smeerolie van onze mobiliteit* (Den Haag: Ministerie van Verkeer en Waterstaat, Kennisinstuut voor Mobiliteitsbeleid, 2015), 26. Pedestrian kilometers have increased even more (13 percent).

118 See Oldenziel and Albert de la Bruhèze, "Amsterdam: World Bicycle Capital, By Chance," 24.

119 Olaf Jonkeren, Hans Wust, and Mathijs de Haas, *Mobiliteit in stedelijk Nederland* (Den Haag: Ministerie van Infrastructuur en Waterstaat, Kennisinstituut voor Mobiliteitsbeleid, 2019), 16-7.

120 Ibid., 36.

121 Bicycle Sale Statistics, 2011-2018: https://www.bovag.nl/BovagWebsite/media/BovagMediaFiles/Cijfers/2019/Fietsverkoopcijfers-2011-2018.pdf [accessed February 10, 2020]. On the shift to e-bikes, see Qi Sun, Tao Feng, Astrid Kemperman, and Andreas Spahn, "Modal shift implications of e-bike use in the Netherlands: Moving towards sustainability?," *Transportation Research Part D: Transport and Environment* 78, no? (2020): 1-10.

This chapter has shown that Dutch cycling governance still functions relatively well despite suffering problems likely to occur in a decentralized governance arrangement, where the state has only a minor role. The state's historical role in cycling governance involves constantly questioning whether cycling is as much a public good as automobility, and whether state engineers are obliged to provide facilities for cyclists. Local government's involvement in building cycling paths dates from the 1930s. The state only tends to take an interest in cycling when it is linked with bigger issues like traffic safety, congestion, or environmental concerns. As soon as the political urgency of these issues wanes slightly, national state agencies come up with arguments to devolve responsibility for cycling governance to local authorities. Local governments often push back. This is no different in the period since 2000. However, as local government now has more control over the spending of mobility budgets, it can and does shape local cycling conditions, in close cooperation with the highly professionalized world of Dutch cycling activists.

Conclusion Part III

The sharp rise of motorization in the Netherlands in the 1960s created the same problems as it had elsewhere in the industrialized West. But a still-sizeable bicycle culture meant that combatting the negative effects of automobility played out differently in the Netherlands. Many policymakers still considered cycling a practical and typically Dutch mode of transport around 1970. Activists then added the claims that promoting cycling was a way to address pollution issues and the scarcity of (urban) space. What is more, the staggering numbers of fatal road accidents, particularly among cyclists and children, stirred widely shared outrage and protest. Consequently, cycling became a political priority in the mid-1970s. The facilitation of suburban commuter cycling, moped riding, and recreational cycling promotion were already policy goals, and now urban cycling joined the list of priorities.

In a wider context of environmental and anti-car activism, cycling experienced a revival in the 1970s and beyond. In the Dutch case, this activism was extraordinarily successful. Policies promoting cycling had been in place for decades and as policymakers were more positive about cycling in the Netherlands than elsewhere, activists could work with the state to win back street space from cars. Added to the mix was the bicycle-train combination, a popular alternative to the car. Combining bicycle parking, and more recently bicycle sharing with the train, was a convenient option for long-distance commuting. As congestion and car parking fees rose, driving by car to urban centers lost its appeal. Culturally, the always popular but unremarkable Dutch bicycle acquired new positive connotations such as health and environmental benefits, making the humble two-wheeler even more attractive.

Since the 1970s, the major change in cycling's status as Dutch public good has been the more positive evaluation of urban cycling. The 1950s and 1960s saw a mixed discourse: certain urban planners, engineers, and police officials complained that cyclists were unruly and disobedient road users. Others realized that bicycles were more space-efficient than cars and deserved a place in urban mobility. They received little or no protection. This changed around 1970. With cyclists and urban policymakers daily experiencing the congestion, air pollution, and hazards caused by automobility, central policy shifted to believing that cycling's place had to be expanded at the cost of driving.

More recently, the value of recreational and suburban commuter cycling has come under the spotlight. Throughout the twentieth century, first private actors and then regional state actors worked on providing infrastructure for recreational cycling networks. With mounting concerns over the importance of an active lifestyle, as well as the ever-growing popularity of cycling as a sport, facilitating recreational cycling is as important as ever. As to suburban cycling, the existing network of separate cycling paths alongside Dutch roads is even more important now that the increasing use of e-bikes has extended the distances cyclists can cover. Municipalities and provinces are promoting this form of mobility as alternative to car journeys.

These shifts in the cycling discourse have had consequences for cycling governance. The belief that cycling should be governed locally has been the key governance theme since the 1970s. At the same time, many activists and politicians demanded that national government supported this local policy financially and with expert advice. Acknowledging activists' demands, which came out of broader social unrest, national politicians called on the Ministry of Public Works to take on cycling policymaking. This call scored success in the general political climate of the 1970s and up to 1985. However, at national level, the underlying policy belief remained that cycling was the provinces and municipalities' responsibility. Criticism that limiting automobility required national support for cycling as an alternative, evoked very little response.

Without a clear, responsible institution, cycling governance remained dispersed over the various national departments and government layers. This obviously created friction and disagreements. Some national ministries wanted Public Works to take over the coordination of all cycling policies, but this never happened. The subsidies arranged in the 1970s and 1980s also led to infighting between municipal and national policymakers. Nevertheless, overall investments in cycling at all government levels rose sharply and have led to more and safer cycling infrastructure. In addition, the autonomy that municipalities enjoyed allowed them to initiate traffic-calming measures and ban cars from city centers regardless of the level of national support.

We cannot overestimate the role of activists in creating these pro-cycling policies. Like the tourist organization ANWB five decades earlier, the Cyclists' Union has been the bedrock for (urban) cycling policies. Armed with detailed knowledge of local road and traffic conditions, and thanks to good media contacts, these activists successfully put pressure on local policymakers. Belief in the value of the expertise everyday cyclists possessed has paid off: this knowledge of local road conditions and mobility patterns was indeed useful for local policymakers and helped the Cyclists' Union gain a place in the cycling policy coalition.

In the Netherlands, 1970s activism found an attentive audience because popular opinion, political and engineering developments, and a wave of activism all converged. Firstly, the wider public was experiencing a changing mentality and cultural perception of traffic safety and environmental degradation. Parents – mothers in particular – organized many spontaneous and local protests against unsafe cycling routes for children, demonstrating the widespread support for better cycling infrastructure. It is a sign of a functioning democracy that local officials listened to these calls. Crucially important too was cycling's broad-based modal share: many Dutch men, women, and children still used the bicycle for everyday mobility in the 1960s and 1970s. Consequently, activism did not aim to create better conditions for *future* cyclists, but for safer, possibly life-saving conditions for people who were already cycling. It was about maintaining an existing and vibrant culture of cycling under threat. This lent cycling activism a much greater sense of urgency and made the claims on local policymakers so much more powerful. Environmental awareness was also growing: the Club of Rome 1972 report *Limits to Growth* had a huge impact in the Netherlands, more so than elsewhere, and followed by the 1973 Oil Crisis, increased many people's resistance to the car.

Secondly, government engineers traditionally already built cycling infrastructure outside cities: believing in the model of separating traffic streams as the best solution, engineers started constructing separated cycling paths along major roads in the 1920s and continued this model throughout the 1950s and 1960s. There were some but not many cycling lanes in cities. Nevertheless, the start-up costs of constructing urban cycling infrastructure were relatively low because provincial engineers could extend a tradition of rural cycling path construction to the city, with national subsidies. The Dutch government was also used to working on cycling policies with non-governmental partners because of the major role the ANWB played historically. While the character of the new action groups differed initially, their pragmatic and constructive approach worked well with the urban Public Works administrations. The Dutch political culture of compromise and consensus helped the action groups. It was not difficult to access advisory boards or local policymakers. Again, the ANWB provided a precedent thanks to its presence in numerous traffic-related governmental institutions since the 1920s. Finally, there have always been politicians and engineers interested in cycling – this is inevitable given the high numbers of cyclists in the Netherlands. In the early 1970s, many local governments were making cycling policies. These were perhaps not all great policies, nor informed by extensive knowledge of cyclists' behavior and preferences, but they demonstrate an intention to foster cycling. And they greatly increased activists' chances of finding a willing ear at their local city hall.

Over the past fifty years, Dutch activists, engineers, and planners have worked towards implementing a vision of mobility that dates from the late 1960s. Central to this vision are aspects such as improving traffic safety (especially for vulnerable cyclists), enhancing urban livability through the redistribution of space, and combating pollution by substituting car trips with bicycle trips. This process requires time, money, and effort. Implementation cannot happen overnight as urban streets and neighborhoods must be transformed in turn. Over a fifty-year time span however, with policymakers committed to this vision, and activists continually guarding its progress, this ongoing process has achieved a remarkable transformation in how Dutch public and traffic spaces are designed.

Conclusion

In the 1930s, an accountant by the name of Duyts from the village Loosdrecht decided to build a cycling commuter path himself, because the government had failed to provide one. Today's Dutch cyclists do not think twice about having a state-built network of infrastructure at their disposal, allowing them to travel safely through and between cities and tour nature over a dense network of cycling infrastructure. Few cyclists stop to wonder how this network came about. What made the government realize that providing cycling infrastructure was a public task? *Cycling Pathways* argues that contested political compromises, unintended consequences, and path-dependent developments all played a role. As with any material infrastructure, creating a network depended on significant investments and political effort. Prior to the 1920s, non-state actors invested in dedicated cycling infrastructure before making the case for cycling infrastructure as a public good. By putting cycling on the agenda – and providing road engineers with user-based and lay expertise – cycling citizens played a key role. Within state circles, much energy and manpower have gone into creating this system – even though the financial instruments, while larger than elsewhere in the world, are dwarfed by public investments in automobility.

In *Cycling Pathways*, I set out to address a wider question: why the Netherlands has become one of the world's leading cycling nations. While there are multiple explanatory factors, this project focuses on the governance processes behind cycling, asking: to what extent has Dutch cycling governance since 1880 contributed to the Netherlands' success as a cycling country? By taking a long-term view covering over a hundred years, as well as a multi-level view of governance, this book adds to our knowledge of Dutch cycling history from an international perspective. It shows how provincial and national policymakers, as well as non-state actors and advocacy groups, played a crucial role. The study also refines our insight into the key role of urban politics. Some aspects of these findings are more novel than others. The research here builds on established scholarship, confirming many of its conclusions with new evidence for some of its claims. It also offers several original arguments.

Explaining Dutch Cycling Success

Remarkably perhaps, there never was a masterplan for building Dutch cycling infrastructure. It is the product of a general positive attitude towards cycling and a gradual accumulation of material infrastructure, built over a period of decades by several different, partially overlapping coalitions. *Cycling Pathways* examines the role of politics and social movements in creating this habit among Dutch engineers of constructing cycling infrastructure. Still, explaining the historical trajectory of Dutch cycling from an international perspective is no easy task. There is no single factor that explains everything. While this book's focus is politics and social movements, other causal factors proposed by Albert de la Bruhèze, Oldenziel, and Veraart are also crucial: the availability of mobility alternatives, the (urban) landscape, and the cultural status of cycling. The late adoption of automobility, the compact nature of the Dutch (urban) landscape, and the bicycle's positive cultural connotations are major reasons why the Dutch have always cycled so much. However, each factor matters, and each shapes cycling policy. The way these different factors have reinforced one another goes a long way toward explaining cycling's success as a mode of transport and recreation in the Netherlands. Compared to other countries, there were few obstacles and many factors favoring cycling.

A potentially vibrant cycling culture needs a material infrastructure consisting of roads, cycling paths, and other facilities. Together, these should make it possible to ride a bicycle in a way that cyclists perceive as sufficiently safe and comfortable to make it preferable to alternatives such as the car or public transit. In other words, the governance of space distribution, risk, and traffic safety, and of cycling as a public good, needs to be a positive factor. This infrastructure – a key topic in this project – is a necessary condition for achieving high cycling levels, but not the only one. I trace how stakeholders shaped this process of governance over more than a century. The central theme running through all the chapters is that the Netherlands has benefited from an evolving pro-cycling coalition, consisting of influential interest groups from all levels of society on the one hand, and on the other, state officials who have taken the bicycle seriously as a mode of transport and recreational vehicle.

The politics of infrastructure discussed here are hard to disentangle from other factors. Cycling advocates' successful claims for cycling infrastructure as a public good have a cultural component: the high status of cycling – not framed as working-class, outdated, for children, or purely recreational – partly explains why these claims were successful. The normalization of cycling and its framing as a typically Dutch activity made the reasons engineers would study it, formulate design norms, and build infrastructure around it self-evident. As we have seen, for much of the twentieth century, this attention was also limited to certain groups like recreational cyclists or suburban commuters. At no point, however, did a majority of cycling policymakers come to see cycling as an outlandish activity unworthy of government attention. We could say that cultural factors underlie the political and governance outcomes and are a fundamental explanatory factor.

There is also a positive feedback loop connecting cycling practices to infrastructure provision in the Netherlands, a process we could term path dependent. The growing network of cycling infrastructure supported cycling as a practice by making it a safer and more comfortable option, even when mobility alternatives became more widely available. As a result, more people continued cycling and forced policymakers to keep taking cycling seriously. This also meant that by 1970, social movements had a large power base in the Netherlands. Again, this process of infrastructure provision was uneven, and engineers implemented it slowly over decades. A combination of separate infrastructure outside cities and relatively low car ownership levels within cities made cycling during the mid-twentieth century such a practical option that the Dutch did not need to think twice about it. The subsequent rise in car ownership, coupled with the neglect of urban cyclists, led to waves of protest in the 1970s resulting in policies which have once again made cities more welcoming to cyclists. Cycling is now often faster, cheaper, and more pleasant than driving or taking public transit. This is partly an outcome of infrastructure: cycling routes creating shortcuts where cars need to take detours, bicycle parking facilities next to stations and shopping centers, and so on. The bike-train combination kept cycling relevant as commuter distances grew and more citizens kept on cycling. The interlocking and reinforcing factors of politics, culture, and the practice of cycling itself created a positive spiral in the Netherlands but led to the rapid decline of cycling in other countries.

The Dutch have incrementally built a country-wide cycling path network that we can consider – uniquely in an international context – a national network. Up until the 1970s, this network ended at the urban fringe. Since then, urban policymakers have also gradually implemented it in cities. The many interactions between local, provincial, and national policymakers, stimulated by activists and lobby groups, have connected networks created by a variety of actors. Journeys between cities, along rural roads, or in nature, are all possible in a network that cyclists experience as one and the same national network, which has in fact been built with very different sources of funding by many different government and non-government actors.

Historians have studied the politics of urban cycling, the role of the tourist organization ANWB, and the early history of bicycle paths and bicycle taxation to find answers to the question: what has made cycling in the Netherlands so unusual? Missing from this history is the part national government and state engineers played in Dutch cycling governance and the provinces' crucial role in mediating between the city and national government. And while the authors of the comparative research project *Cycling Cities* underline the significance of the Cyclists' Union for the urban revival of cycling in the 1970s, what is lacking is an analysis of this movement's origins, its strategies, and its relationship with policymakers at different governmental levels. Drawing for the first time on rich archival material to fill this gap, my multi-level analysis of cycling governance, with a specific focus on non-governmental actors, constitutes a step forward as we seek to explain the trajectory of Dutch cycling.

Theoretically, the core concepts of the policy arrangement approach and path dependency ideas informed my work. I operationalized the concept of cycling governance by asking how the provision of cycling infrastructure came to be considered a public task. This involves a discursive move: framing specific types of cycling infrastructure as public good. It also concerns policy: how were these discursive beliefs translated into material realities by specific actors, operating with certain power sources according to specific rules of the game? The concepts of the policy arrangement approach have guided my research questions and chapters. To recap, this approach defines a policy arrangement as "the temporary stabilization of the organization and substance of a policy domain at a specific level of policy making."[1] According to this political science theory, such an arrangement includes policy discourse, policy coalitions, power and resources, and rules of the game. In what follows, I tease out the main findings regarding these aspects of policy arrangement, focusing in the next section on the discourse on cycling over time (cycling politics) and then on the implementation of cycling policies, to obtain a bird's eye perspective of the policy coalitions, their power and interactions. After some final reflections on the path-dependent elements of this process, I then place these findings within the larger historiography on cycling and note the book's contributions to different fields.

Making the Case for Cycling Infrastructure

Dutch cyclists were a heterogeneous group. They came from urban, suburban, and rural backgrounds, with social differences between commuting

1 Bas Arts, Jan van Tatenhove, and Pieter Leroy, "Policy Arrangements," in *Political Modernisation and the Environment: The Renewal of Environmental Policy Arrangements*, eds. Jan van Tatenhove, Bas Arts, and Pieter Leroy, 53-69. (Dordrecht: Kluwer, 2000), 54.

and leisure. Throughout its policy history, the government treated them differently. Only recently has the state seen and taken all or most cyclists into account. The struggle to get cycling recognized as a public good has been a political contest: conflicting visions of the right to mobility are a source of contention. Starting from the idea that traffic should be separated, driving, cycling, walking, and public transit require different infrastructures and therefore compete for public resources: often for the same road space and always for funding and politicians' or planners' attention. Traditionally, recreational cycling managed to lay a potent political claim – something historians have not yet sufficiently recognized. Later, utilitarian cycling – that is, commuters and children cycling to school – were even more powerful sources of political claims. For this group, historians have not yet systematically studied national and provincial politicians' attitudes towards cycling, nor the records of ministerial and provincial engineering committees and working groups. Based on this new primary material, my research refines and expands our knowledge of cycling's political status since the nineteenth century. Compared to the rapid marginalization of cyclists elsewhere, Dutch policymakers supported cycling enough to keep it on the agenda in one way or another for over a century, and to ensure it would survive until the 1970s came around and reenergized political support for cycling. For policymakers, cycling was – and is – rarely a goal in itself, but a means to different policy ends. Dutch cycling advocates' success in linking cycling to what were salient policy goals at certain times have led to larger investment in cycling in the Netherlands than elsewhere. Ultimately, the alignment of social movements with cycling policymakers led to a pro-cycling coalition more powerful and effective than it was anywhere else, with the possible exception of Denmark. From the perspective of cycling advocates, this has been a long and hard struggle. *Cycling Pathways* tells the story of that historical contest.

In Part I, "Roots," I showed how linking cycling financially to roadbuilding for cars led to a model of traffic separation, with engineers providing separate cycling paths for rural or suburban cyclists. At a time when public transit was either not available or too expensive, and before the car broke through, Dutch politicians and engineers acknowledged that utilitarian cycling should be a public good: they recognized that cycling gave access to employment, schools, and social gatherings. Non-governmental groups like the tourist organization ANWB – having transformed from a cycling organization to one that, by the 1920s, was also for cars – actively promoted this discourse and presented cycling as an activity for the entire nation.

Infrastructure that facilitated this was something the state should consider funding. Still, the argument only became politically relevant in the context of car-centered policies. The introduction of the bicycle tax – and the public statistics it generated – meant politicians felt compelled to also acknowledge the millions of cyclists, who formed a much larger group than drivers. This discourse around giving cyclists something in return for their tax money was powerful. The vast number of cyclists brought in substantial revenue, unlike car drivers who existed mostly as a future projection and in the imagination of policymakers and engineers than on the actual road. It is hard to say how positively other countries viewed cycling. What historians have shown, however, is that car infrastructure became a more highly valued public good in other countries. Economic and cultural arguments helped prioritize automobility: motorized vehicles were the engines of economic growth and symbols of modernity. In the ensuing conflict over road space and funding allocation, cyclists often lost out. When supplying car infrastructure became a public concern for the Netherlands in the 1920s, commuter cyclists' place on major roads came into question. As a compromise, Dutch politicians and engineers created a system of separation that provided both car and cycling infrastructure on these roads. The tax-paying status of cycling citizens reinforced the general perception among engineers and politicians that cyclists had some rights.

Advocates of recreational cycling also made the case for the public status of a cycling path network independent of the country's road network. The ANWB promoted a national culture of tourism and recreation. The tourist organization helped to make cycling popular and lobbied for a separate and different type of cycling infrastructure. The national government, however, refused to become involved – at first. Non-governmental organizations filled the gap. While these organizations' ability to obtain subsidies from local policymakers represented recognition of their public service, before the 1930s most of their funding came from private sources. Reacting to public funding of utilitarian cycling, advocates of the tourist club and affiliated cycling path organizations emphasized that both commuters and tourists used their paths – particularly during the economic crisis of the 1930s, when funds were more precarious and the government created public works projects to solve unemployment. While clearly exaggerated for strategic reasons, the claim that both tourists and commuters used the non-urban roads was nevertheless based on truth. National policymakers refused, however, to respond to the demands for investment in the rural, interurban, and suburban network.

Part II, "Divergence," discussing the 1950s and 1960s, showed how engineers viewed cycling in an era when policymakers facilitated automobility. Not all engineers, but certainly the key group of Public Works engineers considered the car a greater public good than public transit, let alone cycling. The engineers' cycling discourse no longer rested on the bicycle tax logic of giving cyclists something in return. Instead, realizing that cyclists might solve the urban congestion caused by cars – engineers introduced a new element, albeit with few practical results, in the form of facilities for urban cyclists. Rather than a negative attitude towards cycling, there was a certain "benign neglect": (national) policymakers and engineers no longer engaged in public discussions about cycling because they were preoccupied with the vision (or nightmare) of mass car use. Despite the ambivalence in policy and political circles, the traffic separation compromise of the 1920s, acknowledging that the provision of car and commuter cycling infrastructure was a public task, still held sway in the 1950s and 1960s. It was not contested and was even reinforced by the decision to consign mopeds to cycling paths. After that decision in the early 1950s, engineers became even more committed to separating traffic on the nation's major roads – and in the process provided some infrastructure for cyclists and moped riders. The rhetoric of fairness and giving the numerous cyclists at least some protection still weighed sufficiently on engineers' minds to make a difference in this period, even if they now only paid lip service to this ideal.

This crucial period also saw an important development in recreational cycling. Where there was previously little (local) to no (national) interest in this category, around 1950, provincial and municipal policymakers started to see recreation in general, including cycling, as a core element of their mandate. Provincial policymakers publicly coupled cycling to public health, recreation, and tourism, depending on the province's background and political color at the time. Even as cycling's modal share declined in the Netherlands (as elsewhere) due to increasing competition from cars, public transit, and mopeds, the Dutch cycling path network kept on expanding. Engineers and politicians were much less convinced than their international counterparts that cycling was about to disappear and still discussed cycling in more positive terms than public officials elsewhere. Indeed, Part II revealed that this is when the direction of Dutch cycling started to diverge, although the roots of this shift date back even further, to the 1920s.

Part III, "The Dutch Model," showed how activists and politicians successfully – and in line with developments outside the Netherlands – linked cycling discursively to pressing issues: safety, pollution, and urban livability.

From the mid-1960s, urban cycling became a focal point, not just in urban politics but also on a provincial and national level, importantly among activists. Policymakers discussed cycling much more than they had from the late 1940s onwards, and tirelessly stressed the many advantages they saw resulting from higher cycling levels. Activist citizens made the case that the state had allowed automobility to invade cities unchecked. The government had a responsibility for traffic safety, which it ignored. Urban cyclists, or cycling schoolchildren throughout the country, suffered as a result. The argument for giving cyclists their due – in this case a minimum level of protection – struck a chord with people because it aligned with a decades-old discourse on the bicycle tax that went back to the 1920s. Cyclists had always been acknowledged at some level as citizens with certain rights – which policymakers had come to neglect. This fact also struck a particularly raw nerve when activist parents rallied to argue that children had a right to be, play, and feel safe on the street. Under pressure from social groups, many politicians at all levels of government, as well as a new generation of urban planners and engineers, took up this call.

Activists also argued that livability and quality of life, along with environmental protection, were public goods the state should provide, and all required support for cycling (infrastructure). As Adri Albert de la Bruhèze and Frank Veraart have already shown, political support for cycling among key aldermen or city council members helped to revive cycling in the 1970s. My research shows that something similar applies to provincial and national policymakers, whose involvement in cycling governance has not been studied. MPs often demanded in plenary and special meetings that the Ministry of Public Works invest more in utilitarian cycling. I have also shown that national and provincial engineers were sometimes slow to fulfil these demands or react to information inquiries. No doubt without this political pressure, investments in cycling would have been significantly lower. Politicians in turn translated (perceived) social demands for better and safer cycling infrastructure. In short, this is a successful example of democracy in action.

This chronological – diachronic – treatment of cycling governance thus shows the historical process of how policy came about. *Cycling Pathways* also reveals some enduring – synchronic – themes. Throughout the book, particularly in Parts I and II, I show that recreational cycling, both as a practice and in governance terms, has thus far not received the attention it deserves. Partly overlapping with governance coalitions for utilitarian cycling, it nevertheless formed a largely separate tradition. Throughout the twentieth century, recreational cyclists could count on political support and funding, mainly at a provincial and regional level. This is somewhat speculative: the continued support for recreational cycling might also have to do with its enduring popularity among elites and groups who switched to commuting by car but could still receive political support for cycling through this avenue. More important, perhaps, was the recognition of the economic value of bicycle tourism and ideas about public health. Historians have amply studied the recreational bourgeois cycling culture of the late nineteenth century; studies of various national (professional) sport cycling cultures are also increasing. However, scholars have neglected recreational cycling as a key element of governance arrangements for the twentieth century. *Cycling Pathways* shows that recreational cycling, as contributing factor to cycling's persistent place on the political agenda, remained popular as a practice, became a source of engineering expertise, and merits more attention.

That Dutch engineers built so much cycling infrastructure during the twentieth century without strong national state governance provides an important lesson. While there were guidelines and soft policies "nudging" engineers in a certain direction, policymakers could have taken cycling less seriously. The institutional arrangements for mobility allowed this – and in many European countries, that is exactly what happened. The difference in cycling's development in the Netherlands has to do with its more positive status there. This fluctuated over time and differed according to government level and type of cycling. Still, the general trend is clear. Cycling infrastructure had the status of a public good, whether utilitarian or recreational. Even when engineering norms for cycling infrastructure were more like guidelines than formal requirements, the construction of separate cycling paths alongside roads never stopped. When national traffic funding for provinces and municipalities was no longer earmarked, cycling advocates feared investment in cycling would drop, which did not happen.

Understanding the mentality, beliefs, and even cycling practices of key policymakers – politicians and engineers – is crucial for explaining cycling policies and therefore cycling's success in the Netherlands. Many of them had a familiarity with and affinity for the bicycle, which explains their susceptibility to cycling advocacy. The cycling discourse I have reconstructed based on archival material, trade journals, and other publications, provides insight into why policymakers and engineers have consistently seen it as their task to "do something" for cyclists. Which group of cyclists, and what this "something" had to be, were subjects for discussion, as was the question of who should provide that something.

Turning Beliefs into Infrastructure

Turning cycling beliefs into a network of infrastructure has been a piecemeal, gradual process that has taken well over a century. A large, decentralized policy coalition of social movements, politicians, and engineers has contributed to this. How did they do it? Governance scholar Rod Rhodes argues that network governance "evokes a world in which state power is dispersed among a vast array of spatially and functionally distinct networks composed of all kinds of public, voluntary, and private organizations with which the center now interacts."[2] This is also a fair description of the Dutch situation. In terms of the policy arrangement approach – policy coalitions, power and resources, and rules of the game – Dutch cycling governance had several key characteristics: it involved a policy coalition that had a crucial role for non-governmental organizations, in particular the ANWB (since the 1890s) and the activist group Cyclists' Union (since the 1970s), plus a dynamic relationship with multi-layer policymakers; it involved a powerful role for cyclists both through sheer numbers and also via their representation in citizen organizations employing engineering or user expertise; and it involved a political tradition where the rules of the game revolve around accommodation, consensus-seeking, and interest trade-offs in the polder model.

In *Cycling Pathways,* I drew attention to some major governance themes. First, I indicated throughout how the proper place for cycling was not just contentious when it came to the distribution of road space. It also involved the issue of governance. The national government's role was particularly contested. Second, jumping in the gap created by the lack of national coordination, interest and activist groups played a crucial role – not least in producing and coordinating cycling expertise. Third, the success of these groups in cycling advocacy is partly thanks to Dutch political traditions of accommodating outside groups in the country's governance. Fourth, their use of expertise (engineering expertise in the case of the ANWB, lay expertise in the case of the Cyclists' Union) was a highly successful strategy. Finally, while there are path-dependent elements to this story, it also should be considered as the outcome of a process of contestation, negotiation, and compromise.

The policy coalition for cycling in the first half of the twentieth century consisted of national engineers, politicians, and the national tourist and

2 R.A.W. Rhodes, "Waves of Governance," in *The Oxford Handbook of Governance*, ed. David Levi-Faur (Oxford: Oxford University Press, 2012), 34.

lobby organization ANWB, which put recreational cycling on the political map. In the 1950s and 1960s, the ANWB's role remained significant, while the center of gravity within the government shifted from national to provincial policymakers and engineers. When the focus on cycling governance moved more towards urban cycling in the 1970s, a new urban coalition emerged to put utilitarian cycling on the political agenda: the new cycling activist group Cyclists' Union worked with urban engineers, planners, and politicians. The role for national and provincial engineers and politicians also continued, making the policy coalition larger and more complex than before.

The examination of the national government's role in cycling governance shows how national policymakers were pulled into the cycling governance policy coalition time and again, despite their reluctance. This reluctance had as much to do with political ideas about decentralization and local government as with the national government giving priority to automobility, a recurring theme. National policymakers considered the bicycle a mode of transport for shorter distances. This is a myopic governance view: once combined with the train or bus, the bicycle becomes a viable alternative to long-distance car traffic. Still, the perception of the bicycle as a vehicle for short trips has affected its place in Dutch mobility governance over the last century. Opponents of national involvement in cycling governance continually insisted that a local transport mode like the bicycle should be governed locally. Meanwhile, car transport supposedly transcended the local and caught the interest of the national government. These attempts at making cycling governance the exclusive domain of local governance did not entirely succeed, for two main reasons. First, activists and local politicians demanded that national government address major safety issues with cycling. Second, as national policymakers recognized at times, the bicycle was a solution to congestion, air quality, and climate change problems, which are not the sole responsibility of local government but also key concerns for national government.

Cycling Pathways thus shows that a cycling history of the Netherlands needs to consider national and provincial policymakers (both politicians and engineers). Due to the lack of historical international scholarship – with the exception of Maxime Huré's work on the French national government's role in cycling governance since the 1970s – it is hard to determine whether this applies to other countries. Dutch provinces played a major role, and not just as the authorities responsible for major provincial roads. They supplied a large part of the Dutch road network with cycling infrastructure from the 1920s onward, initiated recreational cycling plans, and, after 1970, coordinated similar plans for utilitarian cycling. Provincial authorities also served as intermediaries between municipal and national policymakers in subsidy procedures and other multi-level issues. Together with the recent creation of urban areas and other regional forms of governance, provincial policymakers have become key actors because their operational level is well-suited to cycling's action radius.

Other historians have described the importance of the tourist organization ANWB and the city-based Cyclists' Union. My main contribution is to take our knowledge of the Cyclists' Union's strategies a step further. Using new source material, the analysis of the Cyclists' Union presents new insights and arguments on its governance role. The role of private initiatives in Dutch cycling governance has fluctuated but has always been strong. Governments around the world have not been very proactive in launching cycling policies. The efforts by activists and lobby groups to put cycling on the agenda seem significant everywhere. Groups in the Netherlands, however, were more successful than anywhere else in gaining access to government circles and contributing to cycling governance. Until the 1920s, the ANWB played an even more important role in national cycling governance than the state. Thereafter, it was on equal footing with state engineers. In the 1950s and 1960s, the ANWB continued to fulfil a less visible but still crucial (gatekeeper) role in the production and dissemination of traffic engineering knowledge. In this role, while advocating car use as much if not more than cycling, the organization remained significant in the cycling governance network. In the 1970s, a new group of urban cyclists entered the frame and quickly gathered momentum. Dutch cycling governance coalitions have always been a mixture of state and civil society actors.

This close cooperation is a choice: writing in the late 1980s, Jacqueline Cramer warned that environmental action groups increasingly carried out (free) work for the government. In line with the international trend to reduce governmental bureaucracies, outsourcing tasks to volunteer organizations operating with small subsidies was an attractive option. The access to policymaking gave action groups like the Cyclists' Union the opportunity to work with the government while remaining outside it. Radical critique of the system becomes harder in such a close co-governance model, as some Cyclists' Union branches observed and deplored. However, the majority chose to go down this route. Pragmatism and compromise produced enough concrete results and were a more attractive option than standing on the sidelines.

Cycling Pathways argues that both the ANWB and the Cyclists' Union were able to achieve success because they functioned as experts presenting themselves as knowledgeable about cycling. The ANWB did this by employing its own civil engineers, who knew as much if not more about cycling infrastructure as state engineers. In contrast, the Cyclists' Union claimed that its members, everyday cyclists, provided a different kind of user or lay expertise which should be taken into account for transforming cities. We could say these non-state actors functioned as what transport historian Marcus Popkema calls a "free consultancy agency" for state agencies.[3] Knowledge served as a bargaining chip for these organizations. Without costing the state anything, they provided valuable input for cycling policymaking. In return, the organizations had a seat at the discussion table and received certain concessions.

Cycling Pathways offers a novel, multi-level perspective on state agencies. Earlier works tend to focus on urban governance. Without detracting from this perspective – urban governance is indeed crucial in cycling governance everywhere – the provincial and national levels also deserve attention. They are responsible for the policy frameworks within urban policymakers' operations, and crucially steer these through financial instruments.

The recurring analysis of the ongoing struggle over cycling governance should not be read as an indictment or evaluation of its efficacy. If anything, the analysis shows how widespread the concerns for cycling have been at different levels. The ever-present debate over the extent of the national government's involvement, the struggles to coordinate policies, and the attempts to shift the responsibility for cycling onto other actors, seems ineffective and negative regarding cycling. Yet cycling's current position in the Netherlands remains undeniably strong. It is firmly embedded in the material infrastructure, engineering norms and practices, and culture. Has this been achieved because of or despite Dutch cycling's form of governance? Without engaging in too much counterfactual speculation, the centralization of cycling governance probably did more harm than good. A 1997 report on municipal cycling policies notes that besides Utrecht and Amsterdam, only one other municipality (Arnhem) had a designated bicycle coordinator, which had "the logical negative effect: the others forget about the bicycle."[4] The dispersal of responsibility for cycling among diverse actors may have contributed to the survival of cycling policies throughout the twentieth century. If knowledge, funding, and advocacy are present in many different decision-making bodies, their loss in one location does not endanger the commitment to cycling as a whole.

Once they were represented by these civil society actors in the Dutch policy coalition, cyclists gained a powerful position. What were the sources

3 Marcus Popkema, "Tussen techniek en planning," 79.

4 Ligtermoet et al., *Gemeentelijk fietsbeleid*, 18.

of this power? Engineers' focus on cyclists may have fluctuated, as it did in traffic counts, but for most of the twentieth century they were aware of cycling's popularity. The often-repeated assertion that cycling was typically Dutch, which also appeared in engineering trade journals, reinforced this perception. Consequently, the sheer number of cyclists was a constant source of pressure. Perhaps even more importantly, cyclists had lobby and action groups on their side: first the ANWB and later the Cyclists' Union. Both used expertise as a source of power to justify their position in governance, allowing them to lobby more effectively for cyclists' interests.

There are important differences and similarities between the national tourist ANWB and the urban-based Cyclists' Union. A key difference is that, while the ANWB had power through its board members' prominent social standing, the organization had no knowledge of traffic; however, it developed this over time. Initially the Cyclists' Union did not have this power, but from the start it possessed traffic knowledge. The Union gained power over time. Both had sprung from a fine-meshed network of local organizations: the ANWB from its system of local representatives or consuls, the Cyclists' Union through its organizational structure based around local chapters. Both are rooted locally, although the ANWB has become an interest group that operates nationally. The Cyclists' Union also lobbies nationally, however. What sets both organizations apart from many cycling groups outside the Netherlands is their activism in a context with a high proportion of cyclists. Much non-Dutch activism concentrates on getting non-cyclists on bicycles by sharing knowledge and tips about everyday cycling. The Dutch organizations offer their members this kind of practical advice, but more important is their advocacy for the interests of existing cyclists.

How well were everyday cyclists represented by the ANWB and Cyclists' Union? While these organizations were founded by active cyclists, over time they professionalized and day-to-day leadership was in the hands of a smaller group of activists. Nevertheless, their claim to represent all or most cyclists was certainly acknowledged in their access to politics and policymaking on behalf of the cycling public. Cyclists came closest to directly participating in politics in the early stages of the Cyclists' Union's activities in the 1970s and 1980s. By contributing information about local blackspots, bottlenecks, and other barriers to cycling, relayed by the chapters to local policymakers, the Cyclists' Union made sure cyclists were heard. Later, as the organization professionalized, a small core of volunteers established more permanent relationships with local policymakers. Policymakers increasingly recognized leading activists as experts in their own right. As cyclists, they represented users, but the broad social movement character of early activism has declined over the past twenty to thirty years. How these activists influenced cycling policies in the Netherlands requires further research.

The Dutch cycling policy coalition, consisting of multiple actors with a prominent role for citizen groups, knows the rules of the game within the Dutch polder model of governance. The relationship between national, provincial, and municipal government – part of the "rules of the game" – is not top-down or coercive. At no point in history has the national government forced lower-level authorities to construct cycling infrastructure, for example. Engineering guidelines, sometimes a precondition for subsidies, allow opportunities to steer policy quite strongly. This "golden strings" approach allowed the national government to ensure local authorities spent money on cycling in light of MPs' demands to create these budgets. After decentralization of this funding in the late 1990s, cities still invested significantly in cycling, proving that national suspicions of local spending patterns were unfounded. In recent decades, the national government has decided that producing and disseminating cycling expertise is its primary role in cycling governance. Nevertheless, as MPs push for cycling investments, they still reserve budgets for (local) cycling policies.

The polder model is an important factor in understanding negotiations over public space in the Netherlands. According to Carstensen and Ebert, "[t]he avoidance of antagonisms between the bicycle and the car is an important explanation of the successful mass cycling cultures in the Netherlands and Denmark."[5] Cycling and mobility historians – Veraart, Albert de la Bruhèze, Mom – have previously stated that Dutch politicians do not firmly choose or reject the car. Dutch politicians try to do it all: promote a high-capacity railway network, a dense road network, *and* cycling all at once, even though some of these are easier to combine (bicycle and train travel) than others. With its dense (car) road network, car ownership and use, the Netherlands is just as much a car country as any other in Western Europe. Despite ongoing discussions since the 1960s, road pricing measures to fight congestion have continually failed to pass parliament. Out of fear that they are accused of bullying car drivers, many political parties eschew any overt anti-car rhetoric.[6]

The concepts of policy coalitions, resources, and rules all have to do with the way stakeholders employ their agency to shape cycling governance. There are also more structural, path-dependent elements to the story of

5 Carstensen and Ebert, "Cycling Cultures in Northern Europe," 53.

6 Smaal, *Politieke strijd om de prijs van automobiliteit: de geschiedenis van een langlopend discours, 1895-2010*, 581, 96, 618, 720.

cycling infrastructure construction. *Cycling Pathways* shows how many individuals and organizations crucially shaped cycling policies. All the same, infrastructural systems often create lock-in patterns that limit the possibility to introduce change at a later stage. Critics usually cast such systems in a negative light, as obstacles preventing the transition from car-based to more sustainable mobility. In the case of Dutch cycling, certain lock-in patterns may have contributed to cycling's long-term perseverance. The interaction of cycling policies and practice can trigger feedback processes that put cycling on either a downward or upward trajectory in the long run. Fortunately for Dutch cyclists, it was an upward spiral: in the first half of the twentieth century, users, interest groups, and state officials all in their own way contributed to a strong cycling culture and an extensive infrastructure network. That material network enabled a large proportion of Dutch citizens to continue cycling in the 1950s and 1960s – engineers and traffic experts could no longer ignore them. Path dependency and crucial policy choices played a key role: the traffic separation model had created safe cycling conditions but also high cycling numbers. We can consider this a positive feedback process for cyclists. It was hard for planners to stop focusing on cycling infrastructure, since cyclists were not going anywhere, and engineers certainly did not want them on roads where cars were supposed to have an uninterrupted flow. Combined with the important decision to consign mopeds to cycling paths, there was plenty of incentive for Dutch traffic experts to keep on planning cycling infrastructure. Dutch bicycle planning is as old, if not older, than planning for cars, and the two have always progressed in tandem.

For my final analysis, I do not intend to present an overly deterministic account of cycling governance in the Netherlands. Within the structural limitations, individual actors and organizations were able to operate with a degree of agency. In terms of path dependency, there were many critical junctures when this agency made a difference. Indeed, there was no grand design for Dutch cycling infrastructure. Instead, different elements were built at different times, for different reasons, and by different actors. Cycling lobbyists and advocates like ANWB officials, (social democrat) politicians, (local) engineers, and activists managed to carve out a space for cycling throughout the twentieth century. That there is currently a comprehensive system of cycling infrastructure tells us that cycling did not just have a higher cultural status in the Netherlands than in many other countries; it also carried weight with policymakers. As in any large-scale historical process, structural constraints nevertheless leave some scope for agency.

Contributions, Limitations, and Further Research

First of all, *Cycling Pathways* contributes to the growing international historiography on cycling. What other cycling historians could take from this study is timescale: few cycling studies have as broad a scope. By covering such a large period, this study can observe long-term developments. While this approach certainly has its limitations and drawbacks – as I discuss below – it suggests that we can only understand the fate of attempts to revive cycling since the 1970s in the context of earlier decades. Many studies of the cycling revival since the 1970s would benefit from closely considering the fifty years or so of road planning that went before. Decisions made then and cycling levels as a result are key factors that explain cycling activists' success or lack thereof in the 1970s and beyond. Cycling did not reappear in Europe *ex nihilo* some fifty years ago. If anything, the 1950s and 1960s "car decades" are crucial for understanding the diverging national trajectories in cycling. Too often, the actual process of cycling's decline is not studied in any detail but just ascribed to the car's ascendancy and planners' single-minded focus on cars. This study has shown there is a lot more of the story to tell, at least with regard to the Netherlands. Studying these disparate trajectories of disappearance is an important avenue for future research.

In addition to the broad timescale, the novel contribution of this book also stems from the use of a large body of new archival sources, supplemented with trade journals, newspaper articles, and interviews, allowing triangulation. This approach was fruitful: provincial and national archives had not been used much (if at all) in previous studies. The exploration of other provincial archives might offer an even richer picture. It turns out that engineers discussed cycling regularly in committees, wrote about it in trade journals, and left a large paper trail on the many cycling paths in the Netherlands. It will be very interesting to see if future historians can uncover similar sources that tell the story of cycling in other countries in more detail.

This observation brings us to one of the major limitations of the present study. A study with such a broad approach necessarily sacrifices depth for breadth, in the process perhaps raising as many questions as it answers. The suggested positive political developments in the Netherlands are cautiously framed in an international context. The historical scholarship on automobility and cycling in other European countries suggests pro-cycling policies were much stronger in the Netherlands. While this is plausible, given what we know about cycling's central place in Dutch culture and mobility, it is only a suggestion. Further research in other state archives might well nuance this picture. A comparative approach to cycling activism since the 1970s could enrich our understanding of Dutch cycling activism. It is necessary to put this broad research project in an even broader context. At the opposite end of the spectrum, a very specific analysis of the material transformation on one or more Dutch city streets, while taking into account the role of activists and policymakers, would add a new level of detail that I cannot provide due to space constraints.

This study is relevant for scholars of contemporary cycling and practitioners as well as historians of mobility and cycling. In popular narratives and accounts of cycling promotion, the Dutch story since the 1970s is often invoked as the ultimate example. The transformation of Dutch cities in this period from car-centered to bicycle-friendly is often presented as proof that "it can be done elsewhere." However, as noted earlier, how to achieve this is not clear. Do we really understand how cities transformed over the past fifty years? And to explain this, do we not also need to understand what happened before the 1970s? The historical perspective is not something that otherwise interdisciplinary cycling studies take sufficiently on board.[7] For scholars studying contemporary cycling politics and activism, this study offers insight into one pathway to a cycling society. The Dutch case demonstrates that the historical constraints at play limit the range of pathways to the (near) future. At the same time, it shows us how a vibrant civil society can achieve success if it manages to establish good working relationships with an interested (local) government.

It is important to focus on policymaking processes and their history, since a growing body of policy-oriented research is attempting to come up with strategies that will increase cycling levels. The challenge is to connect this practice-oriented body of scholarship to historical bicycle research. A good strategy is to focus on the tenacity and persistence of long-term developments, such as the evolution of certain mobility cultures and the long-lasting impact of completely forgotten policy choices.[8] Very few people know about the debates regarding cycling and motorization in the Dutch political world of the 1920s and 1930s, yet the decisions taken then have changed the course of mobility ever since, as have the choices other countries made. On the one hand, historical work can make us more aware of the weight and importance of decisions in infrastructure policy now, because they set us on a certain course that will be hard to change in the future. On the other

7 See Peter Cox and Till Koglin, eds., *The Politics of Cycling Infrastructure: Spaces and (In) Equality* (Bristol: Polity Press, 2020).

8 Oosterhuis, "Bicycle Research between Bicycle Policies and Bicycle Culture," 35.

hand, understanding choices in the past and the particular road taken to arrive at the present moment might also offer us ideas and inspiration as to how to "unmake" these decisions and create more sustainable and equitable mobility systems. Contemporary policymakers can learn from historical examples, as Bert Toussaint has argued for mobility policy.[9] For instance, as a governance process, cycling policymakers are exploring the use of living labs to study user experiences.[10] Chapter 6 showed how social movements pioneered bottom-up approaches to cycling governance where user knowledge formed a source of counter expertise.

In analyzing Dutch cycling policy, I have borrowed concepts and ideas from governance scholarship and Dutch political history, a third body of scholarship to which this project contributes. Scholars in this field could draw on my work as a rich empirical case study of how coalitions, networks, and the Dutch "polder model" function. Contemporary governance scholarship acknowledges that non-hierarchical forms of governance have a long history, yet tend to focus on the past fifty years. This study goes back further in time to show how, without a clear center of power, a varied stakeholder coalition produced Dutch cycling infrastructure. This understudied case supports Dutch political history ideas about the "polder model." The role of the Cyclists' Union and ANWB on many government committees, but also the efforts to accommodate and balance the interests of drivers versus cyclists in policymaking, are themes of scholarship that recur in my work. Dutch historiography on social movements and politics has not engaged with the politics of mobility. This is a real surprise and oversight, since it represents a successful and influential form of activism.

From a wider perspective, this work deals with the relationship between technologies, politics and users, themes familiar to historians of technology. Here as well, we learn about the governance of everyday technologies. The way users shape technologies and technological systems is a well-established

9 Bert Toussaint, "Using the Usable Past: Reflections and Practices in the Netherlands," 15-27; and "(Transport) History as Policy Lab for Democratic Governance," *The Journal of Transport History* 40, no. 2 (2019): 270-80.

10 Yuliya Voytenko et al., "Urban Living Labs for Sustainability and Low Carbon Cities in Europe: Towards a Research Agenda," *Journal of Cleaner Production* 123 (2016): 45-54; Matthias Mück, Christoph Helf, and Miriam Lindenau, "Urban Living Labs Fostering Sustainable Mobility Planning in Munich," *Transportation Research Procedia* 41 (2019): 741-44; Matthew Bruno, "The Challenge of the Bicycle Street: Applying Collaborative Governance Processes while Protecting User Centered Innovations," *Transportation Research Interdisciplinary Perspectives* 7 (2020); Julian Alexandrakis, "Cycling towards Sustainability: The Transformative Potential of Urban Design Thinking in a Sustainable Living Lab," *Transportation Research Interdisciplinary Perspectives* 9 (2021): 1-9.

theme in the literature. This work showcases this process, how users (can claim to) possess counter-expertise to dominant engineering knowledge, making them indispensable partners in shaping mobility systems. Mobility "system-builders" should take note: engineers tend to simplify systems for their own purposes until a certain stakeholder comes along to complicate this view.[11] Conceptualizing the city as a flow of vehicles could not be the norm for very long in Dutch cities because of citizens' and activists' efforts. The scholarship on the life and/or decline of technologies discarded by engineers and elites as obsolete, exemplified by Edgerton's *Shock of the Old*, also resonates with this book. The ongoing practice of cycling requires further study from the user perspective.

Here ends our story of more than a century of Dutch cycling politics, policies, and governance. Where does this leave Dutch cyclists, benefiting from all these years of cycling policies and governance, in 2021? In March 2021, the Dutch Ministry of Public Works announced its National Future Vision for the Bicycle (*Nationaal Toekomstbeeld Fiets*) in the period 2020 to 2040.[12] This is the Ministry of Public Works' response to parliamentary requests for national government to be more proactive in cycling governance.[13] The minister announced that the "primary responsibility" lies with "decentral governments." She also emphasized that a state coalition with "provinces, transport regions, municipalities," as well as the intergovernmental body on cycling (its so-called Tour de Force) together shape cycling policy in the Netherlands.[14] This is a symbolic moment: governmental cycling policies were established roughly a century ago. In the meantime, much has changed, yet much has stayed the same. The bicycle is enduringly popular, both for recreation and commuting. The governance process remains a coalition without a clear center. But perhaps this is not such a problem – after all, regional and local policymakers believe in cycling, and activists make sure they keep the government on its toes. More than ever before, cycling enjoys

11 See Michel Callon, "Society in the Making: The Study of Technology as a Tool for Sociological Analysis," in *The Social Construction of Technological Systems: New Directions in the Sociology and History of Technology*, eds. Wiebe E. Bijker, Thomas P. Hughes and Trevor Pinch (Cambridge, MA: MIT Press, 1987), 83-103.

12 https://www.rijksoverheid.nl/documenten/rapporten/2021/03/08/bijlage-nationaal-toekomstbeeld-fiets-op-hoofdlijnen

13 *Kamerstukken II* 2019/20, no. 35300 A, sub-no. 109; no. 35300 XII, sub-no. 104; no. 35426, sub-no. 16.

14 Letter Ministry of Public Works to Lower House, March 8, 2021. https://www.rijksoverheid.nl/documenten/rapporten/2021/03/08/bijlage-nationaal-toekomstbeeld-fiets-op-hoofdlijnen

policymakers' support. The urgent action required on climate change, the attempts to reduce traffic deaths to zero, along with many other areas of policy, all point in one direction: we need to make cycling an even more attractive and safe option for everyone. The National Future Vision for the Bicycle promises large investments in cycling. Whether it signals a truly new chapter in cycling governance remains to be seen.

Appendix 1

Cycling Paths Subsidized by the Ministry of Culture and Recreation (Per Province), 1960-1970

	Awarded	Built
Groningen		
Along Leekstermeer (Lettelbert)	1961-1962	1966
Friesland		
3 cycling paths Ameland	1963	Not (yet)
From youth hostel 'Us Blau Hiem' to Appelscha	1965	Not (yet)
In land consolidation (*ruilverkaveling*) Haskerveenpolder	1963	1963
In 'de Kleine Wielen'	1965	1965
To 'de Kleine Wielen'	1965	1966
In land consolidation Koningsdiep	1962	1963
'West a/Zee-Midsland a/Zee'	1966	Not (yet)
Along Slotermeer near Balk	1966	Under construction
Drenthe		
Assen-Zuidlaren	1964	1964
Bunne-Frisian border	1965	1964
Assen-Spier	1966	23 percent in 1966
Assen-Roden (Nietap)	1965	Contracted out in 1967
Near Zuidwolde	1965-1966	Not (yet)
Near Sleen-Odoorn	1966	Not (yet)
Hoogeveen-Pesse	1967	Not (yet)
Near Assen	1967	Not (yet)
Gelderland		
Barneveld-Putten (6.5 km)	1966	Not (yet)
Alongside Hessenweg (reconstruction)	1965	1967
Utrecht		
In land consolidation Harmelen-Kockengen	1963	1964
Noord-Holland		
Huizen-Naarden	1962	1963
Texel (to Westerslag)	1963	1964
Texel (Waalenburgerdijk)	1961-1963	Unknown
Texel (in Eyerland dunes)	1964	Not (yet)

	Awarded	Built
Noord-Brabant		
Recreation area Markiezaat Bergen op Zoom (10 km)	1967	Not (yet)
Recreation area De Kempen (phase one, 26 km)	1965-1966	Unknown
Pannenhoef-Bremberg	1963	1964
Breda-Dongen	1963	1964
Forestry St. Anthonis	1965	Unknown
Recreation area Brabants Centrum (27 km in Loonse dunes)	1964-1965	70 percent in 1966
Recreation area Brabants Centrum (3 paths)	1966	Not (yet)
Recreation area Baronie van Breda (17 km)	1966	Not (yet)

List of cycling paths the Ministry of Culture and Recreation subsidized in the 1960s. Not all subsidized paths were constructed, possibly because the local government could not or would not pay its share, which they had to since national cycling subsidies never awarded the full 100 percent of costs. Some paths, like those in Noord-Brabant and Drenthe, were part of the provincial recreational plans discussed in Chapter 3. Source NA 2.27.5111, inv. no. 60.

Appendix 2

Provinces' Cycling Projects Subsidized with National Investments, 1975-1980

Groningen	**Urban**	Groningen	Eikenlaan	Cycling path
		Groningen	Korreweg	Cycling path
	Non-urban	Secondary road 5	Westervalge at Warffum	Cycling path
		Secondary road 10	Jan Bronsweg at Appingedam	Cycling path
		Secondary road 18	Oostwedde-Vlagtwedde	Cycling path + bridges
		Tertiary road 26	Eemshaven (Loppersum-Garrelsweer)	Cycling path
		Tertiary road 29	Near Holwierde	Cycling tunnel
		Tertiary road 30	Ruischerbrug-Klein Harkstede	Cycling path
		Tertiary road 33	Ten Post-Wittewierum	Cycling path
		Tertiary road 34	Siddeburen-Wagenborgen	Cycling path
		Tertiary road 52	Graaf Adolfbrug-Heiligerlee	Cycling path
Friesland	**Urban**	Dokkum	Harddraversdijk	Cycling path
		Franeker	Burg. Dijkstraweg-Loréstraat	Cycling path connection with bridge Dongjumervaart
		Franeker	Noordergracht	Cycling/pedestrian bridge
		Sneek	Groenedijk	Cycling path
	Non-urban	Secondary road 9	Galamadammen	Cycling tunnel
		Tertiary road 8	Kievitshorne-Witmarssum-Kimswerd-national road 9	Cycling path
		Tertiary road 21	Hoeve De Driesprong-Haulerwijk	Cycling path
		Tertiary road 22	Oosterwolde-Hoeve De Driesprong	Cycling path
		Tertiary road 33	Nes-Rallum-Hollum on Ameland island	Cycling path
		Tertiary road 35	Kromten-Koaibosk	Cycling path
		Tertiary road 35	Lindenbrug-Noordwolde	Cycling path

Drenthe	**Urban**	No projects		
	Non-urban	Secondary road 4	Rolde-Borger	Cycling path
Overijssel	**Urban**	Almelo	Agglomeratieweg-Zuid	Cycling tunnel
		Deventer	Mensinksdijkje (section)	Cycling path
		Diepenveen	Mensinksdijkje (section)	Cycling path
		Hengelo	Deldenerstraat (south side)	Cycling path
		Noordoostpolder	Routes by Harmen Visserplein and Meldestraat	Traffic-calmed route with cycling bridge
		Oldenzaal	Rijksrondweg (section)	Cycling path with tunnels
		Zwolle	Junction Zwartewaterallee/ Biesbosch	Traffic light installation
	Non-urban	Secondary road 6	Zwolle-Vollenhove	Cycling path
		Secondary road 19	Den Ham-Vroomshoop	Cycling path
		Secondary road 24	Mariaparochie-Fleringen	Cycling path including Robbenschaarsbridge
		Secondary road 69	Losser-Overdinkel	Cycling path including bridge over Dinkel
Gelderland	**Urban**	Arnhem	Thomas à Kempislaan	Cycling path
		Nijmegen	Energieweg	Cycling path
		Nijmegen	Grootstalselaan	Cycling path
		Nijmegen	Scheidingsweg	Cycling path
		Wijk bij Duurstede	Groene Woud-Vikinghof	Cycling connection
		Zevenaar	De Methen	Cycling path
	Non-urban	Secondary road 104	Buren-tertiary road 110	Cycling path
		Tertiary road 7	Valkseweg at Barneveld	Cycling path
		Tertiary road 17	Halfweg-Vierhouten at Nunspeet	Cycling path
		Tertiary road 19a	Garderenseweg at Uddel (Appeldoorn)	Cycling path
		Tertiary road 22	Tertiary road 23-Rheden (Brummen)	Cycling path
		Tertiary road 63	Aalten-Dinxperlo	Cycling path
		Tertiary road 64	Ulft-Varsselder	Cycling path
		Tertiary road 65	Laag Keppel-Rondweg Hummelo	Cycling path

		Tertiary road 103	Bemmel-Gendt	Cycling path
		Tertiary road 109	Ingensesteeg-secondary road 101	Cycling path
Utrecht	**Urban**	Mijdrecht	Industrieweg	Cycling path
		Zeist	De Dreef	Cycling path
		Zeist	Geroplein	Cycling path
		Zeist	Dijnselburgerlaan	Cycling path
	Non-urban	Secondary road 9	Maarten Maartenshuis-Doorn	Cycling path
		Tertiary road 2	Zandheuvelweg at Baarn	Cycling path
		Tertiary road 27	Kockengen-Breukelen	Cycling path
		Tertiary road 27	Along De Dreef in Kockengen	Cycling path
		Tertiary road 28	Lange Meentweg (Wilnis and Kamerik)	Cycling path
Noord-Holland	**Urban**	Amstelveen	Vierlingsbeeklaan	Cycling path
		Amsterdam	Kamperbrug at Central Station	Underpass
		Amsterdam	Weesperstraat/ Nieuwe Herengracht	Underpass
		Amsterdam	Weesperstraat/ Nieuwe Kerkstraat	Underpass
		Amsterdam	Weesperplein/ Spinozastraat	Underpass
		Amsterdam	Sections route Slotervaart-Duivendrecht; Parnassusweg-Amstel trajectory	Cycling path + 3 tunnels + bridge
		Anna Paulowna	Kleine Sluis-Spoorbuurt	Cycling connection
		Haarlem	Pr. Bernhardlaan/ Pr. Beatrixlaan/V. Zeggelenplein	Reconstructed junction
		Haarlem	Kinderhuisvest	Cycling path
		Haarlemmermeer	Lijnden-Badhoeve-dorp	Section cycling connection
		Heerhugowaard	Westtangentweg and Zuidtangentweg	Cycling paths
		Den Helder	Lange Vlieg and Nieuwe Weg	Cycling paths
		Den Helder	Waddenzeestraat-Texelstroomstraat	Cycling connection
		Purmerend	Hoornselaan and Van IJssendijkstraat	Cycling paths
		Velsen	Schoterkerkpad	Cycling path

		Zaanstad	Vincent van Goghweg	Cycling path
	Non-urban	Secondary road 6	Niedorperverlaat-national road 7	Cycling path
		Tertiary road 1	Nieuwlanderweg and Kogerweg on Texel island	Cycling path
		Tertiary road 5	Doggersvaart-Julianadorp	Cycling path
		Tertiary road 12	Nieuwesluizerweg at Wieringermeer	Cycling path
		Tertiary road 12	Medemblikkerweg at Wieringermeer	Cycling path
		Tertiary road 21	Kanaalweg at Schoorldam	Cycling path
		Tertiary road 24	Middenweg at Heerhugowaard	Cycling path
		Tertiary road 33	Dijkgr. Grootweg/ Esdoornlaan Andijk and Grootebroek)	Cycling path
		Tertiary road 49	Zuiderweg at Zuidoostbeemster	Cycling path
Zuid-Holland	**Urban**	Delft	Brasserskade-Aan 't Verlaat	Cycling path
		Delft	Componistenpad until Martinus Nijhofflaan	Widened and extended cycling path
		Delft	Hambrug over Schie river	Cycling/pedestrian bridge
		Delft	Hambrug-area	Cycling path
		Delft	Prinses Irenetunnel-Engelsestraat	Cycling path
		Dordrecht	Sterrenbrug-Noordendijk	Cycling path + bridge + 2 underpasses
		Gouda	Access northern side station	Cycling path + bridge
		's-Gravenhage	Demonstration project	Cycling route
		Krimpen a/d IJssel	Nieuwe Tiendweg	Cycling path
		Krimpen a/d IJssel	Groenendaal	Cycling path
		Leiden	Haagsche Schouw-weg-Rijnsburgerweg (section)	Cycling path
		Leiden	Merenwijk-Schuttersveld	Cycling path + bridge
		Oegstgeest	Rijnsburgerweg-Haagsche Schouw (section	Cycling path

		Rotterdam	Langenhorst, section Slinge-railway crossing	Cycling path
		Rotterdam	Cliostraat	Cycling path
		Rotterdam	Rozenlaan	Cycling path
		Rotterdam	Schieweg	Cycling path
		Rotterdam	Mathenesserweg and square	Cycling path
		Rotterdam	Van Cittersstraat	Cycling path
		Rotterdam	Bunschotenweg	Cycling path
		Rotterdam	Eemhavenweg	Cycling path
		Rotterdam	Spaanseweg-Spaansebocht	Cycling path
		Rijswijk	Adm. Helfrichsingel and Weth. Brederodelaan	Cycling paths
		Voorburg	Rodelaan-Prinses Margrietplantsoen	Cycling path
		Voorschoten	Trompweg (section)	Cycling path
	Non-urban	Secondary road 19	Pijnacker	Cycling path
		Secondary road 21	Berkel en Rodenrijs	Cycling path
		Secondary road 43	Alongside Merwedebrug	Cycling path
		Tertiary road 11	Voorschoterweg at Valkenburg	Cycling path
		Tertiary road 16	Second section Oud Adeselaan Alkemade	Cycling path
		Tertiary road 20	Hoogmade-Woubrugge	Cycling path
		Tertiary road 21	Schilkerweg (section) Ter Aar	Cycling path
		Tertiary road 24	Nieuwkoop-Noorden	Cycling path
		Tertiary road 65	Noordweg and Oudeweg (section) Pijnacker	Cycling paths
		Tertiary road 65	Oudeweg (section) at Nootdorp	Cycling path
		Tertiary road 107	Langeweg (Watersch. De Groote Waard)	Cycling path
Zeeland	**Urban**	No projects		
	Non-urban	Tertiary road 23	Wolphaartsdijk-National road 258	Cycling path
		Tertiary road 46	Aagtekerke-Westkapelle	Cycling path
		Tertiary road 73	Graauw-Zandberg	Cycling path
Noord-Brabant	**Urban**	Bergen op Zoom	Zuidzijde Zoom	Cycling path + bridge
		Berkel-Enschot	Section route Tilburg-Berkel-Enschot-Oisterwijk	Cycling path

		Geldrop	Dommeldal	Cycling paths
		's-Hertogenbosch	Orthenseweg/ Gordelweg	Cycling tunnel
		Oss	Phase 1 and 2 routeplan	Cycling paths
		Roosendaal en Nispen	Rucphensebaan	Cycling path
		Tilburg	Demonstration project	Cycling route
		Tilburg	Section route Tilburg-Berkel-Enschot-Oisterwijk	Cycling path
		Valkenswaard	Nieuwe Waalreseweg	Cycling path
		Veldhoven	Noordelijke Pendelweg	Cycling path
	Non-urban area	Secondary road 8	Nieuwe Molen-Stampersgat	Cycling path
		Secondary road 9	Hoogerheide-Putte	Cycling path
		Secondary road 12	Alphen-Baarle Nassau	Cycling path
		Secondary road 21	Schijndel-St. Oedenrode	Cycling path
		Secondary road 23	Nuenen-Gerwen	Cycling path
		Secondary road 23	Gemert-Elsendorp	Cycling path
		Secondary road 25	Veghel-Erp	Cycling path
		Secondary road 30	Nistelrode-Uden	Cycling path
		Tertiary road 12	Lieshout-Aarle Rixtel	Cycling path
		Tertiary road 13	Aarle Rixtel-Helmond	Cycling path
		Tertiary road 15	Rips-Oploo	Cycling path
		Tertiary road 17	Bakel-Helmond	Cycling path
		Tertiary road 25	Bergeijk-Luyksgestel	Cycling path
		Tertiary road 31	Oisterwijk-Moergestel	Cycling path
Limburg	Urban area	No projects		
	Non-urban area	Secondary road 1	Horst-Venraij	Cycling path
		Secondary road 9	Echterbosch-Putbroek	Cycling path
		Tertiary road	Bergen-German border	Cycling path
		Tertiary road	Venraij-Overloon	Cycling path
		Tertiary road	Haelen-Roggel	Cycling path
		Tertiary road	Helden-Neer-Roggel	Cycling path
		Tertiary road	Meijelsedijk at Nederweert	Cycling path
		Tertiary road	Budel-Weert	Cycling path
Flevoland	**Urban**	Dronten	Bisonweg	Cycling path
	Non-urban	No projects		

Source: *Kamerstukken II* 1979/80, 15 885 (*Meerjarenplan Personenvervoer* [Multi-Year Plan Personal Traffic]), no. 4, 19-22.

Appendix 3

National Public Works Subsidies Awarded to Municipalities, c. 1975-1980

Awarded to	**Amount** (guilders)	**Purpose**
Geldrop[a]	350,000	Cycling paths in Kleine Dommeldal
Delft[b]	67,000	Extension Componistenpad as cycling and pedestrian path
Dokkum[b]	216,000	Cycling path along Harddraversdijk
Krimpen aan de IJssel[b]	418,000	Cycling paths along Nieuwe Tiendweg
Amsterdam[c]	2,340,000	Cycling route Slotervaart-Duivendrecht
Hengelo[c]	400,000	Cycling path along Deldenerstraat
Berkel-Enschot[d]	1,660,000	Connections and extensions to cycling path demonstration project Tilburg
Oisterwijk[d]	1,320,000	Ibid.
Tilburg[d]	380,000	Ibid.
Rotterdam[e]	55,000	Cycling path Slinge-harbor railway Rotterdam-Zuid
Rijswijk[e]	60,000	Cycling path along Admiraal Helfrichsingel and Hammarskjöldlaan
Den Bosch[f]	800,000	Cycling tunnel in Orthenseweg/Grondelweg
Roosendaal and Nispen[g]	252,000	Separate cycling path along Rucphensebaan
Haarlemmermeer[g]	96,000	Cycling path Lijnden-Badhoevedorp (between Veldweg and Keizersweg)

Examples of urban cycling infrastructure construction costs in the 1970s and 1980s Sources: (a) *Verkeerskunde* 27, no. 1 (1976): 40-41; (b) *Verkeerskunde* 27, no. 12 (1976): 608; (c) *Verkeerskunde* 28, no. 1 (1977): 10-11; (d) *Verkeerskunde* 28, no. 2 (1977): 62; (e) *Verkeerskunde* 28, no. 4 (1977): 162; (f) *Verkeerskunde* 28, no. 7 (1977): 308-10; (g) *Verkeerskunde* 28, no. 9 (1977): 402-3.

Bibliography

Archival Sources

Digitized Parliamentary Records (*Handelingen*) and Documents (*Kamerstukken*) of the Netherlands (1814-1995). https://zoek.officielebekendmakingen.nl/uitgebreidzoeken/historisch

NATIONAL ARCHIVES OF THE NETHERLANDS, THE HAGUE (NA)
Ministry of Public Works (V&W)
Inspecteurs van de Waterstaat, 1849-1930 (2.16.07)
Raad van de Waterstaat, 1892-1992 (2.16.11)
Commissie inzake Wegbelasting, 1924-1925 (2.16.19.10)
Commissie Motor- en Rijwielverkeer (2.16.19.31)
Commissie van Overleg voor de Wegen, 1915-1940 (2.16.59)
Ministerie van Waterstaat: Directie van de Waterstaat (DW), 1930-1949 (NA 2.16.22.01)
Centrale Commissie van Advies en Bijstaand voor het Verkeersfonds, 1935-1939; Subcommissie voor de Kosten van de Weg, 1935-1938 (2.16.51.02)
Ministerie van Verkeer en Waterstaat: Stuurgroep Verkeer en Vervoer, 1941-1983 (2.16.5113)
Ministerie van Verkeer en Waterstaat: Directoraat-Generaal van het Verkeer, 1947-1980 (2.16.5117)
Ministerie van Verkeer en Waterstaat: Dienst Weg- en Waterbouwkunde (DWW), 1961-1994 (2.16.5285)
Ministerie van Verkeer en Waterstaat: Permanente Internationale Commissie van de Permanent International Association of Road Congresses (PIARC), 1908-1989 (2.16.5287)

Ministry of Culture, Recreation and Social Work (CRM)
Interdepartementale Coördinatiecommissie voor de Openluchtrecreatie en het Toerisme (INCORET), 1958-1976 (2.27.5111)
Ministerie van Cultuur, Recreatie en Maatschappelijk Werk, Beleidsterrein Recreatie, 1949-1982 (2.27.5215)

Ministry of Public Housing, Spatial Planning and the Environment (VROM)
Centrale Sector van het Ministerie van VROM, 1940-1981 (2.17.03)
Rijksdienst voor het Nationale Plan (RNP), 1941-1965 (2.17.04)

Ministry of Agriculture and Fisheries (LNV)
Cultuurtechnische Dienst/Landinrichtingsdienst, 1935-1995 (2.11.5006)
Coördinatiecommissie Openluchtrecreatie (2.11.5164)

Other National Governmental Bodies
Tweede Kamer der Staten-Generaal, 1945-1989 (2.02.28)
Ministerie van Binnenlandse Zaken: Afdeling Binnenlands Bestuur, 1879-1950 (2.04.57)
Rijkscommissie Werkverruiming, 1922-1946 (2.06.081)
Interprovinciaal Overleg (IPO), 1947-2003 (2.19.063)
Vereniging van Nederlandse Gemeenten (VNG), 1912-1989 (2.19.185)

Action and Interest Groups
Veilig Verkeer Nederland, 1932-2000 (2.19.197)
Pressiegroep Stop de Kindermoord/Stichting Pressiegroep Kinderen Voorrang, 1973-2000 (2.19.198)

Industry
Kamer van Koophandel en Fabrieken voor Rotterdam, 1922-2013 (3.17.17.04)

Provincial Government Zuid-Holland
Rijkswaterstaat in Zuid-Holland, 1849-1980 (3.07.11)
Rijkswaterstaat: Directie Zuid-Holland, 1981-1989 (3.07.5018)
Rijkswaterstaat: Directie Zuid-Holland: Dienstkring Leiden, 1911-1998 (3.07.5069)
Provinciale Waterstaat in Zuid-Holland, 1945-1986 (3.02.27)
Griffie van de provincie Zuid-Holland, 1946-1975 (3.02.22)
Griffie van de Provincie Zuid-Holland, 1976-1986 (3.02.46)

Others
Personal archive Mr. H.J. Smidt, 1851-1921 (2.21.026.08)

PROVINCIAL ARCHIVES DRENTHE, ASSEN (DRENTS ARCHIEF; DA)
Bestuursarchief provincie Drenthe, 1950-1987 (0923)
Provinciale Staten, 1814-1951 (0025)
Gedeputeerde Staten, 1814-1951 (0031)
Provinciale Waterstaat, 1845-1945 (0029)
Provinciale Waterstaat Drenthe, 1945- (0924)
Centrale Vereeniging voor den Opbouw van Drenthe, 1925-1970 (0196)
Collectie Zwinderman, 1693-1847 (0635)

PROVINCIAL ARCHIVES GELDERLAND, ARNHEM (GELDERS ARCHIEF; GA)
Gedeputeerde Staten, 1814-1950 (0039)
Provinciale Staten en Gedeputeerde Staten 1950-1979 (3204)
Provinciale Waterstaat, 1864-1955 (0244)

PROVINCIAL ARCHIVES NOORD-HOLLAND, HAARLEM (NOORD-HOLLANDS ARCHIEF, NHA)
Provinciaal Bestuur van Noord-Holland, 1851-1943 (18)
Rijkswaterstaat Directie Noord-Holland te Haarlem, 1931-1951 (228)
Provinciale Waterstaat van Noord-Holland te Haarlem, 1911-1942 (553)
Provinciaal Bestuur van Noord-Holland, Wegen, Verkeer en Waterstaat en bijbehorende commissies te Haarlem, 1943-1989 (639)
Digitized Council Minutes Heemstede (https://nha.courant.nu/periodicals/RNH)

PROVINCIAL AND MUNICIPAL ARCHIVES UTRECHT, UTRECHT (HET UTRECHTS ARCHIEF, HUA)
Gedeputeerde Staten van Utrecht, 1920-1954 (1201)
Gedeputeerde Staten van Utrecht, 1955-1988 (1205)
Provinciale Staten van Utrecht, 1954-1987 (1204-1)
Commissies uit de Provinciale Staten Utrecht, 1956-1988 (1204-2)
Provinciale Waterstaat van Utrecht, 1830-1925 (1208)
Dienst van de Provinciale Waterstaat van Utrecht, 1955-1988 (1210)
Provinciale planologische dienst van Utrecht, 1943-1989 (1211)
Gemeentebestuur van Utrecht, 1813-1969, deel 2, 1813-1910 (1007-2)
Gemeentebestuur van Utrecht, 1813-1969, deel 3: 1911-1969 (1007-3)
Gemeentebestuur van Utrecht, 1970-1989 (1338)
Gemeente Nieuwegein, secretarie 1971-1990 (1651)
Verzamelingen betreffende spoorwegen, 1839-2000 (971)

Regional Archives Amersfoort, Amersfoort (Archief Eemland, AE)
Eerste Enige Echte Wielrijdersbond (E.N.W.B.), afdeling Amersfoort, 1975-1996 (0242)

Regional Archives Eindhoven, Eindhoven (Regionaal Historisch Centrum Eindhoven, RHCE)
Gemeenschappelijke regeling Agglomeratie Eindhoven, 1977-1985 (10884)
Intergemeentelijke organisatie voor rijwielpaden in de Kempen, 1953-1985 (10091)

Municipal Archives Amsterdam, Amsterdam (Gemeentearchief Amsterdam; GAA)
R.A.I., 1894-1994 (1302)
Collectie Stadsarchief Amsterdam: persdocumentatie op onderwerp (30486)

Municipal Archives The Hague, The Hague (Gemeentearchief Den Haag; GADH)
Werkgroep Dooievaar, 1970-1985 (0772-01)

Municipal Archives Zeist, Zeist (Gemeentearchief Zeist; GAZ)
Rijwielpadvereniging Utrecht met omstreken Umo, 1918-1968 (8020)

International Institute for Social History, Amsterdam (IISG)
Archief ENFB (ARCH01969)
Archief Amsterdam fietst (ARCH02118)

ANWB Foundation Archives, The Hague (ANWB)

Interviews

January 8, 2018, André Pettinga – Former traffic engineer Delft 1970s; currently cycling consultant
January 15, 2019, Peter Plantinga – Founding member of Eindhoven chapter of Cyclists' Union
February 5, 2019, Hans van Beek and Leo Hamer – Founders of Dooievaar
August 12, 2019, Koos Louwerse – Involved in Masterplan Fiets; currently cycling consultant
December 12, 2018, Jan Ploeger – Former traffic engineer at multiple levels
July 16, 2019, Jan Ploeger
November 13, 2019, Steven Schepel – Former chair of Stop the Child Murder
June 5, 2020, Maartje van der Putten – Founding member of Stop the Child Murder
June 9, 2020, Wim Bot – Policy advisor/lobbyist for Cyclists' Union
May 18, 2021, Eisse Kalk and Annemarie Wijn – Founding members of Werkgroep 2000
June 2, 2021, Tom Godefrooij and Jaap Rijnsburger – Leading members of Cyclists' Union, 1980s/1990s

Audio Sources

NTR/NPO Radio 1, *De Spaak*, podcast, Stop de Kindermoord, August 1, 2019. https://www.nporadio1.nl/podcasts/de-spaak [accessed September 2, 2019].

Newspaper and Journal Articles

"Afdeling Amersfoort: 'Bericht over het ontstaan van een fietsrapport'." *De Ketting* 1 (1976): 2-3.
"Ambtenaar in Deurne, actievoerder in Venray." *De Vogelvrije Fietser* 11, no. 1 (1985): 12-14.
"Amsterdam publiceert handboek fietsbeleid." *NRC Handelsblad*, July 23, 1991, 2.
"Amsterdams Dagboek: Notities." *Het Parool*, September 7, 1970, 2.
"ANWB: Hoe zit dat?" *Bromfietskampioen* 1, no. 1 (1951): 14.
"ANWB: Hoe zit dat?" *Bromfietskampioen* 1, no. 7 (1951): 123.
"ANWB: Hoe zit dat?" *Bromfietskampioen* 1, no. 11 (1951): 195.
"ANWB: Hoe zit dat?" *Bromfietskampioen* 1, no. 12 (1951): 224.
"Arnhems plan voor fietspaden." *Verkeerskunde* 26, no. 6 (1975): 296-97.
"Auto kapot rameien voor 'veilig verkeer'." *Rotterdamsch Parool/De Schiedammer*, May 22, 1965, 2.
"Belangen voetganger in het verkeer." *Arnhemsche Courant*, January 8, 1954.
"Belangrijke rede van den Rijkscommissaris." *Nieuwsblad van Friesland*, March 14, 1941, 1.
"Brabant is onveiligste provincie." *Het Vrije Volk*, April 19, 1975, 7.
"(Brom)fietser niet vergeten." *Verkeerskunde* 26, no. 6 (1975): 285.
"Bromfiets zal rijwiel niet overvleugelen." *Bromfietskampioen* 5, no. 4 (1955): 284-85.
"Bromfiets: op naar het miljoen...!". *Trouw*, February 25, 1959.
"'Burger moet echt kunnen meespreken'." *Het Parool*, March 24, 1981, 7.
"Burger wordt dupe van groeibeleid." *Trouw*, November ember 27, 1972, T12.
"Bus-stop, school-stop." *Algemeen Dagblad*, November 1, 1972, 14.
"Compacte massa auto's en fietsers." *Utrechtsch Nieuwsblad*, May 31, 1955, 2.
"Coördineren en stimuleren: Het fietsbeleid van de provincie Zeeland." *Fietsverkeer* 2, no. 5 (2003): 8-10.
"De A.N.W.B. werkt door!" *De Kampioen* 57, no. 12 (1940): 177.
"De bromfiets: het snelverkeer van het rijwielpad: Nationale Verkeerstechnische Leergang 1954." *Bromfietskampioen* 4, no. 3 (1954): 142-44.
"De correcte fietser." *De Kampioen* 57, no. 12 (1940): 179-80.
"De lange adem van Zwolle: Perfecte fietsroutes." *Fietsverkeer* 4, no. 11 (2005): 1-7.
"De Noord heeft schoon genoeg van de onveilige weg naar school." *Het Parool*, April 23, 1974, 9.
"De opbouw van een goed wegennet, in het bijzonder in Friesland." *Wegen* 36, no. 6 (1962): 126-33.
"De plek van de fiets in gemeentelijk verkeersbeleid." *Fietsberaad* 3, no. 9 (2004): 15-17.
"De Rijkswegen in 1956." *Berichten en beschouwingen van de verkeersafdeling van de ANWB* 8, no. 1 (1957): 9-20.
"De rijwielbelasting verdwijnt." *Haagsche Courant*, April 24, 1941, 1.
"De Rijwielpadvereniging Twente: Haar werk en haar vele moeilijkheden," *Twentsch dagblad Tubantia*, April 30, 1949, 2.
"De verkeerskundige moet eenvoudig durven zijn! Gesprek met ir. C.A. Kuysten bij zijn afscheid als Directeur Verkeer en Recreatie van de ANWB." *Verkeerskunde* 28, no. 11 (1977): 506-7.
"De verschoppeling der steden." *De Kampioen* 61, no. 5 (1946): 3-4.
"Den Haag neemt proef met fietsroute." *Trouw*, June 17, 1975, 6.
"Directeur Amsterdams verkeerbureau: Geen doorbraken in binnenstad." *Algemeen Handelsblad*, April 25, 1968, 2.
"Dooievaar krijgt steun voor alternatief: PvdA vraagt raad uitstel 'Koekamp'." *Trouw*, April 17, 1975, 6.
"Dooievaar wil Hagenaar weer laten fietsen." *De Volkskrant*, February 24, 1973, 39.
"Een jaar ANWB geschiedenis." *Bromfietskampioen* 5, no. 7 (1955): 514-21.
"Eerstkomende 15 jaar: bijna 250 fietspaden erbij in Friesland." *Leeuwarder Courant*, December 1, 1979, 13.

"Eisse Kalk: 'In dit werk is een brede rug nodig'." *Het Parool*, April 13, 1983, 6.
"ENFB in actie voor onderhoud fietspaden." *Nederlands Dagblad*, May 8, 1984.
"ENFB organiseert actie 'Fiets en trein'." *Het Vrije Volk*, October 21, 1982.
"ENFB wil meer stallingsgelegenheid bij leeszaal." *Nieuwsblad van het Noorden*, April 10, 1982.
"ENFB wil speciale wegzijzers voor fietsers." *Nieuwsblad van het Noorden*, October 6, 1983.
"ENWB afd. Maastricht." *De Ketting* 1 (1976): 5-6.
"f 500.000 gevoteerd voor rijwielpadennet." *Nieuwsblad van het Noorden*, July 12, 1955, 3.
"Fietsdemonstratie naar Hoofddorp." *Trouw*, May 16, 1978, 12.
"Fietsen op de paden van de Zuid-Veluwe." *De Kampioen* 75, no. 6 (1960): 176-77.
"Fietsenmaker H. Kind Jr.: 'Ze zeggen weer u tegen me'." *Het Vrije Volk*, November 8, 1973, 7.
"Fietser in de berm van het verkeersbeleid. Fietsen is een volwassen verkeersvorm." *De Kampioen* 89, no. 4 (1974): 246-48.
"Fietsers hebben weinig aan de politiek." *Trouw*, June 10, 1980.
"Fietsers willen paden." *De Telegraaf*, May 18, 1978, 7.
"Fietsersbond ENFB keurt fietsroutes in Flevoland." *Fietsverkeer* no. 4 (1994): 3.
"'Fietspad langs Stadsweg vertoont tal van gebreken'." *Nieuwsblad van het Noorden*, October 20, 1983.
"Fietsverkeer trein te weinig gestimuleerd." *Leeuwarder Courant*, November 1, 1980.
"Fietsvriendelijke stad is nog toekomstmuziek." *Trouw*, May 14, 1982.
"G.A.M. v. d. Muyzenberg promotor Drents fietspad vertrekt op 13 december." *Nieuwsblad van het Noorden*, November 30, 1974, 15.
"Geen gefundeerde visie op verkeer: Secretaris RAI kraakt verkeerscirculatieplan." *NRC Handelsblad*, April 21, 1972, 2.
"Gemengd nieuws. N.S.B. contra rijwielbelasting." *Nieuwsblad van Friesland*, February 24, 1941, 2.
"Generaal van een roemrijk leger, in een land vol vredestrofeeën: Ir. A.G. Maris ten afscheid." *Algemeen Handelsblad*, October 19, 1961, 4.
"Gevaarlijk punt in Bergschenhoek eist wéér mensenleven." *Het Vrije Volk*, July 9, 1963, 1.
"Gouden tijden voor rijwielindustrie." *NRC Handelsblad* November 8, 1973, 14.
"Groene Eland en Dooievaar vormen reddend tandem." *De Posthoorn*, January 10, 1974.
"Groningen ziet wel iets in fietsbruggen." *Nieuwsblad van het Noorden*, January 16, 1981.
"Happening voor veilig verkeer Schiedam." *Het Vrije Volk*, May 22, 1965, 7.
"Harlinger haven was nog niet 'haalbaar' daarom uitdieping van grote kanelen. Fietspaden: gemeenten voeren plan uit en GS coördineren." *Leeuwarder Courant*, July 4, 1962, 7.
"Heden en toekomst van de rijwielpaden." *De Kampioen* 63, no. 10 (1948): 266-67.
"Het 53ste jaarverslag (Vervolg)." *De Kampioen* 53, no. 16 (1936): 346.
"Het comfortabele netwerk van Veenendaal: Verwennerij per 300 meter." *Fietsverkeer* 4, no. 12 (2005): 4-7.
"Het probleem van de fiets: het opbergen." *De Tijd*, February 24, 1973, 4.
"Hysterische 'kunstrakkers' slopen Nieuwlandse galerie." *Het Nieuwe Stadsblad*, May 25, 1965, 1.
"In hart Rotterdam tóch bewaakte fietsenstalling." *Algemeen Dagblad*, August 25, 1982.
"In het kort." *Verkeerskunde* 26, no. 6 (1975): 318-19.
"In het kort." *Verkeerskunde* 29, no. 6 (1978): 274-75.
"In Nederland komt een miljoen auto's." *Nieuwsblad van het Zuiden*, February 28, 1956.
"Ir. Hoogenboom: 'Bloed, zweet en tranen'." *NRC Handelsblad*, January 19, 1972, 8.
"Ir. L.T. van der Wal †." *De Kampioen* 79, no. 10 (1964): 616.
"Jaarlijkse GDU-bestedingen vanaf 1996: Aandacht voor fietsvoorzieningen houdt stand." *Fietsverkeer* 2, no. 5 (2003): 11.
"Johan Leferink, provincie Gelderland: 'Behoefte aan uitwisseling met andere provinces'," *Fietsverkeer* 1, no. 2 (2002): 8.

"Kind moet veilig kunnen leven." *Het Parool*, January 20, 1973, 5.
"Kinderen blijven thuis: Telegram aan minister, motie aan de raad." *Tubantia*, February 22, 1968, 5.
"Lezers-tribune: Ook bromfietser moet aan z'n trek komen." *Utrechtsch Nieuwsblad*, June 10, 1953, 4.
"Linda de Klein-Bevers, provincie Brabant: 'Wees realistisch: de wereld verandert niet in één keer'," *Fietsverkeer* 1, no 2 (2002): 9.
"Met GDU (fiets)verkeersbeleid regisseren." *Fietsverkeer* 2, no. 5 (2003): 10-11.
"Miljoen bromfietsen in 1960?" *Algemeen Handelsblad*, March 27, 1956, 7.
"Misvattingen over het werk van den A.N.W.B.". *De Kampioen* 55, no. 23 (1938): 413.
"Neerlands rijwielpaden en hun bewijzering." *De Kampioen* 59, no. 9 (1942): 65-66.
"Nieuw pleidooi Wiegel voor snelweg Midlum naar Zurich." *Leeuwarder Courant*, October 25, 1984, 19.
"Nieuwe rijwielpaden in Zuid-Holland." *Trouw*, August 13, 1959, 2.
"Nieuwe verkeerswet sluit duizenden kilometers rijwielpad voor bromfietsen." *Berichten en beschouwingen van de verkeersafdeling van de ANWB* 1, no. 5 (1950): 1-2.
"Nieuwe wegen zullen moeten worden aangelegd." *Het Parool*, February 28, 1956, 5.
"Notulen van de jaarlijksche Algemeene Vergadering." *De Kampioen* 56, no. 18 (1939): 345-47.
"Notulen van de vergadering van het Algemeen Bestuur ... op Zondag 22 december 1935." *De Kampioen* 53, no. 6 (1936): 131-33.
"NS inventariseert knelpunten voor fiets bij stations." *Nederlands Dagblad*, March 12, 1980.
"Onzeker of provincie aanleg paden met geld kan steunen." *Leeuwarder Courant*, June 11, 1978.
"Open deze paden voor ons!". *Bromfietskampioen* 2, no. 5 (1952): 170-72.
"Ouders houden hun kinderen van school uit protest." *Tubantia*, September 27, 1972, 7.
"Plan om fietsendieven te remmen." *Algemeen Dagblad*, January 31, 1981.
"Politie lost maar 1pct fietsendiefstallen op." *Het Vrije Volk*, November 9, 1984.
"Politie steekt stokje voor blokkade Marnezijl door schooljeugd." *Leeuwarder Courant*, September 3, 1976, 15.
"Protest tegen slechte klimaat voor fietsers." *NRC Handelsblad*, June 12, 1973, 10.
"Provincies willen meer grip op besteding gemeentelijke subsidies." *Fietsverkeer* 6, no. 17 (2007): 23.
"Rapport over het probleem van voetgangers in het verkeer: Wettelijke bescherming gevraagd van oversteekplaatsen." *Het Vaderland*, September 15, 1954, 1.
"Rijk legt weg aan naast E 8 bij Holten." *Volkskrant*, February 22, 1968, 11.
"Rijwielpaden in de Geldersche Achterhoek." *De Kampioen* 62, no. 11 (1947): 329.
"Rijwielpadennet in het centrum van Noord-Brabant." *De Kampioen* 67, no. 12 (1952): 348-49.
"Rotterdam moet kiezen: auto of mens." *NRC Handelsblad*, October 31, 1970, 20.
"Steden laten de fietsers versukkelen." *Trouw*, February 28, 1964, 11.
"Tienduizenden trokken naar bos en water." *Utrechtsch Nieuwsblad*, May 26, 1953, 2.
"Tussentijdse balans." *Bromfietskampioen* 2, no. 1 (1952): 6-9.
"Verkeerstechnische leergangen 1976: Samenvattingen lezingen." *Verkeerskunde* 26, no. 12 (1975): 614-15.
"Verkort Verslag van de Congresdag op 14 december 1956 te Utrecht: Maatregelen ter bevordering van de Vlotheid van het verkeer en van de parkeer-mogelijkheden in de binnenstad." *Wegen* 31, no. 7 (1957): 192-202.
"Vlakke rijwielpaden." *Berichten en beschouwingen van de verkeersafdeling van de ANWB* 5, no. 2 (1954): 6.
"Voetganger in actie." *Algemeen Dagblad*, November 3, 1979, 35.
"Voorzieningen voor fietsers en bromfietsers." *Verkeerskunde* 26, no. 6 (1975): 286-95.
"Vrijliggende rijwielpaden zijn noodzakelijk." *De Kampioen* 56, no. 11 (1939): 184.
"Waar blijft het geld voor onze 'gewone' fietspaden?" *Leeuwarder Courant*, June 26, 1954, 5.

"Walcheren bewegwijzerd voor fietsers." *Fietsverkeer* 5 (1994): 1-2.
"Wanneer een auto een fietser inhaalt." *De Kampioen* 53, no. 4 (1936): 83.
"'Weg met auto's': 100 kleuters demonstreren." *De Telegraaf*, April 21, 1970, 7.
"Wij spraken met Ir. M. le Cosquino de Bussy, Hoofdingenieur-directeur Rijkswaterstaat." *Het Vaderland*, February 28, 1958, 3.
"Zieke kabouter weggebracht: In drie steden acties." *De Volkskrant*, April 13, 1970, 3.
"Zo wil Amsterdam zijn verkeersknooppunten doorhakken." *Het Parool*, February 23, 1974, 3.

[Ankersmit], J.W.A. "Divergerende lijnen." *Bromfietskampioen* 6, no. 8 (1956): 524-25.
Arendshorst, Dick. "Janus Meulendijks regelt het wel met de hoge heren." *De Vogelvrije Fietser* 1, no. 2 (1976): 8-9.
Arlman, Hugo, and Rudie van Meurs. "'De adem van de ANWB stinkt eeuwig naar benzine.' De macht van het vijftiende ministerie." *Vrij Nederland* (Bijvoegsel) 17 (1980): 3-30.
Bakker, H.B. "Opnieuw over rijwielpaden." *Wegen* 38, no. 3 (1964): 70-76.
Bakker, H.B. "Over het saneren in het algemeen en over het saneren van een oud Rijkswegvak in het bijzonder." *Wegen* 25, no. 1 (1951): 1-4.
Bakker, H.B. "Rijwielpaden langs Rijkswegen." *Wegen* 24, no. 5 (1950): 137-44.
Bakker, Henk, and William Rothuizen. "'Elke dode die er niet valt is er voor ons één'." *Haagse Post* 59, no. 51 (1972): 6-7.
Bakker, J.P., and R. v.d. Hout. "Herziening provinciaal fietspadenplan Zuid-Holland." *Verkeerskunde* 26, no. 7 (1975): 356-61.
Beekmans, Kees. "Amsterdam werkt aan 'fietssnelwegen'." *NRC Handelsblad*, March 13, 1990, 7.
Been, H., and G. van der Kolk. "Fietsverkeer in steden." *Tijdschrift voor Verkeerstechniek* 25, no. 4 (1974): 164-68.
Berg, Cees van den. "Op de barricaden voor de fietser." *Algemeen Dagblad*, April 17, 1986.
Bergsma, E. "Middelen om verkeersonveiligheid te bestrijden." *De Kampioen* 53, no. 3 (1936): 49-50.
Bergstra, Piter, and Martin de Bruin. "Een goede CdK opent Haagse deuren." *Nieuwsblad van Friesland*, August 27, 1994, 23.
Bommels, Bert. "ANWB, kampioen op snelwegen en zijpaden." *Het Parool*, April 19, 1980, 25-26.
Bont, Wim. "Fiets met bus en trein mee." *Het Vrije Volk*, December 24, 1980.
Boost, A.G.M. "Rijwielverkeer en toerisme." *Wegen* 24, no. 5 (1950): 158-59.
Bouma, Joop. "Fooi voor fiets, maar tòch een omslag." *Trouw*, September 27, 1991.
Broecke, A.A.J. ten, and J. Hoekwater. "Kwalitatieve evaluatie van de demonstratie fietsroutes in Den Haag en Tilburg (1)." *Verkeerskunde* 30, no. 9 (1979): 417-21.
Broecke, A.A.J. ten, and J. Hoekwater. "Kwalitatieve evaluatie van de demonstratie fietsroutes in Den Haag en Tilburg (2)." *Verkeerskunde* 30, no. 10 (1979): 492-96.
Broek, Marc van den. "Pad voor fietser is hobbelig en slecht aangegeven." *De Volkskrant*, August 28, 1984.
Broer, Karin. "Praten, praten en praten." *De Vogelvrije Fietser* 19, no. 3 (1994): 8-10.
Bruggeman, Marnix. "Handboek laat wegbeheerder door een fietsbril kijken." *Trouw*, October 9, 1987, 13.
Bruijn, N.H.J. de. "Wie mee wil praten moet het nu vertellen." *NRC Handelsblad*, February 5, 1973, 8.
Bruin, Willem de. "Nationaal verkeersveiligheidscongres: Belangen fietsers en voetgangers voorop." *De Waarheid*, April 23, 1980, 4.
Bussy, M. Le Cosquino de. "Verkeer – heden en toekomst." *De Kampioen* Special issue '75 jaar ANWB' 1 July i 1958 (1958): 37-39.
D.H.G. "ANWB Hoe zit dat?" *Bromfietskampioen* 1, no. 4 (1951): 71.
Daalmeijer, Joop. "We moeten bewuster leven ook met het verkeer." *Het Nieuwe Stadsblad*, May 19, 1965, 9.

Deltrap, Ien. "Er is nog hoop voor Zuid-Holland." *De Vogelvrije Fietser* 3, no. 1 (1977): 9.

Deltrap, Ien. "Waar blijven de dames?" *De Vogelvrije Fietser* 4, no. 2 (1979): 4.

Dersjant, A., and C. Wentink. "Het gebruik van de demonstratie fietsroutes in Den Haag en Tilburg." *Verkeerskunde* 30, no. 8 (1978): 358-61.

Dragtstra, Dirk, and Paulien Osse. "Het ENWB-bestuur doet een boekje open: 'De fiets hoort juist in de stad!'." *De Vogelvrije Fietser* 1, no. 5 (1976): 18-19.

Engelen, Gert van. "'Actie is hier een vies woord'." *De Vogelvrije Fietser* 8, no. 6 (1983): 10-11.

Engelen, Gert van. "Beschaafd, constructief en tóch lastig." *De Vogelvrije Fietser* 10, no. 2 (1984/85): 10-11.

Engelen, Gert van. "'Je ziet hier geen dikke schoolkinderen'." *De Vogelvrije Fietser* 6, no. 1 (1980): 16-17, 37.

Engelen, Gert van. "Overleggen is niet fout, ingepakt worden wel." *De Vogelvrije Fietser* 16, no. 1 (1990): 24-25.

Fekkes, J.C. "Provinciaal Rijwielpadenplan voor Zuid-Holland." *Wegen* 30, no. 1 (1956): 1-5.

Fekkes, J.C. "Provinciaal rijwielpadenplan voor Zuid-Holland." *Tijdschrift voor Verkeerstechniek* 15, no. 8 (1964): 360-63.

Godefrooij, Paul, and Toni Niël. "'Rijkswegen, dat is de grootste prioriteit in Brabant'." *De Vogelvrije Fietser* 9, no. 3 (1984): 18-19.

Goudriaan, Huib. "Een kind blijft ook in het verkeer een kind." *Trouw*, October 21, 1972, 21.

Gurp, P.B. van, and R.A. Sangen. "De fietsroute Tilburg." *Verkeerskunde* 28, no. 4 (1977): 170-71, 91.

Gurp, P.B. van, and R.A. Sangen. "Tilburgers op de fiets: een demonstratieproject." *Verkeerskunde* 26, no. 11 (1975): 550-56.

Hielkema, Haro. "Zuid-Hollandse fietsers krijgen meer de ruimte: Gedeputeerde Borgman kan nog wel paar miljoen gebruiken." *Trouw*, November 10, 1978, 6.

Hoogenboom, G.H.A. "Actie is zinvol maar vereist veel geduld." *De Tijd*, October 24, 1971, 3.

Huitema, T. "Rijwielpaden." *Wegen* 5, no. 11 (1929): 270-74.

Hupkes, Henk Dikker. "ENFB heeft inhoudelijk net zoveel in te brengen als ambtenaren." *De Vogelvrije Fietser* 10, no. 2 (1984/85): 8-9.

Illich, Ivan. "Gradaties in het voortbewegen op eigen kracht." *De Vogelvrije Fietser* 2, no. 6 (1977): 18-19.

Jong, Jacques de. "'Aangepaste wegwijzers nodig voor fietsers'." *De Volkskrant*, September 27, 1982.

Jong, Theo de. "Knelpunten-nota's grootste knelpunt." *De Vogelvrije Fietser* 3, no. 6/4, no. 1 (1978): 14-15.

Kanstein, F. "Rijwielpaden." *Wegen* 5, no. 7 (1929): 161-74.

Kee, Peter. "Amsterdam moet de fiets nog ontdekken." *Het Parool*, February 20, 1990, 12.

Koolhaas Revers, J. "Bromfiets de 'Veroveraar': Een beschouwing over het ontstaan van een nieuwe phase in ons gemotoriseerde verkeer." *Bromfietskampioen* 1, no. 1 (1951): 4-6.

Koolhaas Revers, J. "Bromfietsers van Nederland, de A.N.W.B. heet u welkom." *Bromfietskampioen* 1, no. 1 (1951): 3.

Kouwenberg, Paul. "Inspraak-begeleider Eisse Kalk: Democratie verbeteren kost tijd." *Volkskrant*, March 8, 1975, 19.

Kouwenberg, Paul. "Veilig Verkeer Nederland houdt niet van kinderen." *De Volkskrant*, May 12, 1973, 27.

Krijn, C. "Rijwielpaden langs Provinciale Wegen." *Wegen* 24, no. 5 (1950): 145-48.

Kuysten, C.A. "Rijwielpaden in Zuid-Holland." *De Kampioen* 62, no. 9 (1947): 274-75.

Kuysten, C.A. "Rijwielpaden ten Plattelande." *Wegen* 24, no. 5 (1950): 154-57.

L., Th.C.M. "Bromfietsenwel." *Utrechtsch Nieuwsblad*, June 6, 1953, 3.

Langenhoff, Vic. "'Pressiegroep Stop de kindermoord'." *De Tijd*, September 20, 1972, 9.

Langenhoff, Vic. "Waarom laten wij kinderen bij honderden van de weg rijden?" *De Tijd*, September 20, 1972, 9.

Lewin, Lisette. "Amsterdammer is weer lastig: metro geen comfort." *NRC Handelsblad*, January 26, 1973, 2.

Lewin, Lisette. "Fielt." *NRC Handelsblad*, November 11, 1972.

Lewin, Lisette. "Henk Bakker: metro is Lammers' struikelblok." *NRC Handelsblad*, November 20, 1972, 2.

Lewin, Lisette. "Lastige Amsterdammer Henk Bakker: 'Alles is te regelen als eerst die auto maar weg is'." *Algemeen Handelsblad*, March 18, 1970, 2.

Ligtermoet, Dirk. "De geleidelijke weg naar een prettig fietsklimaat." *Fietsverkeer* 9, no. 26 (2010): 24-28.

Ligtermoet, Dirk. "Het fietsklimaat van Culemborg: waarom Nederland zo'n fietsland is." *Fietsverkeer* 10, no. 27 (2011): 10-14.

Lohuizen, Th. K. van. "Het schetsontwerp voor het streekplan 'Utrecht-Oost'." *Tijdschrift voor Volkshuivesting en Stedebouw* 16, no. 4 (1935): 68-74.

Lowe, Patricia Penrhyn. "Weg fiets? ... Fietsweg. Oprichting van de Eerste, Enige, Echte Nederlandse Wielrijders Bond." *Jonas: tijdschrift voor sociale en kulturele bewustwording* 6, no. 5 (1975): 10-11.

Mak, Geert. "Anatomie van een romantische stadsrebellie." *NRC Handelsblad*, October 3, 1992.

Maurits, Henk. "Energiecrisis leidt tot recordverkoop van fietsen." *De Tijd*, January 3, 1974, 3.

Mees, B. "Het rijwielverkeer in de grote stad." *Wegen* 24, no. 5 (1950): 149-54.

[Meijer, Jan.] "Spreeuwenpraat." *Het Vrije Volk*, July 31, 1950, 5.

Metselaar, J.W.E. "De vuist van de gemangelde fietsers." *NRC Handelsblad*, December 3, 1977, 5.

Muyzenberg, G.A.M. van den. "De fiets en het verkeer: Pleidooi voor het maken van voorzieningen voor de fietser." *Tijdschrift voor Verkeerstechniek* 24, no. 5 (1973): 239-42.

Osse, Paulien. "'Beroepskrachten vind ik een gevaarlijk element'." *De Vogelvrije Fietser* 6, no. 1 (1980): 4-5.

Osse, Paulien. "De lange mars door de instituties." *De Vogelvrije Fietser* 3, no. 5 (1978): 8-9.

Osse, Paulien. "'Ieder van ons kan zich voor de geest halen hoe het fietsparadijs eruitziet'." *De Vogelvrije Fietser* 3, no. 4 (1978): 9.

Osse, Paulien, and Theo de Jong. "'Als je in de commissie zit, verwacht de gemeente dat je geen harde acties meer voert'." *Vogelvrije Fietser* 5, no. 4 (1980): 18-19.

Ouwerkerk, Hendrik J. van. "Fietspaden of RONA-banen? Norm over de schreef." *Natuur en milieu* 1979, no. 7-8 (1979): 18-23.

Oversteegen, George. "Fietspaden zijn veel te smal." *Utrechtsch Nieuwsblad*, May 16, 1956, 7.

Peppink, H.J. "Aparte rijwielpaden ... òf: wielrijders op den hoofdweg?" *Wegen* 5, no. 9 (1929): 220-22.

[Peppink, H.J.] "Er zijn 43.297.587 automobielen op de wereld." *De Kampioen* 56, no. 17 (1939): 319-20.

Ploeger, Jan. "De komende tien jaar zouden wel eens oogstjaren kunnen worden." *De Vogelvrije Fietser* 11, no. 1 (1985): 6-7.

Ploeger, Jan. "Eindelijk normen voor onderhoud fietspaden." *De Vogelvrije Fietser* 12, no. 2 (1986/87): 18-19.

Ploeger, Jan. "Tekenen voor de fiets." *Verkeerskunde* 44, no. 11 (1993): 22-25.

Pouderoyen, G. "Van de lezer voor de lezers: De bromfietser heeft 't gedaan." *Bromfietskampioen* 5, no. 9 (1955): 661.

Raukema, Anne-Mariken. "Meter voor meter fietspaden bevechten." *Op eigen kracht: Blad van de afdeling Amsterdam van de Fietsersbond*, no. 77 (2009): 12-13.

Ringlever, Dick. "De deskundige is de fietser zélf." *Trouw*, October 18, 1980.

[Rivière], R. [P.F.J.H.] van de. "De fiets in het nauw." *Provinciale Geldersche en Nijmeegsche Courant*, September 9, 1923, 1.
Rodenburg, R.H.J. "Enkele voorbeelden van fiets- en voetgangersvoorzieningen in en rond woonwijken." *Verkeerskunde* 26, no. 5 (1975): 256-58.
Schelven, F. van, and G. Bakker. "Overzicht belangrijkste fietspadenplannen in Nederland." *Recreatievoorzieningen* 12, no. 8 (1980): 373-77.
Schenk, Niek. "De fiets als redding." *Algemeen Dagblad*, April 15, 1989.
Schenk, Niek. "Fietser wil meetellen." *Algemeen Dagblad*, May 6, 1989.
Schie, Joop van. "Stop kindermoord op de wegen." *De Volkskrant*, October 11, 1972, 11.
Schoemaker, Jenny. "Uitzicht op succes is geen voorwaarde om door te zetten." *De Vogelvrije Fietser* 11, no. 1 (1885): 9-11.
Smits, H. "'Inspraak': nog minder dan meedenken." *NRC Handelsblad*, June 12, 1971, 7.
Sniekers, Vic. "Bekende auto-journalist J. Koolhaas Revers (75 jaar): Heimwee naar motorfiets." *De Tijd*, February 19, 1966, 7.
Ster, Elly van der. "Hoe sterk is de eenzame fietser." *De Volkskrant*, March 29, 1975, 35.
Stoovelaar, F., and W. van Tilburg. "De fietsroute Den Haag (1) – verkeerstechnische aspecten." *Verkeerskunde* 30, no. 11 (1979): 544-48.
Stoovelaar, F., and W. van Tilburg. "De fietsroute Den Haag (2) – verkeerstechnische aspecten." *Verkeerskunde* 30, no. 12 (1979): 597-601.
Tilburg, W. van, and F. Stoovelaar. "De fietsroute Tilburg: verkeerstechnische aspecten." *Verkeerskunde* 28, no. 4 (1977): 163-69.
Van den Berg, Han. "50 is teveel, vindt comité." *Het Vrije Volk*, August 29, 1979.
Vries, D. de. "Over fietsen gesproken." *De Kampioen* 94, no. 12 (1979): 1158.
Werkgroep Dooievaar. "Weg fiets? Nee fietswegen!". *Tijdschrift voor Verkeerstechniek* 24, no. 11 (1973): 548-53.
Wilmink, A. "Onderzoek betere richtlijnen fietsvoorzieningen." *Verkeerskunde* 28, no. 2 (1976): 76-77.

Published Documentation of Government and Action Groups

Een halve eeuw A.N.W.B. 1883-1933. Gedenkboek, uitgegeven ter gelegenheid van het 50-jarig jubileum van den A.N.W.B. – Toeristenbond voor Nederland, en aan dezen Bond en zijn bestuurders aangeboden door J. Jonker Roelants, Mr. H.A.M. Roelants en H.L. Jonker Roelants. Schiedam: Roelants, 1934.
Eindrapport Masterplan Fiets. Samenvatting, evaluatie en overzicht van de projecten in het kader van het Masterplan Fiets, 1990-1997. Den Haag: Ministerie van Verkeer en Waterstaat, Directoraat-Generaal Personenvervoer, 1998.
"Maak van het fietspad geen moordkuil!": Protest van de Fietsersbond ENFB tegen de bezuinigingen op het onderhoud van fietsverbindingen. Woerden: Fietsersbond ENFB, 1984.
MIRT Overzicht. Meerjarenprogramma Infrastructuur, Ruimte en Transport 2019. Den Haag: Ministerie van Infrastructuur en Waterstaat, mede namens het Ministerie van Economische Zaken en Klimaat, het Ministerie van Binnenlandse Zaken en Koninkrijksrelaties, 2019.
Onderhoud fietspaden: Voorlopige normen en richtlijnen. Mededeling van de werkgroep R4 'Onderhoudsnormen van fietspaden' van het Studie Centrum Wegenbouw en het Studiecentrum Verkeerstechniek. Studiecentrum Wegenbouw/Studiecentrum Verkeerstechniek, 1986.
Staatsblad van het Koningrijk der Nederlanden, 1953, no. 301-399, 01-01-1953. 's-Gravenhage: Algemeene Landsdrukkerij 's-Gravenhage, 1953.

Veertig jaar. A.N.W.B., Toeristenbond voor Nederland, 1883-1923. S.l.: S.n., 1923.
Verslag betreffende de verkeerswaarnemingen in 1926. I. Rijkswegen. 's-Gravenhage: Algemeene Landsdrukkerij, 1928.
Verslag betreffende de verkeerswaarnemingen in 1932. I. Rijkswegen. 's-Gravenhage: Algemeene Landsdrukkerij, 1934.
Verslag betreffende verkeerswaarnemingen in 1923 op de rijkswegen in Nederland. 's-Gravenhage: Algemeene Landsdrukkerij, 1925.

Bike-minded Design Consulting, M. Lopes Cardozo, and M. Bielderman. *Voorbeeldenboek fiets: toewerken naar het structureler meenemen van de fiets in rijksprojecten*. s.l.: Ministerie van Infrastructuur & Milieu, Rijkswaterstaat, 2020.
Broek, Jan van den. "Op de fiets de provincie in!": Fietsersbond ENFB, 1990.
Campagne, Niek, ed. *Fietsbewegwijzering: Welke kant op?* Woerden: Fietsersbond ENFB, 1985.
Centraal Bureau voor de Statistiek. *Statistisch zakboek 1974*. 's-Gravenhage: Trio, 1974.
Dienst Gemeentewerken 's-Gravenhage. *Demonstratie-fietsroute Den Haag 1975-1979*. Den Haag: Dienst Gemeentewerken 's-Gravenhage, 1978.
Fietsersbond Landelijk Bureau & afdelingen Noord-Holland,. *Top 10 knelpunten op de hoofdfietsroutes per regio in Noord-Holland*. Haarlem: Fietsersbond, 2008.
Godefrooij, Tom. *Snelheid remmen: Over weginrichting en aangepaste rijsnelheden*. Woerden/Amsterdam/Den Haag: Fietsersbond ENFB/Stop de Kindermoord/Vereniging Bescherming Voetgangers, 1982.
Guit, André. *Hoe krijg ik mijn fiets terug? Het nationale plan van de fietsersbond ENFB ter bestrijding van fietsdiefstal*. Woerden: Fietsersbond ENFB, 1984.
Gurp, P.B. van. "De (brom)fietser: zijn plaats in de plannen en planvorming." In *Congres fiets- en bromfietsvoorzieningen Tilburg 1977*, 18-20. Den Haag: ANWB, 1978.
Holland, Henk, Reinoud de Haan, and Jac de Jonge. *Naar een veilig fietsverkeer op het platteland: Plattelands wegen zónder sluipverkeer, mét snelheidsbeperking*. Woerden: Fietsersbond ENFB, 1993.
Kalk, E. "Communicatie tussen burgerij en overheid." In *Congresverslag: de veiligheid van voetgangers, fietsers en bromfietsers binnen de bebouwde kom. Verslag van het Eerste Nationale Verkeersveiligheidscongres, gehouden op 19 en 20 april 1978 in het RAI-congrescentrum te Amsterdam*, edited by Stichting Wetenschappelijk Onderzoek Verkeersveiligheid SWOV, 40-41. 's-Gravenhage: Koninklijke Nederlandse Toeristenbond ANWB, 1978.
Kalk, E. "Verkeerd verkeer en ruimtelijke wanorde." In *Milieu-aktie in Nederland*, edited by E. Tellegen and J. Willems, 62-77. Amsterdam: De Trommel/Vereniging Milieudefensie, 1978.
Kennisinstituut voor Mobiliteitsbeleid. *Fietsen en lopen: de smeerolie van onze mobiliteit*. Den Haag: Ministerie van Verkeer en Waterstaat, Kennisinstuut voor Mobiliteitsbeleid, 2015.
Kennisinstuut voor Mobiliteitsbeleid. *Mobiliteitsbeeld 2019*. Den Haag: Ministerie van Infrastructuur en Waterstaat, Kennisinstutuut voor Mobiliteitsbeleid, 2019.
Koninklijke Nederlandsche Toeristenbond ANWB. *Congres fiets- en bromfietsvoorzieningen Tilburg 1977*. Den Haag: Koninklijke Nederlandsche Toeristenbond ANWB, 1978.
Koninklijke Nederlandsche Toeristenbond ANWB. *Fietspaden en -oversteekplaatsen*. [Den Haag]: Koninklijke Nederlandsche Toeristenbond ANWB, 1966.
Koninklijke Nederlandsche Toeristenbond ANWB. *Verkeersstructuurplan en dwarsprofielen van stadsstraten*. Verkeersmemorandum, vol. 16. [Den Haag]: Koninklijke Nederlandsche Toeristenbond ANWB, 1961.
Ligtermoet, Dirk, Tonny Bosch, Edo Okkema, André Pettinga, and Henk Tromp. *Gemeentelijk fietsbeleid: stand van zaken*. Rotterdam: Ministerie van Verkeer en Waterstaat, Directoraat Generaal Rijkswaterstaat, Adviesdienst Verkeer en Vervoer, Hoofdafdeling Vervoer en Maatschappelijke Ontwikkelingen, Afdeling Personenvervoer, 1997.

Ligtermoet, Dirk Maarten. *Beleid en planning in de wegenbouw: De relatie tussen beleidsvorming en planning in de geschiedenis van de aanleg en verbetering van rijkswegen* Rijkswaterstaat-serie, vol. 51. Den Haag: Rijkswaterstaat, 1990.

Linden, Piet van der. *Knelpunten in het hoofdfietsnetwerk in Noord-Holland 2016*. Utrecht: Fietsersbond, 2016.

Lunsingh Tonckens, N.A. *Bekostiging van wegen*. Hengelo: H.L. Smit & Zn., 1929.

Ministry of Transport, Public Works and Water Management: Directorate-General for Passenger Transport. *The Dutch Bicycle Master Plan: Description and Evaluation in an Historical Context*. The Hague: Ministry of Transport, Public Works and Water Management: Directorate-General for Passenger Transport, 1999.

Pieters, Ingrid. *Mee op de fiets, van baby tot kleuter*. Woerden: Fietsersbond ENFB, 1992.

Ploeger, Jan. *Fiets en trein op één lijn: Een ENFB-schrift over de integratie van langzaam verkeer en openbaar vervoer*. Woerden: Fietsersbond ENFB, 1982.

Ploeger, Jan. *Wegwijzer fietsvoorzieningen: Uitgangspunten en planvorming*. Ede: Stichting CROW, 1990.

Poelstra, H. "De (brom)fietser: zijn plaats in de plannen en planvorming. 4e preadvies: Plannen maken voor de fiets; enkele kanttekeningen." In *Congres fiets- en bromfietsvoorzieningen Tilburg 1977*, 27-28. Den Haag: ANWB, 1978.

Post, Menno. *Decentralisatie van het fietsbeleid: Stuurloos naar de toekomst?* Amsterdam: Adviesburo voor Mobiliteit, 1996.

Raad voor Verkeer en Waterstaat: Commissie Personenvervoer & Commissie Verkeersinfrastructuur,. *Advies Rijk op de fiets: advies over de rol van de rijksoverheid in het fietsbeleid*. Den Haag: Raad voor Verkeer en Waterstaat, 1995.

Reitsma, S.A. *De rijwielbelasting als bestemmingsheffing voor het motorsnelverkeer*. Assen: Van Gorcum, 1938.

Reitsma, S.A. *De weg naar coördinatie van het verkeer in Nederland*. Den Haag: Moorman's Periodieke Pers, 1933.

Reitsma, S.A. *Herwaardeering van verkeerseconomische waarden*. Den Haag: Moorman's Periodieke Pers, 1942.

Reitsma, S.A. *Laat den weg vrij! Een ernstig woord tot het denkende deel der natie*. Den Haag Moorman's Periodieke Pers, 1933.

Reitsma, S.A. *Open brief aan de leden der Tweede Kamer van de Staten-Generaal n.a.v. den Nederlandschen verkeerschaos*. Utrecht: Drukkerij J. van Broekhoven, 1932.

Rijksinstituut voor Volksgezondheid en Milieuhygiene. *Zorgen voor morgen: Nationale Milieuverkenning 1985-2010*, edited by F. Langeweg. Alphen aan den Rijn: Samson H.D. Tjeenk Willink, 1988.

Rijkswaterstaat: Centrale Dienst voor de Wegen en de Bruggen: Bureau voor de Wegen- en Verkeersstatistiek. *Atlas van de Rijkswegen 1949*. 's-Gravenhage: Rijkswaterstaat, 1950.

Rijkswaterstaat Directie Algemene Dienst. *Atlas Rijkswegen 1960*. 's-Gravenhage: Staatsdrukkerij- en Uitgeverijbedrijf, 1962.

Schaap, D. *Een eeuw wijzer 1883-1983. 100 jaar Koninklijke Nederlandse Toeristenbond ANWB*. Den Haag: ANWB, 1983.

Schouten, Marian. *Veilig schoolverkeer: Een doe-het-zelf-zaak: een handleiding voor ouders die het verkeer rond school veiliger willen maken*. Woerden: Fietsersbond ENFB, 1996.

Slop, M. "De (brom)fietser: zijn plaats in de plannen en planvorming. 5e preadvies: Waarom niet meer groene golven voor (brom)fietsers?" In *Congres fiets- en bromfietsvoorzieningen Tilburg 1977*, 30. Den Haag: ANWB, 1978.

Snelderwaard, Ad, Eric Nijland, and Anita Bakker, eds. *Landelijk Fietsplatform 1987-2012: 25 jaar in beeld*. Amersfoort: Stichting Landelijk Fietsplatform, 2012.

Stichting: fiets!, *Fit door fietsen: hoe fiets je voor je gezondheid?: wat zegt de dokter?: conditietraining, trimmen, trainingsschema's.* Amsterdam: Stichting: fiets!, 1974.

Stichting: fiets!, *Meer fietsen ... meer fietspaden: Over de noodzaak van het aanleggen van meer fietspaden voor verkeersveiligheid en recreatie* Amsterdam: Stichting: fiets!, 1967.

Stichting: fiets!, *Leefruimte voor de fiets: Over de noodzaak van het treffen van voorzieningen met betrekking tot het stallen en parkeren van fietsen.* Amsterdam: Stichting: fiets!, 1969.

Stichting: fiets!, *Optimale aandacht voor fietsers: 27 stellingen over voorzieningen voor het fietsgebruik.* Amsterdam: Stichting: fiets!, 1972.

Stichting Wetenschappelijk Onderzoek Verkeersveiligheid SWOV. *De bromfietser en de verkeersveiligheid: Een beschrijving van de groep bromfietsbezitters en van de onveiligheid van bromfietsers* Voorburg: SWOV, 1973.

Stichting Wetenschappelijk Onderzoek Verkeersveiligheid SWOV. *Snorfiets, veilig of niet? Te verwachten consequenties van de eventuele invoering van de snorfiets voor de verkeersveiligheid.* Voorburg: SWOV, 1976.

Stop de Kindermoord. *Handleiding voor Inspraak en Aktie.* Amsterdam: Pressiegroep Stop de Kindermoord, 1975.

United Nations. "Convention on Road Traffic." Geneva: United Nations, 1949.

United Nations. "Convention on Road Traffic." Vienna: United Nations, 1968.

Verein Deutscher Fahrrad-Industrieller. *Festschrift zum Vierzigjährigen Bestehen des Vereins Deutscher Fahrrad-Industrieller e.V. 1888-1928.* Berlin: Verein Deutscher Fahrrad-Industrieller, 1928.

Vereeniging Het Nederlandsche Wegen-Congres. *Ontwerp voor eene wegbelastingwet.* [Den Haag]: Haagsche Drukkerij, 1924.

Vereniging Het Nederlandsche Wegencongres. *Prae-adviezen te behandelen op de Congresdag 1908 [van de] Vereniging Het Nederlandsche Wegencongres [te] Utrecht, 19 december 1958: Het Nederlandse Wegennet in verband met de toekomstige verkeersbehoeften.* 's-Gravenhage: Nederlandsche Wegencongres, 1958.

Vries, F. de, and S.A. Reitsma. *Twee lezingen over het verkeerswezen.* Arnhem: G.W. van der Wiel & Co., 1933.

Vries, Marten de, and Tom Godefrooij. *Fietsen in winkelgebieden: Een pleidooi voor het toelaten van fietsers in autovrije winkelgebieden.* Woerden: Fietsersbond ENFB, 1990.

Wal, L.T. van der. *Het wegennet en de wegenverbetering in de provincie Zuid-Holland.* Middelharnis: Flakkeesche Boek- en Handelsdrukkerij, 1916.

Welleman, Ton. *The Dutch Bicycle Master Plan: Description and evaluation in an historical context.* The Hague: Ministry of Transport, Public Works and Water Management, Directorate-General for Passenger Transport, 1999.

Werkgroep 2000 in association with Werkgroep Dooievaar. "Fietser... kop op!". *Katernen 2000* 1974, no. 9-10 (1974).

Scholarly Publications

Aerts, Remieg, Herman de Liagre Böhl, Piet de Rooy and Henk te Velde. *Land van kleine gebaren: Een politieke geschiedenis van Nederland 1780-1990.* Nijmegen/Amsterdam: Uitgeverij SUN, 1999.

Albert de la Bruhèze, Adri A. "Enschede: An Experiment in Cycling." In *Cycling Cities: The European Experience. Hundred Years of Policy and Practice,* edited by Ruth Oldenziel, Martin Emanuel, Adri Albert de la Bruhèze and Frank Veraart, 41-51. Eindhoven: Foundation for the History of Technology, 2016.

Albert de la Bruhèze, Adri A., and Martin Emanuel. "Hannover: Sidelining the Bicyclist." In *Cycling Cities: The European Experience. A Hundred Years of Policy and Practice*, edited by Ruth Oldenziel, Martin Emanuel, Adri Albert de la Bruhèze and Frank Veraart, 113-23. Eindhoven: Foundation for the History of Technology, 2016.

Albert de la Bruhèze, Adri A., and Ruth Oldenziel. *Cycling Cities: The Munich Experience*. Eindhoven: Foundation for the History of Technology, 2018.

Albert de la Bruhèze, Adri A., and Ruth Oldenziel. *Manufacturing Technology, Manufacturing Consumer: The Making of Dutch Consumer Society*. Amsterdam: Aksant, 2009.

Albert de la Bruhèze, Adri A., and Ruth Oldenziel. "Who Pays, Who Benefits? Bicycle Taxes as Policy Tool, 1890-2012." In *Cycling and Recycling: Histories of Sustainable Practices* edited by Ruth Oldenziel and Helmuth Trischler, 73-100. New York/London: Berghahn, 2015.

Albert de la Bruhèze, Adri A., and Frank Veraart. *Fietsverkeer in praktijk en beleid in de twintigste eeuw: overeenkomsten en verschillen in fietsgebruik in Amsterdam, Eindhoven, Enschede, Zuidoost-Limburg, Antwerpen, Manchester, Kopenhagen, Hannover er en Basel*. Den Haag: Ministerie van Verkeer en Waterstaat, Directoraat-Generaal Rijkswaterstaat, Hoofdkantoor van de Waterstaat, Directie Kennis/ Stichting Historie der Techniek, 1999.

Alexandrakis, Julian. "Cycling Towards Sustainability: The Transformative Potential of Urban Design Thinking in a Sustainable Living Lab." *Transportation Research Interdisciplinary Perspectives* 9 (2021): 100269.

Altena, H.W. van, and B. Rakic. *GDU subsidies: Inventarisatie van probleemgebieden en maatregelcategorieën, nu en in de toekomst*. Veenendaal: Traffic Test, 2001.

Andeweg, Rudy B. "Consociational Democracy." *Annual Review of Political Science* 3 (2000): 509-36.

Andeweg, Rudy B., and Galen A. Irwin. *Governance and Politics of the Netherlands*. 2nd ed. Basingstoke: Palgrave Macmillan, 2005.

Arnold, David. "The Problem of Traffic: The Street-Life of Modernity in Late-Colonial India." *Modern Asian Studies* 46, no. 1 (2012): 119-41.

Arnold, David, and Erich DeWald. "Cycles of Empowerment? The Bicycle and Everyday Technology in Colonial India and Vietnam." *Comparative Studies in Society and History* 53, no. 4 (2011): 971-96.

Arnouts, Rikke. "Regional Nature Governance in the Netherlands: Four Decades of Governance Modes and Shifts in the Utrechtse Heuvelrug and Midden-Brabant." PhD diss., Wageningen University, 2010.

Arnouts, Rikke, Mariëlle van der Zouwen, and Bas Arts. "Analysing Governance Modes and Shifts – Governance Arrangements in Dutch Nature Policy." *Forest Policy and Economics* 16 (2012): 43-50.

Arts, Bas, and Pieter Leroy, eds. *Institutional Dynamics in Environmental Governance*, Environment & Policy, vol. 47. Dordrecht: Springer, 2006.

Arts, Bas, Jan van Tatenhove, and Pieter Leroy. "Policy Arrangements." In *Political Modernisation and the Environment: The Renewal of Environmental Policy Arrangements*, edited by Jan van Tatenhove, Bas Arts and Pieter Leroy. Environment & Policy, vol. 24, 53-69. Dordrecht: Kluwer, 2000.

Assmann, Carla. "The Emergence of the Car-Oriented City: Entanglements and Transfer Agents in West-Berlin, East-Berlin and Lyon, 1945-75." *The Journal of Transport History* 41, no. 3 (2020): 328-25.

Balkmar, Dag. "Cycling Politics: Imagining Sustainable Cycling Futures in Sweden." *Applied Mobilities* 5, no. 3 (2020): 324-40.

Balkmar, Dag, and Jane Summerton. "Contested Mobilities: Politics, Strategies and Visions in Swedish Bicycle Activism." *Applied Mobilities* 2, no. 2 (2017): 151-65.

Bax, Charlotte, Pieter Leroy, and Marjan P. Hagenzieker. "Road Safety Knowledge and Policy: A Historical Institutional Analysis of the Netherlands." *Transportation Research Part F* 25 (2014): 127-36.

Beaujon, Otto. "De Franse fietsbelasting." *De oude fiets* 1997, no. 3 (1997): 4-6.

Berg, Caspar F. van den. "Dynamics in the Dutch Policy Advisory System: Externalization, Politicization and the Legacy of Pillarization." *Policy Sciences* 50, no. 1 (2017): 63-84.

Berkers, Eric. "Fietsgebruik en -beleid in Maastricht en Parkstad in historisch perspectief." Eindhoven: TU Eindhoven/Stichting Historie der Techniek, 2017.

Berkers, Eric. *Technocraten en bureaucraten: Ontwikkeling van organisatie en personeel van de Rijkswaterstaat, 1848-1930*. Zaltbommel: Europese Bibliotheek, 2002.

Berkers, Eric, Frans Botma, and Ruth Oldenziel. *Cycling Cities: The Hague Experience*. Eindhoven: Foundation for the History of Technology, 2018.

Berkers, Eric, and Ruth Oldenziel. *Cycling Cities: The Arnhem and Nijmegen Experience*. Eindhoven: Foundation for the History of Technology, 2017.

Berkers, Eric, Frank Schipper, Patrick Bek, and Ruth Oldenziel. *Cycling Cities: The Rotterdam Experience*. Eindhoven: Foundation for the History of Technology, 2019.

Bertho-Lavenir, Catherine. "How the Motor Car Conquered the Road." In *Cultures of Control*, edited by Miriam R. Levin, 113-34. Amsterdam: Harwood Academic Publishers, 2000.

Bevir, Mark, and R.A.W. Rhodes. *The State as Cultural Practice*. Oxford: Oxford University Press, 2010.

Bianconi, Marco, and Mark Tewdwr-Jones. "The Form and Organisation of Urban Areas: Colin Buchanan and *Traffic in Towns* 50 Years On." *Town Planning Review* 84, no. 3 (2013): 313-36.

Bijker, Wiebe E. *Of Bicycles, Bakelites, and Bulbs: Toward a Theory of Sociotechnical Change*. Cambridge, MA: MIT Press, 1995.

Block, Greet de, David de Kool, and Bruno de Meulder. "Paradise Regained? Crossing Borders Between Planning Concepts in the Netherlands and Belgium (1830-2012)." In *Builders and Planners: A History of Land-Use and Infrastructue Planning in the Netherlands*, edited by Jos Arts, Ruud Filarski, Hans Jeekel and Bert Toussaint, 17-67. Delft: Uitgeverij Eburon, 2016.

Blomkvist, Pär, and Martin Emanuel. "Regulating Innovation: Utility vs. Leisure in Swedish Moped History, 1952-1961." *Technology and Culture* 61, no. 3 (2020): 815-42.

Bogaard, A. van den. "De geplande stad, 1914-1945." In *Techniek in Nederland in de twintigste eeuw. Deel 6. Stad, bouw, industriële productie*, edited by J.W. Schot, H.W. Lintsen, A. Rip and Adri A. Albert de la Bruhèze, 51-73. Eindhoven: Stichting Historie der Techniek, 2003.

Bonham, Jennifer, and Peter Cox. "The Disruptive Traveller? A Foucauldian Analysis of Cycleways." *Road and Transport Research* 19, no. 2 (2010): 42-54.

Bontje, Marco. "A 'Planner's Paradise' Lost? Past, Present and Future of Dutch National Urbanization Policy." *European Urban and Regional Studies* 10, no. 2 (2003): 135-51.

Boot, O.J., and J. Ploeger. *De economische waarde van het fietsverkeer in het bijzonder met motief onderwijs: Eindrapportage*. Den Haag: Adviesbureau Van Roon, 1987.

Borrás, Susana, and Jakob Edler. "Introduction: On Governance, Systems and Change." In *The Governance of Socio-Technical Systems: Explaining Change*, edited by Susana Borrás and Jakob Edler, 1-22. Cheltenham: Edward Elgar Publishing, 2014.

Bosch, A., W. van der Ham, E. Berkers, M.L. ten Horn-Van Nispen, W.H.P.M. van Hooff, H.W. Lintsen (editor), H.C. Toussaint, and G.P. van de Ven. *Twee eeuwen Rijkswaterstaat: 1789-2015*. 2nd revised and expanded ed. Asten: Nieuwe Uitgevers, 2015.

Bosma, Jitske, J. Bosma, H. Bijnen, S. Davidson, J. Eissens, K. van Harn, and T. Nijenhuis. *"... de beste aktiegroep ter wereld ...": 40 dorpsverhalen uit de Nieuwmarkt* Amsterdam: Stichting Uitgeverij De Oude Stadt, 1984.

Briese, Volker. *Besondere Wege für Radfahrer. Zur Geschichte des Radwegebaus in Deutschland von den Anfangen bis 1940*. Unpublished manuscript, Paderborn, 1993.

Brouwer, Petra. *Van stad naar stedelijkheid: Planning en planconceptie van Lelystad en Almere 1959-1974*. Rotterdam: NAi Uitgevers, 1997.

Bruno, Matthew. "The Challenge of the Bicycle Street: Applying Collaborative Governance Processes while Protecting User Centered Innovations." *Transportation Research Interdisciplinary Perspectives* 7 (2020).

Bruno, Matthew, Henk-Jan Dekker, and Letícia Lemos Lindenberg. "Mobility protests in the Netherlands of the 1970s: Activism, innovation, and transitions," *Environmental Innovation and Societal Transitions* (forthcoming).

Bruno, Matthew, and Anna Nikolaeva. "Towards a Maintenance-Based Approach to Mode Shift: Comparing Two Cases of Dutch Cycling Policy Using Social Practice Theory." *Journal of Transport Geography* 86 (2020).

Buchanan, Colin. *Traffic in Towns: A Study of the Long Term Problems of Traffic in Urban Areas*. London: HM Stationery Office, 1963.

Buiter, Hans. "Als een spin in het web: de ANWB en de Nederlandse rijwielpadenverenigingen." *Tijdschrift voor historische geografie* 1, no. 2 (2016): 76-86.

Buiter, Hans. "'Hoogviadukten in het polderland?' De introductie van de autosnelweg in Nederland." *NEHA-jaarboek voor economische, bedrijfs- en techniekgeschiedenis* 60 (1997): 285-306.

Buiter, Hans. "Utrecht: Bicycles Rule – Again." In *Cycling Cities: The European Experience. Hundred Years of Policy and Practice*, edited by Ruth Oldenziel, Martin Emanuel, Adri Albert de la Bruhèze and Frank Veraart, 29-39. Eindhoven: Foundation for the History of Technology, 2016.

Buiter, Hans, and Peter Staal. *Het avontuur van de ANWB: 135 jaar onderweg*. Bussum: Uitgeverij THOTH, 2018.

Buiter, Hans, and Kees Volkers. *Oudenrijn: geschiedenis van een verkeersknooppunt*. Utrecht: Matrijs, 1996.

Buizer, Marleen. "Worlds Apart: Interactions Between Local Policy Initiatives and Established Policy." PhD diss., Wageningen University, 2008.

Burg, Arjen van der, Odette van de Riet, and Bert Toussaint. "Contested Spatial Planning: Ca. 1965-1985." In *Builders and Planners: A History of Land-Use and Infrastructure Planning in the Netherlands*, edited by Jos Arts, Ruud Filarski, Hans Jeekel and Bert Toussaint, 209-69. Delft: Uitgeverij Eburon, 2016.

Burri, Monika. *Das Fahrrad: Wegbereiter oder überrolltes Leitbild?: eine Fussnote zur Technikgeschichte des Automobils*. Zürich: Eidgenössische Technische Hochschule, Institut für Geschichte, Technikgeschichte, 1998.

Callon, Michel. "The Role of Lay People in the Production and Dissemination of Scientific Knowledge." *Science, Technology and Society* 4, no. 1 (1999): 81-94.

Callon, Michel. "Society in the Making: The Study of Technology as a Tool for Sociological Analysis." In *The Social Construction of Technological Systems: New Directions in the Sociology and History of Technology*, edited by Wiebe E. Bijker, Thomas P. Hughes and Trevor Pinch, 83-103. Cambridge, MA: MIT Press, 1987.

Cammen, Hans van der, and Len de Klerk. *Ruimtelijke ordening: van grachtengordel tot Vinex-wijk*. Utrecht: Spectrum, 2003.

Caracciolo, Carlos Héctor. "Bicicleta, circulación vial y espacio público en la Italia Fascista." *Historia Critica* 39 (2009): 20-42.

Carstensen, T.A., A.S. Olafsson, N.M. Bech, T.S. Poulsen, and C. Zhao. "The Spatio-Temporal Development of Copenhagen's Bicycle Infrastructure, 1912-2013." *Geografisk Tidsskrift-Danish Journal of Geography* 115, no. 2 (2015): 142-56.

Carstensen, Trine Agervig, and Anne-Katrin Ebert. "Cycling Cultures in Northern Europe: From 'Golden Age' to 'Renaissance'." In *Cycling and Sustainability*, edited by John Parkin, 23-58. Bingley: Emerald, 2012.

Conekin, Becky E. "Fashioning Mod Twiggy and the moped in 'swinging' London." *History and Technology* 28, no. 2 (2012): 209-15.

Cox, Peter. "'A Denial of our Boasted Civilization': Cyclists' View on Conflicts over Road Use in Britain, 1926-1935." *Transfers* 2, no. 3 (2012): 4-30.

Cox, Peter, and Till Koglin, eds. *The Politics of Cycling Infrastructure: Spaces and (In)Equality*, Bristol. Polity Press, 2020.

Cramer, Jacqueline. *De groene golf: geschiedenis en toekomst van de Nederlandse milieubeweging* Utrecht: Van Arkel, 1989.

Dauncey, Hugh. *French Cycling: A Social and Cultural History*. Contemporary French and Francophone Cultures, vol. 23. Liverpool: Liverpool University Press, 2012.

Dauncey, Hugh, and Geoff Hare, eds. *The Tour de France 1903-2003: A Century of Sporting Structures, Meanings and Values*. London: Routledge/Cass, 2003.

Dauncey, Hugh, and Geoff Hare. "The Tour de France: a Pre-Modern Contest in a Post-Modern Context." *The International Journal of the History of Sport* 20, no. 2 (2003): 1-29.

David, Paul A. "Clio and the Economics of QWERTY." *American Economic Review* 75, no. 2 (1985): 332-37.

De Jong, Wim, and Bas Van Kleef. *De Puch-story en andere brommerverhalen*. Bussum: Uitgeverij Thoth, 1994.

De Liagre Böhl, Herman. *Amsterdam op de helling: De strijd om stadsvernieuwing*. Amsterdam: Boom, 2010.

Dekker, Henk-Jan. "An Accident of History? How Mopeds Boosted Dutch Cycling Infrastructure (1950-1970)," *Journal of Transport History* (forthcoming).

Dekker, Henk-Jan. "Jeremiades over de Jeremiebrug: Fietsfiles bij spoorwegovergangen in Utrecht, 1915-1940." *Oud-Utrecht* 91, no. 2 (2018): 32-38.

Derksen, J.B.D., and A. Rombouts. "The Demand for Bicycles in the Netherlands." *Econometrica* 5, no. 3 (1937): 295-300.

Disco, Nil. "De bewogen stad, 1945-1980." In *Techniek in Nederland in de twintigste eeuw. Deel 6. Stad, bouw, industriële productie.*, edited by J.W. Schot, H.W. Lintsen, A. Rip and A. Albert de la Bruhèze, 74-99. Eindhoven: Stichting Historie der Techniek, 2003.

Disco, Nil. "Made in Delft: Professional Engineering in the Netherlands 1880-1940." PhD diss., University of Amsterdam, 1990.

Disco, Nil, and Harry Lintsen. "De vervlechting van ingenieursberoep en industrie: 1890-1925." *Tijdschrift voor Sociale Geschiedenis* 9 (1983): 343-69.

Divall, Colin. "Mobilizing the History of Technology." *Technology and Culture* 51, no. 4 (2010): 938-60.

Divall, Colin Michael, Peter Lyth, and Gijs Mom. "Towards a Paradigm Shift? A Decade of Transport and Mobility History." In *Mobility in History: The State of the Art in the History of Transport, Traffic and Mobility*, edited by Gijs Mom, Gordon Pirie and Laurent Tissot, 13-40. Neuchatel: Alphil, 2009.

Dokkum, J.D.C. van, and D.J. van der Ven. *Ons mooie Nederland: Het Gooi*. Amsterdam: J.M. Meulenhoff, 1921.

Duizer, Bram. *"In het nut van actie moet je geloven": Dertig jaar actievoeren door de Fietsersbond.* Utrecht: Wetenschapswinkel Letteren Universiteit Utrecht, 2005.

Duyn, Roel van. *Provo. De geschiedenis van de provotarische beweging 1965-1967.* Amsterdam: Meulenhoff, 1985.

Duyvendak, J.W., and R. Koopmans. "Protest in een pacificatiedemocratie. Nieuwe sociale bewegingen en het Nederlandse politieke systeem." In *Tussen verbeelding en macht: 25 jaar nieuwe sociale bewegingen in Nederland*, edited by J.W. Duyvendak, H.A. van der Heijden, R. Koopmans and L. Wijmans, 39-57. Amsterdam: Sua, 1992.

Ebert, Anne-Katrin. "Cycling Towards the Nation: The Use of the Bicycle in Germany and the Netherlands, 1880-1940." *European Review of History: Revue europeenne d'histoire* 11, no. 3 (2004): 347-64.

Ebert, Anne-Katrin. *Radelnde Nationen: Die Geschichte des Fahrrads in Deutschland und den Niederlanden bis 1940.* Frankfurt: Campus Verlag, 2010.

Ebert, Anne-Katrin. "Review of: Ruth Oldenziel, Martin Emanuel, Adri Albert de la Bruhèze, Frank Veraart (eds), Cycling Cities: The European Experience." *The Journal of Transport History* 39, no. 1 (2018): 128-30.

Ebert, Anne-Katrin. "When Cycling Gets Political: Building Cycling Paths in Germany and the Netherlands, 1910-40." *Journal of Transport History* 33, no. 1 (2012): 115-37.

Edgerton, David. *The Shock of the Old: Technology and Global History since 1900.* London: Profile Books, 2006.

Emanuel, Martin. "Constructing the Cyclist: Ideology and Representations in Urban Traffic Planning in Stockholm, 1930-70." *The Journal of Transport History* 33, no. 1 (2012): 67-91.

Emanuel, Martin. "Controlling walking in Stockholm during the inter-war period." *Urban History* 48, no. 2 (2021): 248-265.

Emanuel, Martin. "Copenhagen: Branding the Cycling City." In *Cycling Cities: The European Experience. A Hundred Years of Policy and Practice*, edited by Ruth Oldenziel, Martin Emanuel, Adri Albert de la Bruhèze and Frank Veraart, 89-99. Eindhoven: Foundation for the History of Technology, 2016.

Emanuel, Martin. "Making a Bicycle City: Infrastructure and Cycling in Copenhagen since 1880." *Urban History* 46, no. 3 (2019): 493-517.

Emanuel, Martin. "Malmö: A Center of Cycling Innovation." In *Cycling Cities: The European Experience. A Hundred Years of Policy and Practice*, edited by Ruth Oldenziel, Martin Emanuel, Adri Albert de la Bruhèze and Frank Veraart, 137-47. Eindhoven: Foundation for the History of Technology, 2016.

Emanuel, Martin. "Moments of Unsustainability: Planning, Path Dependence, and Cycling in Stockholm." In *Cycling and Recycling: Histories of Sustainable Practices*, edited by Ruth Oldenziel and Helmuth Trischler, 101-21. New York/Oxford: Berghahn, 2016.

Emanuel, Martin. "Stockholm: Where Public Transit Eclipses Cycling." In *Cycling Cities: The European Experience. A Hundred Years of Policy and Practice*, edited by Ruth Oldenziel, Martin Emanuel, Adri Albert de la Bruhèze and Frank Veraart, 149-59. Eindhoven: Foundation for the History of Technology, 2016.

Emanuel, Martin, Frank Schipper, and Ruth Oldenziel, eds. *A U-Turn to the Future: Sustainable Urban Mobility since 1850.* New York: Berghahn Books, 2020.

Emanuel, Martin, Frank Veraart, and Peter Cox. "Manchester: Cycling at a Standstill." In *Cycling Cities: The European Experience. A Hundred Years of Policy and Practice*, edited by Ruth Oldenziel, Martin Emanuel, Adri Albert de la Bruhèze and Frank Veraart, 101-11. Eindhoven: Foundation for the History of Technology, 2016.

Engels, Jens Ivo. *Naturpolitik in der Bundesrepublik: Ideenwelt und politische Verhaltensstile in Naturschutz und Umweltbewegung, 1950-1980.* Paderborn: Ferdinand Schöningh, 2006.

Engler, Harald. "Social Movement and the Failure of Car-Friendly City Projects: East and West Berlin (1970s and 1980s)." *Journal of Transport History* 41, no. 3 (2020): 353-80.

Epperson, Bruce. *Bicycles in American Highway Planning: The Critical Years of Policy-Making 1969-1991*. Jefferson: McFarland & Company, 2014.

Epperson, Bruce. "The Great Schism: Federal Bicycle Safety Regulation and the Unraveling of American Bicycle Planning." *Transportation Law Journal* 37, no. 2 (2010): 73-118.

Epstein, Steven. "The Construction of Lay Expertise: AIDS Activism and the Forging of Credibility in the Reform of Clinical Trials." *Science, Technology and Human Values* 20, no. 4 (1995): 408-37.

Ewalds, Doreen, Ger Moritz, and Michel Sijstermans. *Bromfietsen in Nederland*. Den Haag/ Heerlen: Centraal Bureau voor Statistiek, 2013.

Faludi, Andreas, and Arnold van der Valk. *Rule and Order: Dutch Planning Doctrine in the Twentieth Century*. GeoJournal Library, vol. 28. Dordrecht: Springer, 1994.

Feddes, Fred, and Marjolein De Lange. *Fietsstad Amsterdam: Hoe Amsterdam de fietshoofdstad van de wereld werd*. Amsterdam: Uitgeverij Bas Lubberhuizen, 2019.

Filarski, Ruud. "The Emergence of the Bus Industry: Dutch Transport Policy During the Interwar Years." *Transfers* 1, no. 2 (2011): 61-82.

Filarski, Ruud, in cooperation with Gijs Mom. *Shaping Transport Policy: Two Centuries of Struggle between the Public and Private Sector: A Comparative Perspective*. The Hague: SDU Uitgevers, 2011.

Filarski, Ruud, and Gijs Mom. *Van transport naar mobiliteit: de transportrevolutie [1800-1900]*. Zutphen: Walburg Pers /Stichting Historie der Techniek, 2008.

Finison, Lorenz. *Boston's Cycling Craze, 1880-1900: A Story of Race, Sport, and Society*. Amherst: University of Massachusetts Press, 2014.

Fläschner, Thomas. "Stahlroß auf dem Aussterbe-Etat: zur Geschichte des Fahrrades und seiner Verdrängung in den 50er-Jahren." *Eckstein – Journal für Geschichte* 9 (2000): 4-22.

Fläschner, Thomas. "Zum Gebrauch des Rades gezwungen: Fahrradkultur in Saarbrücken zwischen 1885 und 1939." In *Saarbrücken in Fahrt: 125 Jahre Automobil an der Saar*, edited by Hans-Christian Herrmann and Ruth Bauer. Veröffentlichungen des Stadtarchivs Saarbrücken, vol. 1, 184-221. Marpingen-Alsweiler: Edition Schaumberg, 2011.

Flyvbjerg, Bent. *Rationality and Power: Democracy in Practice*. Chicago: University of Chicago Press, 1998.

Friss, Evan. *The Cycling City: Bicycles and Urban America in the 1890s*. Chicago: University of Chicago Press, 2015.

Friss, Evan. *On Bicycles: A 200-Year History of Cycling in New York City*. New York: Columbia University Press, 2019.

Furness, Zack. "Biketivism and Technology: Historical Reflections and Appropriations." *Social Epistemology* 19, no. 4 (2005): 401-17.

Furness, Zack. *One Less Car: Bicycling and the Politics of Automobility*. Philadelphia: Temple University Press, 2010.

Gaboriau, Philippe. "Les trois âges du vélo en France." *Vingtième Siècle. Revue d'histoire* 29, no. 1 (1991): 17-34.

Gaboriau, Philippe. "The Tour de France and cycling's Belle Epoque." *International Journal of the History of Sport* 20, no. 2 (2003): 57-78.

Gardon, Sébastien. "Gouverner la circulation urbaine: Des villes françaises face à l'automobile (années 1910 – années 1960)." PhD diss., Université Lumière – Lyon II, 2009.

Geyl, Pieter. *Ik die zo weinig in mijn verleden leef: Autobiografie 1887-1940*. Edited and annotated by Wim Berkelaar, Leen Dorsman and Pieter van Hees. Amsterdam: Wereldbibliotheek, 2009.

Giddens, Anthony. *The Constitution of Society: Outline of the Theory of Structuration*. Cambridge: Polity Press, 1984.

Golub, Aaron, Melody L. Hoffmann, Adonia E. Lugo, and Gerardo F. Sandoval. *Bicycle Justice and Urban Transformation: Biking for all?* London/New York: Routledge, 2017.

Gosseye, Janine. "The Janus-Faced Shopping Center: The Low Countries in Search of a Fitting Shopping Paradigm." *Journal of Urban History* 44, no. 5 (2018): 862-86.

Grapperhaus, F.H.M. *Over de loden last van het koperen fietsplaatje. De Nederlandse rijwielbelasting 1924-1941*. Franeker: Uitgeverij Van Wijnen, 2006.

Grundmann, Reiner. "The Problem of Expertise in Knowledge Societies." *Minerva* 55 (2017): 25-48.

Gunn, Simon. "The Buchanan Report, Environment and the Problem of Traffic in 1960s Britain." *Twentieth Century British History* 22, no. 4 (2011): 521-42.

Gunn, Simon. "Introduction to Traffic in Towns." In *Traffic in Towns: A Study of the Long Term Problems of Traffic in Urban Areas*, edited by Simon Gunn, vi-xiv. London: Routledge, 2015.

Gunn, Simon. "People and the Car: The Expansion of Automobility in Urban Britain, c. 1955-70." *Social History* 38, no. 2 (2013): 220-37.

Gunn, Simon. "Ring Road: Birmingham and the Collapse of the Motor City Ideal in 1970s Britain." *The Historical Journal* 61, no. 1 (2018): 227-48.

Haan, Ido de. "Verplaatste democratie? Politieke representatie in functionele organen." In *Omstreden democratie. Over de problemen van een succesverhaal*, edited by Remieg Aerts and Peter de Goede, 89-107. Amsterdam: Boom, 2013.

Haan, Ido de, Jan Willem Duyvendak, and Maarten van Bottenburg. *In het hart van de verzorgingsstaat: Het ministerie van Maatschappelijk Werk en zijn opvolgers (CRM, WVC, VWS), 1952-2002*. Zutphen: Walburg Pers, 2002.

Ham, Willem van der. *Heersen en Beheersen: Rijkswaterstaat in de twintigste eeuw*. PhD diss., Zaltbommel: Europese Bibliotheek, 1999.

Hanna, Erika. "Seeing Like a Cyclist: Visibility and Mobility in Modern Dublin, c. 1930-1980." *Urban History* 42, no. 2 (2015): 273-89.

Hannam, Kevin, Mimi Sheller, and John Urry. "Editorial: Mobilities, Immobilities and Moorings." *Mobilities* 1, no. 1 (2006): 1-22.

Harms, Lucas, Luca Bertolini, and Marco te Brömmelstroet. "Performance of Municipal Cycling Policies in Medium-Sized Cities in the Netherlands since 2000." *Transport Reviews* 36, no. 1 (2016): 134-62.

Harms, Lucas, Luca Bertolini, and Marco te Brömmelstroet. "Spatial and Social Variations in Cycling Patterns in a Mature Cycling Country: Exploring Differences and Trends." *Journal of Transport & Health* 1, no. 4 (2014): 232-242.

Heijden, H.A. van der. "Zandzakken, notenbomen en een taart. Nederlandse sociale bewegingen en mondiale klimaatverandering." *Jaarboek voor Ecologische Geschiedenis* 2008 (2009): 117-43.

Héran, Frédéric. *Le retour de la bicyclette: Une histoire des déplacements urbains en Europe, de 1871 à 2050*. Paris: La Découverte, 2014.

Högselius, Per, Arne Kaijer, and Erik van der Vleuten. *Europe's Infrastructure Transition: Economy, War, Nature*, Making Europe, vol. 3. Basingstoke: Palgrave Macmillan, 2015.

Hommels, Anique. "Studying Obduracy in the City: Toward a Productive Fusion between Technology Studies and Urban Studies." *Science, Technology, & Human Values* 30, no. 3 (2005): 323-51.

Hommels, Anique. *Unbuilding Cities: Obduracy in Urban Sociotechnical Change*. Cambridge, MA: MIT Press, 2005.

Horn-van Nispen, M.L. ten. "Het Rijkswegenplan 1927. De aanpak van het verkeersvraagstuk in de jaren twintig." *Jaarboek voor de geschiedenis van bedrijf en techniek* 9 (1992): 185-208.

Horn, Burkhard. "Vom Niedergang eines Massenverkehrsmittels – Zur Geschichte der Städtischen Radverkehrsplanung." Master Thesis, Gesamthochschule Kassel, 1990.

Horst, Servaas van der. *Berini: Geschiedenis van een Hollandse brommer.* Amsterdam: Uitgeverij Thoth, 1991.

Hughes, Thomas P. "Technological Momentum." In *Does Technology Drive History? The Dilemma of Technological Determinism*, edited by Merritt Roe Smith and Leo Marx, 101-13. Cambridge, MA: MIT Press, 1994.

Huré, Maxime. "Les réseaux transnationaux du vélo. Gouverner les politiques du vélo en ville. De l'utopie associative à la gestion par des grandes firmes urbaines." PhD diss., Université Lyon 2 Lumière, 2013.

Huré, Maxime. "Lyon: The Bicycle Goes Corporate." In *Cycling Cities: The European Experience. A Hundred Years of Policy and Practice*, edited by Ruth Oldenziel, Martin Emanuel, Adri Albert de la Bruhèze and Frank Veraart, 173-83. Eindhoven: Foundation for the History of Technology, 2016.

Iddekinge, P.R.A. van. "Bloemers, Henri Peter Johan (1880-1947)." In *Biografisch Woordenboek van Nederland.* URL:http://resources.huygens.knaw.nl/bwn1880-2000/lemmata/bwn3/bloemers

Ishaque, Muhammad M., and Robert B. Noland. "Making Roads Safe for Pedestrians or Keeping Them Out of the Way?" *The Journal of Transport History* 27, no. 1 (2006): 115-37.

Jaffe, Rivke, and Martijn Koster. "The Myth of Formality in the Global North: Informality-as-Innovation in Dutch Governance." *International Journal of Urban and Regional Research* 43, no. 3 (2019): 563-68.

Jamison, Andrew, Ron Eyerman, Jacqueline Cramer, and Jeppe Læssøe. *The Making of the New Environmental Consciousness: A Comparative Study of the Environmental Movements in Sweden, Denmark, and the Netherlands.* Edinburgh: Edinburgh University Press, 1990.

Jensen, Jens Stissing, Matthew Cashmore, and Morten Elle. "Reinventing the bicycle: how calculative practices shape urban environmental governance." *Environmental Politics* 26, no. 3 (2017): 459-79.

Jonkeren, Olaf, Hans Wust, and Mathijs de Haas. *Mobiliteit in stedelijk Nederland.* Den Haag: Ministerie van Infrastructuur en Waterstaat, Kennisinstituut voor Mobiliteitsbeleid, 2019.

Kaal, Harm. "A Conceptual History of Livability: Dutch Scientists, Politicians, Policy Makers and Citizens and the Quest for a Livable City." *City* 15, no. 5 (2011): 532-47.

Kaal, Harm. "Ruimtelijke ordening." In *Behoedzaam bestuur. Twee eeuwen provincie Zuid-Holland*, edited by Roel Pots and Nico Randeraad, 145-76. Leiden: Primavera Pers, 2014.

Kåstrup, Marie. "Hverdagens beskedne demokrati – Analyser af cyklen som symbol på danskhed." Master thesis, University of Copenhagen, 2007.

Kennedy, James C. *Nieuw Babylon in aanbouw: Nederland in de jaren zestig.* Amsterdam: Boom, 1995.

Kickert, Walter J.M. "Beneath Consensual Corporatism: Traditions of Governance in the Netherlands." *Public Administration* 81, no. 1 (2003): 119-40.

Kickert, Walter J.M. *The History of Governance in the Netherlands: Continuity and Exceptions.* The Hague: Elsevier Overheid, 2004.

Kirchhof, Astrid Mignon. "'For a Decent Quality of Life': Environmental Groups in East and West Berlin." *Journal of Urban History* 41, no. 4 (2015): 625-46.

Kline, Ronald, and Trevor Pinch. "Users as Agents of Technological Change: The Social Construction of the Automobile in the Rural United States." *Technology and Culture* 37, no. 4 (1996): 763-95.

Klühspies, Karl. *München nicht wie geplant: Stadtpolitik, Bürgerwille und die Macht der Medien.* München: Münchner Forum, 2015.

Knudsen, Walther, and Thomas Krag. *På cykel i 100 år – Dansk Cyklist Forbund 1905-2005*. Copenhagen: Dansk Cyklist Forbund, 2005.

Knuts, Stijn, and Pascal Delheye. "Connecting City and Countryside? Faces of Cycling Mobility in Belgium, 1890-1914." *Dutch Crossing* 37, no. 3 (2013): 240-59.

Knuts, Stijn, and Pascal Delheye. "Cycling in the City? Belgian Cyclists Conquering Urban Spaces, 1860-1900." [In English]. *International Journal of the History of Sport* 29, no. 14 (2012): 1942-62.

Knuts, Stijn, and Pascal Delheye. "Identiteiten in koers: Roeselaarse wielrenners als kopmannen van lokale, regionale en (sub)nationale identiteiten, 1900-1960." *Belgisch Tijdschrift voor Nieuwste Geschiedenis/ Revue Belge de Histoire Contemporaine* 41, no. 1-2 (2011): 167-214.

Kohlrausch, Martin, and Helmuth Trischler. *Building Europe on Expertise: Innovators, Organizers, Networkers*. Making Europe, vol. 2, Basingstoke: Palgrave Macmillan, 2014.

Koopmans, Ruud. "Van provo tot RARA. Golfbewegingen in het politiek protest in Nederland." In *Tussen verbeelding en macht: 25 jaar nieuwe sociale bewegingen in Nederland*, edited by J.W. Duyvendak, H.A. van der Heijden, R. Koopmans and L. Wijmans, 59-75. Amsterdam: Sua, 1992.

Koopmans, Ruud, and Jan Willem Duyvendak. "Sociale bewegingen en het primaat van de politiek ". In *Tussen verbeelding en macht: 25 jaar nieuwe sociale bewegingen in Nederland*, 11-38. Amsterdam: Sua, 1992.

Kranakis, Eda. "Writing Technology into History." *Technology and Culture* 62, no. 1 (2021): 212-40.

Kriesi, Hanspeter. "Federalism and Pillarization: The Netherlands and Switzerland Compared." *Acta Politica* XXXI, no. 4 (1996): 539-56.

Kriesi, Hanspeter. "New Social Movements and the New Class in the Netherlands." *American Journal of Sociology* 94, no. 5 (1989): 1078-116.

Kriesi, Hanspeter. "The Political Opportunity Structure of the Dutch Peace Movement." *West European Politics* 12, no. 3 (1989): 295-312.

Kuipers, Giselinde. "De fiets van Hare Majesteit: Over nationale habitus en sociologische vergelijking." *Sociologie* 6, no. 3 (2010): 3-26.

Kuipers, Giselinde. "The Rise and Decline of National Habitus: Dutch Cycling Culture and the Shaping of National Similarity." *European Journal of Social Theory* 16, no. 1 (2013): 17-35.

Laak, Dirk van. *Alles im Fluss: Die Lebensadern unserer Gesellschaft – Geschichte und Zukunft der Infrastruktur*. Frankfurt am Main: S. Fischer, 2018.

Lelieveld, Anja. "Het plaatje met of zonder gaatje." *De oude fiets* 1992, no. 2 (1992): 1-8.

Lente, Dick van, and Johan Schot. "Techniek als politiek: ingenieurs en de vormgeving van de Nederlandse samenleving." In *Techniek in Nederland in de twintigste eeuw – Deel VII – Techniek en modernisering, balans van de twintigste eeuw*, edited by Adri A. Albert de la Bruhèze, H.W. Lintsen, A. Rip and J.W. Schot, 196-231. Zutphen: Walburg Pers/Stichting Historie der Techniek, 2003.

Levels, Annika. "(Re-) Claiming Urban Streets: The Conflicting (Auto)Mobilities of Cycling and Driving in Berlin and New York." *The Journal of Transport History* 41, no. 3 (2020): 381-401.

Lijphart, Arend. *Patterns of Democracy: Government Forms and Performance in Thirty-Six Countries*. New Haven: Yale University Press, 1999.

Lijphart, Arend. *The Politics of Accomodation: Pluralism and Democracy in the Netherlands*. Berkeley: University of California Press, 1968.

Linders, Dick. "Mr. Chr. A. De Ruyter-De Zeeuw, Rotterdam 22 augustus 1907 – IJlst 5 december 1995, Onafhankelijk, sociaal-democraat." *Rotterdams Jaarboekje* Series 10, no. 4 (1996): 146-53.

Linders-Rooijendijk, M.F.A. "Gebaande wegen voor mobiliteit en vrijetijdsbesteding: I. De ANWB als vrijwillige associatie 1883-1937." PhD diss., Katholieke Universiteit Brabant, 1989.

Linders-Rooijendijk, M.F.A. *Gebaande wegen voor mobiliteit en vrijetijdsbesteding: II. De ANWB van vereniging naar instituut, 1937-1983*. Den Haag: ANWB, 1992.

Lintsen, Harry. "De Delftse Polytechnische School als bakermat van het socialisme 1900-1925." In *Het tweede jaarboek voor het democratisch socialisme*, edited by J. Bank, M. Pos and B. Tromp, 81-109. Amsterdam: Wiardi Beckman Stichting/De Arbeiderspers, 1980.

Lintsen, Harry, Frank Veraart, Jan-Pieter Smits, and John Grin, eds. *Well-being, Sustainability and Social Development: The Netherlands 1850-2050*. Cham: Springer International Publishing, 2018.

Lisci-Wessels, Trudy. "Ontwikkeling van het aantal gemeenten sinds 1900." *Bevolkingstrends. Statistisch kwartaalblad over de demografie van Nederland* 52, no. 1 (2004): 56-57.

Liu, George, Marco te Brömmelstroet, Sukanya Krishnamurthy, and Pieter van Wesemael. "Practitioners' perspective on user experience and design of cycle highways." *Transportation Research Interdisciplinary Perspectives* 1 (2019).

Longhurst, James. *Bike Battles: A History of Sharing the American Road*. Seattle/London: University of Washington Press, 2015.

Longhurst, James. "The Sidepath Not Taken: Bicycles, Taxes, and the Rhetoric of the Public Good in the 1890s." *Journal of Policy History* 25, no. 4 (2013): 557-86.

López, Bernat. "The Failed Vuelta Ciclista a España of 1913 and the Launching of the Volta a Catalunya (1911-1913): Centre Versus Periphery in the Struggle for the Governance of Cycling in Early Twentieth-Century Spain." *Sport in History* 30, no. 4 (2010): 547-69.

López, Bernat. "Sport, Media, Politics and Nationalism on the Eve of the Spanish Civil War: The First Vuelta Ciclista a España (1935)." *The International Journal of the History of Sport* 27, no. 4 (2010): 635-57.

Mahoney, James. "Path Dependence in Historical Sociology." *Theory and Society* 29, no. 4 (2000): 507-48.

Mamadouh, Virginie. *De stad in eigen hand: provo's, kabouters en krakers als stedelijke sociale beweging*. Amsterdam: Sua, 1992.

Männistö-Funk, Tiina. "The Crossroads of Technology and Tradition: Vernacular Bicycles in Rural Finland, 1880-1910." *Technology and Culture* 52, no. 4 (2011): 733-56.

Martens, Karel. *Transport Justice: Designing fair transportation systems*. London: Routledge, 2016.

Mauch, Christof, and Thomas Zeller, eds. *The World beyond the Windshield: Roads and Landscapes in the United States and Europe*. Athens, Ohio; Ohio University Press; Stuttgart: Franz Steiner Verlag, 2008.

Meer, Frits van der, Jos Raadschelders, and Toon Kerkhoff. "Van nachtwakersstaat naar waarborgstaat: proliferatie en vervlechting van het Nederlandse openbaar bestuur in de lange twintigste eeuw (1880-2005)." In *Duizend jaar openbaar bestuur in Nederland: van patrimoniaal bestuur naar waarborgstaat*, edited by Pieter Wagenaar, Toon Kerkhoff and Mark Rutgers, 221-90. Bussum: Coutinho, 2011.

Merriman, Peter. "Mobilities, Crises, and Turns: Some Comments on Dissensus, Comparative Studies, and Spatial Histories." *Mobility in History* 6, no. 1 (2015): 20-34.

Mijnssen-Hemmi, Pete, and Florian Boller. *30 Jahre Veloförderung in der Stadt Zürich, 1975 bis 2005*. Zürich: Tiefbauamt der Stadt Zürich, Verkehrsplanung, 2006.

Mil, Yvonne van. "Pieter Bakker Schut (1877-1952), manager in stedenbouw." *Bulletin KNOB* 116, no. 4 (2017): 209-23.

Mom, Gijs. *Atlantic Automobilism: Emergence and Persistence of the Car, 1895-1940*. New York: Berghahn Books, 2015.

Mom, Gijs. "Building an Infrastructure for the Automobile System: PIARC and Road Safety (1908-1938)." In *Proceedings 23th World Road Congress 17-21 September 2007*, 2004-30. Paris: PIARC, 2007.

Mom, Gijs. "The Crisis of Transport History: A Critique, and a Vista." *Mobility in History* 6, no. 1 (2015): 7-19.

Mom, Gijs. "The Emancipation of the Urban View: Dutch Spatial Planning in an International Context (1920-1950)." In *Builders and Planners: A History of Land-Use and Infrastructure Planning in the Netherlands*, edited by Jos Arts, Ruud Filarski, Hans Jeekel and Bert Toussaint, 73-137. Delft: Uitgeverij Eburon, 2016.

Mom, Gijs. "Inter-Artifactual Technology Transfer: Road Building Technology in the Netherlands and the Competition Between Bricks, Macadam, Asphalt and Concrete." *History and Technology* 20, no. 1 (2004): 75-96.

Mom, Gijs. "Mao or Merriman? On Pitjantjatjara and Other Mobilities – A Response." *Mobility in History* 6, no. 1 (2015): 35-39.

Mom, Gijs. "Roads without Rails: European Highway Network-Building and the Desire for Long-Range Motorized Mobility." *Technology and Culture* 46, no. 4 (2005): 745-72.

Mom, Gijs. "The West and the Rest: The Green Heart and the Breakthrough of Spatial Planning (1950-1982)." In *Builders and Planners: A History of Land-Use and Infrastructure Planning in the Netherlands*, edited by Jos Arts, Ruud Filarski, Hans Jeekel and Bert Toussaint, 143-99. Delft: Uitgeverij Eburon, 2016.

Mom, Gijs. "What Kind of Transport History Did We Get? Half a Century of JTH and the Future of the Field." *The Journal of Transport History* 24, no. 2 (2003): 121-38.

Mom, Gijs, and Ruud Filarski. *Van transport naar mobiliteit: de mobiliteitsexplosie [1895-2005]*. Zutphen: Walburg Pers /Stichting Historie der Techniek, 2008.

Mom, Gijs, Johan Schot, and Peter-Eloy Staal. "Civilizing Motorized Adventure: Automotive Technology, User Culture, and the Dutch Touring Club as Mediator." In *Manufacturing Technology, Manufacturing Consumers: The Making of Dutch Consumer Society*, edited by Adri A. Albert de la Bruhèze and Ruth Oldenziel, 139-58 Amsterdam: Aksant, 2009.

Moraglio, Massimo. "Transferring Technology, Shaping Society: Traffic Engineering in PIARC Agenda, in the early 1930s." *Technikgeschichte* 80, no. 1 (2013): 13-32.

Moran, Joe. "Crossing the Road in Britain, 1931-1976." *The Historical Journal* 49, no. 2 (2006): 477-96.

Morgan, Njogu. *Cycling Cities: The Johannesburg Experience*. Eindhoven: Foundation for the History of Technology, 2019.

Morrell, Kevin. "Governance and the Public Good." *Public Administration* 87, no. 3 (2009): 538-56.

Mück, Matthias, Christoph Helf, and Miriam Lindenau. "Urban Living Labs Fostering Sustainable Mobility Planning in Munich." *Transportation Research Procedia* 41 (2019): 741-44.

Norcliffe, Glen. "Associations, Modernity and the Insider-citizens of a Victorian Highwheel Bicycle Club." *Journal of Historical Sociology* 19, no. 2 (2006): 121-50.

Norcliffe, Glen. *The Ride to Modernity: The Bicycle in Canada, 1869-1900*. Toronto: University of Toronto Press, 2001.

Norton, Peter. *Fighting Traffic: The Dawn of the Motor Age in the American City*. Cambridge, MA: MIT Press, 2008.

Norton, Peter. "Four Paradigms: Traffic Safety in the Twentieth-Century United States." *Technology and Culture* 56, no. 2 (2015): 319-34.

Norton, Peter. "Persistent pedestrianism: urban walking in motor age America, 1920s-1960s." *Urban History* 48, no. 2 (2021): 266-89.

Norton, Peter. "Street Rivals: Jaywalking and the Invention of the Motor Age Street." *Technology and Culture* 48, no. 2 (2007): 331-59.

Norton, Peter. "Urban Transport and Mobility in Technology and Culture." *Technology and Culture* 61, no. 4 (2020): 1197-211.

Oldenziel, Ruth. "Accounting Tricks: How Pedestrians & Cyclists were Thrown under the Bus, 1910s-1940s." Unpublished paper, 2016.

Oldenziel, Ruth. "Mode of the Past or Promise for the Future? Cycling in China and the Sustainability Challenge, 1955-Present." *Zeithistorische Forschungen/Studies in Contemporary History* 14, no. 3 (2017): 465-86.

Oldenziel, Ruth, and Adri A. Albert de la Bruhèze. "Theorizing the Mediation Junction for Technology and Consumption." In *Manufacturing Technology, Manufacturing Consumers: The Making of Dutch Consumer Society*, edited by Adri Albert de la Bruhèze and Ruth Oldenziel, 9-39. Amsterdam: Aksant, 2009.

Oldenziel, Ruth, and Adri A. Albert de la Bruhèze. "Amsterdam: World Bicycle Capital, By Chance." In *Cycling Cities: The European Experience. A Hundred Years of Policy and Practice*, edited by Ruth Oldenziel, Martin Emanuel, Adri Albert de la Bruhèze and Frank Veraart, 17-27. Eindhoven: Foundation for the History of Technology, 2016.

Oldenziel, Ruth, and Adri A. Albert de la Bruhèze. "Contested Spaces: Bicycle Lanes in Urban Europe, 1900-1995." *Transfers: Interdisciplinary Journal of Mobility Studies* 1, no. 2 (2011): 29-49.

Oldenziel, Ruth, Adri A. Albert de la Bruhèze, and Onno de Wit. "Europe's Mediation Junction: Technology and Consumer Society in the 20th Century." *History and Technology* 21, no. 1 (2005): 107-39.

Oldenziel, Ruth, Martin Emanuel, Adri A. Albert de la Bruhèze, and Frank Veraart, eds. *Cycling Cities: The European Experience*. Eindhoven: Foundation for the History of Technology and Rachel Carson Center for Environment and Society, 2016.

Oldenziel, Ruth, and Mikael Hård. *Consumers, Tinkerers, Rebels: The People Who Shaped Europe*. Making Europe. Vol. 1, New York/Basingstoke: Palgrave Macmillan, 2013.

Oosterhuis, H., and M. Stoffers. "'Ons populairste vervoersmiddel'. De Nederlandse fietshistoriografie in internationaal perspectief." *BMGN – Low Countries Historical Review* 124, no. 3 (2009): 390-418.

Oosterhuis, Harry. "Bicycle Research between Bicycle Policies and Bicycle Culture." *Mobility in History* 5 (2014): 20-36.

Oosterhuis, Harry. "Cycling, Modernity and National Culture." *Social History* 41, no. 3 (2016): 233-48.

Oosterhuis, Harry. "Ingebakken gewoonte of buitenissige liefhebberij?" *Sociologie* 11, no. 1 (2015): 3-30.

Otten, Marko. *Provo: de ludieke opstand tegen bom en regentendom*. Doetinchem: Het Boekenschap, 2017.

Oudshoorn, Nelly, and Trevor Pinch. *How Users Matter: the Co-Construction of Users and Technologies*. London: MIT Press, 2003.

Parsons, Rorie, and Geoff Vigar. "'Resistance was Futile!' Cycling's Discourse of Resistance to UK Automobile Modernism 1950-1970." *Planning Perspectives* 33, no. 2 (2018): 163-83.

Pas, Niek. *Provo! Mediafenomeen 1965-1967*. Amsterdam: Wereldbibliotheek, 2015.

Pestman, Paul. "Dutch Infrastructure Policies: Changing and Contradictory Policy Arrangements." In *Political Modernisation and the Environment: The Renewal of Environmental Policy Arrangements*, edited by Jan van Tatenhove, Bas Arts and Pieter Leroy. Environment & Policy, vol. 24, 71-96. Dordrecht: Kluwer, 2000.

Petzer, Brett, Anna Wieczorek, and Geert Verbong. "Cycling as a Service Assessed from a Combined Business-Model and Transitions Perspective." *Environmental Innovation and Societal Transitions* 36 (2020): 255-69.

Petzer, Brett, Anna Wieczorek, and Geert Verbong. "Dockless Bikeshare in Amsterdam: A Mobility Justice Perspective on Niche Framing Struggles." *Applied Mobilities* 5, no. 3 (2020): 232-50.

Piersma, Folkert, and Wout Ritzema. *Fietsparkeren bij stations. 20 jaar ontwikkeling, ontwerp en realisatie*. Bussum: THOTH, 2021.

Pierson, Paul. *Politics in Time: History, Institutions, and Social Analysis*. Princeton: Princeton University Press, 2004.

Pivato, Stefano. "The Bicycle as a Political Symbol. Italy, 1885-1955." *International Journal of the History of Sport* 7, no. 2 (1990): 173-87.

Pivato, Stefano. "Italian Cycling and the Creation of a Catholic Hero. The Bartali Myth." *International Journal of the History of Sport* 13, no. 1 (1996): 128-38.

Ploeger, Jan, and Ruth Oldenziel. "The Sociotechnical Roots of Smart Mobility: Bike Sharing since 1965." *Journal of Transport History* 41, no. 2 (2020): 134-59.

Pooley, Colin, Martin Emanuel, Tiina Männistö-Funk, and Peter Norton. "Introduction: historical perspectives on pedestrians and the city." *Urban History* 48, no. 2 (2021): 204-10.Pooley, Colin, and Jean Turnbull. "Modal Choice and Modal Change: The Journey to Work in Britain Since 1890." *Journal of Transport Geography* 8, no. 1 (2000): 11-24.

Pooley, Colin, Jean Turnbull, and Mags Adams. *A Mobile Century? Changes in Everyday Mobility in Britain in the Twentieth Century*. Aldershot: Ashgate, 2005.

Popkema, Marcus. "Tussen techniek en planning: de opkomst van het vak Verkeerskunde in Nederland, 1950-1975." PhD diss., TU Eindhoven, 2014.

Pucher, John, and Ralph Buehler. "Making Cycling Irresistible: Lessons from The Netherlands, Denmark and Germany." *Transport Reviews* 28, no. 4 (2008): 495-528.

Pucher, John, and Ralph Buehler. "Safer Cycling Through Improved Infrastructure." *American Journal of Public Health* 106, no. 12 (2016): 2089-91.

Rautio, Anna-Maria, and Lars Östlund. "'Starvation Strings' and the Public Good: Development of a Swedish Bike Trail Network in the Early Twentieth Century." *Journal of Transport History* 33, no. 1 (2012): 42-63.

Reid, Carlton. *Roads Were not Built for Cars: How Cyclists Were the First to Push for Good Roads & Became the Pioneers of Motoring*. Washington, DC: Island Press, 2015.

Reid, Carlton. *Bike Boom: The Unexpected Resurgence of Cycling*. Washington, DC: Island Press, 2017.

Rhodes, R.A.W. "Waves of Governance." In *The Oxford Handbook of Governance*, edited by David Levi-Faur, 33-48. Oxford: Oxford University Press, 2012.

Richardson, Jeremy. "New Governance or Old Governance? A Policy Style Perspective." In *The Oxford Handbook of Governance*, edited by David Levi-Faur, 311-24. Oxford: Oxford University Press, 2012.

Riet, Odette van de, and Bert Toussaint. "Learning from a Contested Project in the Netherlands: The Clash over the Amelisweerd Forest, 1957-1982." *Transfers* 4, no. 1 (2014): 63-82.

Schepers, P., D. Twisk, E. Fishman, A. Fyhri, and A. Jensen. "The Dutch Road to a High Level of Cycling Safety." *Safety Science* 92 (2017): 264-73.

Schipper, Frank. "Driving Europe: Building Europe on Roads in the Twentieth Century." PhD diss., TU Eindhoven, 2008.

Schmucki, Barbara. *Der Traum vom Verkehrsfluss. Städtische Verkehrsplanung seit 1945 im deutsch-deutschen Vergleich*. Deutsches Museum: Beiträge zur Historischen Verkehrsforschung, vol. 4. Frankfurt: Campus, 2001.

Schot, Johan W., and Vincent Lagendijk. "Technocratic Internationalism in the Interwar Years: Building Europe on Motorways and Electricity Networks." *Journal of Modern European History* 6, no. 2 (2008): 196-217.

Schrijvers, Erik, and Stefan Couperus. "Voorbij verkiezing en parlement. Alternatieve representatie in Nederland na 1870." In *Omstreden democratie. Over de problemen van een succesverhaal*, edited by Remieg Aerts and Peter de Goede, 67-88. Amsterdam: Boom, 2013.

Seely, Bruce E., Donald E. Klingner, and Gary Klein. ""Push" and "Pull" Factors in Technology Transfer: Moving American-Style Highway Engineering to Europe, 1945-1965." *Comparative Technology Transfer & Society* 2, no. 3 (2004): 229-46.

Sheller, Mimi. *Mobility Justice: The Politics of Movement in an Age of Extremes*. London: Verso, 2018.

Sheller, Mimi. "Theorising mobility justice." *Tempo Social* 30, no. 2 (2018): 17-34.

Sheller, Mimi, and John Urry. "The New Mobilities Paradigm." *Environment and Planning A: Economy and Space* 38, no. 2 (2006): 207-26.

Shove, Elizabeth. "The Shadowy Side of Innovation: Unmasking and Sustainability." *Technology Analysis & Strategic Management* 24, no. 4 (2012): 363-75.

Simons, W.J. *Daar fietst men toch zo heerlijk heen: vijftig jaar Provinciale Drentse Rijwielpadvereeniging/Stichting Het Drentse Fietspad*. Assen: Stichting Het Drentse Fietspad, 1990.

Smaal, Maarten. *Politieke strijd om de prijs van automobiliteit: de geschiedenis van een langlopend discours, 1895-2010*. Delft: Eburon, 2012. PhD diss., University of Tilburg.

Smith, Merritt Roe, and Leo Marx, eds. *Does Technology Drive History? The Dilemma of Technological Determinism*. Cambridge, MA/London: MIT Press, 1994.

Steinbeck, Frank. *Das Motorrad: ein deutscher Sonderweg in die automobile Gesellschaft*. Stuttgart: Franz Steiner Verlag, 2012.

Steinbeck, Frank. "Rückenwind für motorisierte Fahrräder." *Kultur & Technik*, no. 2 (2013): 34-39.

Straub, Jan. "Binnenstadsbeleid en verkeers- en vervoersproblematiek. Een onderzoek naar het beleid ten aanzien van de verkeers-, vervoers- en parkeerproblemen in binnensteden in Nederland in de periode 1960-1990." PhD diss., Universiteit van Amsterdam, 1991.

Sun, Qi, Tao Feng, Astrid Kemperman, and Andreas Spahn. "Modal shift implications of e-bike use in the Netherlands: Moving towards sustainability?" *Transportation Research Part D: Transport and Environment* 78 (2020).

Tasman, Coen. *Louter Kabouter: Kroniek van een beweging*. Amsterdam: Babylon-De Geus, 1996.

Tatenhove, Jan van, Bas Arts, and Pieter Leroy, eds. *Political Modernisation and the Environment: The Renewal of Environmental Policy Arrangements*, Environment & Policy, vol. 24. Dordrecht: Kluwer, 2000.

Te Brömmelstroet, Marco. "Framing systemic traffic violence: Media coverage of Dutch traffic crashes." *Transportation Research Interdisciplinary Perspectives* 5 (2020).

Te Brömmelstroet, Marco, Anna Nikolaeva, Meredith Glaser, Morten Skou Nicolaisen, and Carmen Chan. "Travelling together alone and alone together: mobility and potential exposure to diversity." *Applied Mobilities* 2, no. 1 (2017): 1-15.

Tjong Tjin Tai, Sue-Yen, Frank Veraart, and Mila Davids. "How the Netherlands Became a Bicycle Nation: Users, Firms and Intermediaries, 1860-1940." *Business History* 57, no. 2 (2015): 257-89.

Toonen, Theo A.J. "Dutch Provinces and the Struggle for the Meso." In *The Rise of Meso Government in Europe*, edited by L.J. Sharpe, 117-53. London: Sage, 1993.

Toussaint, Bert. "(Transport) History as Policy Lab for Democratic Governance." *The Journal of Transport History* 40, no. 2 (2019): 270-280.

Toussaint, Bert. "Using the Usable Past: Reflections and Practices in the Netherlands." In *Transport Policy: Learning Lessons from History*, edited by Colin Divall, Julian Hine and Colin Pooley, 15-27. Farnham: Ashgate, 2016.

Tsubohara, S. "Democracy through Political Parties and Public Participation: The Case of the Planning History of Groningen, The Netherlands." PhD diss., University of Groningen, 2010.

Udo de Haes, H., and W. ter Keurs. "De Leidse Baan." In *Milieu-aktie in Nederland*, edited by E. Tellegen and J. Willems, 17-23. Amsterdam: De Trommel/Vereniging Milieudefensie, 1978.

Valderpoort, W. *De zelfzuchtige personenauto: Beschouwingen over een onderdeel van het verkeersvraagstuk, met een bijzondere toepassing op de stedebouw*. Amsterdam: Van Saane, 1953.

Valk, Arnold van der. *Het levenswerk van Th. K. van Lohuizen 1890-1956: De eenheid van het stedebouwkundige werk*. Delft: Delftse Universitaire Pers, 1990.

Van Tielhof, Milja. "Op zoek naar het poldermodel in de waterstaatsgeschiedenis." *Tijdschrift voor geschiedenis* 122, no. 2 (2009): 148-61.

Veen, Adriejan van. "From Neo-Corporatism to Regulatory Governance: Interests, Expertise, and Power in Dutch Extraparliamentary Governance, c. 1900-2018." In *Information and Power in History: Towards a Global Approach*, edited by Ida Nijenhuis, Marijke van Faassen, Ronald Sluijter, Joris Gijsenbergh and Wim de Jong, 85-102. London: Routledge, 2020.

Veraart, Frank. "Agriculture and Nutrition: The End of Hunger." In *Well-being, Sustainability and Social Development: The Netherlands 1850-2050*, edited by Harry Lintsen, Frank Veraart, Jan-Pieter Smits and John Grin, 259-92. Cham: Springer International Publishing, 2018.

Veraart, Frank. "Fietspaden. Van gerieflijke paden tot fietsbeleid." *Verkeerskunde* 60, no. 5 (2009): 20-26.

Veraart, Frank. "Geschiedenis van de fiets in Nederland 1870-1940: Van sportmiddel naar massavervoermiddel." Master Thesis, TU Eindhoven, 1995.

Veraart, Frank, and Albert Albert de la Bruhèze. "Fietsen in de Nederlandse bergen. Achterblijvend fietsgebruik in het zuiden van Limburg in historisch perspectief." In *Studies over de sociaaleconomische geschiedenis van Limburg/Jaarboek van het Sociaal Historisch Centrum voor Limburg*, edited by Ad Knotter and Willibrord Rutten, 133-57. Maastricht: Sociaal Historisch Centrum voor Limburg, 2001.

Veraart, Frank, Martin Emanuel, and Ruth Oldenziel. "Eindhoven: Engineering a Path for Bikes?" In *Cycling Cities: The European Experience. A Hundred Years of Policy and Practice*, edited by Ruth Oldenziel, Martin Emanuel, Adri Albert de la Bruhèze and Frank Veraart, 53-63. Eindhoven: Foundation for the History of Technology, 2016.

Veraart, Frank, Stijn Knuts, and Pascal Delheye. "Antwerp: Cycling Claims a Comeback." In *Cycling Cities: The European Experience. A Hundred Years of Policy and Practice*, edited by Ruth Oldenziel, Martin Emanuel, Adri Albert de la Bruhèze and Frank Veraart, 89-99. Eindhoven: Foundation for the History of Technology, 2016.

Verlaan, Tim. "De toekomst van de Nederlandse binnenstad 1960-1978." PhD diss., University of Amsterdam, 2016.

Verlaan, Tim. "Mobilization of the Masses: Dutch Planners, Local Politics, and the Threat of the Motor Age 1960-1980." *Journal of Urban History* 47, no. 1 (2021): 136-56.

Vinne, Vincent van der. *De trage verbreiding van de auto in Nederland 1896-1939: de invloed van ondernemers, gebruikers en overheid*. Amsterdam: De Bataafsche Leeuw, 2007.

Voytenko, Yuliya, Kes McCormick, James Evans, and Gabriele Schliwa. "Urban Living Labs for Sustainability and Low Carbon cities in Europe: Towards a Research Agenda." *Journal of Cleaner Production* 123 (2016): 45-54.

Vries, Michiel S. de "Institutional Fleecing: The Slow Death of Dutch Provinces." *Public Organization Review* 4, no. 4 (2004): 295-315.

Vries, Michiel S. de. "The Rise and Fall of Decentralization: A Comparative Analysis of Arguments and Practices in European Countries." *European Journal of Political Research* 38, no. 2 (2000): 193-224.

Wardlaw, Malcolm J. "History, Risk, Infrastructure: Perspectives on Bicycling in the Netherlands and the UK." *Journal of Transport & Health* 1, no. 4 (2014): 243-50.

Weber, Donald. *De blijde intrede van de automobiel in België 1895-1940*. Gent: Academia Press, 2010.

Wells, Christopher W. *Car Country: An Environmental History*. Seattle: University of Washington Press, 2012.

Williams, Rhys H. "Constructing the Public Good: Social Movements and Cultural Resources." *Social Problems* 42, no. 1 (1995): 124-44.

Windt, Henny van der. "De totstandkoming van 'de natuurbescherming' in Nederland." *Tijdschrift voor geschiedenis* 107, no. 3 (1994): 485-507.

Windt, Henny van der. "Natuurbescherming en landbouw in Nederland 1880-2010, tussen wetenschap en overheid." *Jaarboek voor Ecologische Geschiedenis* 2012/2013 (2014): 119-51.

Wolf, Hanna. "Following America? Dutch geographical car diffusion, 1900 to 1980." PhD diss., Eindhoven University of Technology, 2010.

Woud, Auke van der. *Het landschap, de mensen. Nederland 1850-1940*. Amsterdam: Prometheus, 2020.

Wynne, Brian. "May the Sheep Safely Graze? A Reflexive View of the Expert-Lay Knowledge Divide." In *Risk, Environment and Modernity: Towards a New Ecology*, edited by Scott Lash, Bronislaw Szerszynski and Brian Wynne, 44-83. London: Sage, 1996.

Wynne, Brian. "Sheepfarming after Chernobyl: A Case Study in Communicating Scientific Information." *Environment: Science and Policy for Sustainable Development* 31, no. 2 (1989): 10-39.

Zeller, Thomas. *Driving Germany: The Landscape of the German Autobahn, 1930-1970*. Studies in German history. New York: Berghahn Books, 2007.

Index